Metal Matrix Composites

Metal Matrix Composites

Special Issue Editor

Manoj Gupta

MDPI • Basel • Beijing • Wuhan • Barcelona • Belgrade

MDPI

Special Issue Editor
Manoj Gupta
Materials Group, Department of Mechanical Engineering
Singapore

Editorial Office
MDPI
St. Alban-Anlage 66
Basel, Switzerland

This edition is a reprint of the Special Issue published online in the open access journal *Metals* (ISSN 2075-4701) from 2016–2018 (available at: http://www.mdpi.com/journal/metals/special_issues/matrix_composites).

For citation purposes, cite each article independently as indicated on the article page online and as indicated below:

LastName, A.A.; LastName, B.B.; LastName, C.C. Article title. *Journal Name* **Year**. Article number, Page Range.

ISBN 978-3-03897-087-3 (Pbk)
ISBN 978-3-03897-088-0 (PDF)

Cover photo courtesy of Khin Sandar Tun.

Contents

About the Special Issue Editor ix

Manoj Gupta
Metal Matrix Composites
Reprinted from: *Metals* **2018**, *8*, 379, doi: 10.3390/met8060379 . . . . **1**

Abolfazl Azarniya, Mir Saman Safavi, Saeed Sovizi, Amir Azarniya, Biao Chen, Hamid Reza Madaah Hosseini and Seeram Ramakrishna
Metallurgical Challenges in Carbon Nanotube-Reinforced Metal Matrix Nanocomposites
Reprinted from: *Metals* **2017**, *7*, 384, doi: 10.3390/met7100384 . . . . **3**

Khin Sandar Tun, Yuming Zhang, Gururaj Parande, Vyasaraj Manakari and Manoj Gupta
Enhancing the Hardness and Compressive Response of Magnesium Using Complex Composition Alloy Reinforcement
Reprinted from: *Metals* **2018**, *8*, 276, doi: 10.3390/met8040276 . . . . **47**

Weidong Song, Liansong Dai, Lijun Xiao, Cheng Wang, Xiaonan Mao and Huiping Tang
A Meso-Mechanical Constitutive Model of Particle-Reinforced Titanium Matrix Composites at High Temperatures
Reprinted from: *Metals* **2017**, *7*, 15, doi: 10.3390/met7010015 . . . . **57**

Feng Qiu, Xiang Gao, Jian Tang, Yu-Yang Gao, Shi-Li Shu, Xue Han, Qiang Li and Qi-Chuan Jiang
Microstructures and Tensile Properties of Al–Cu Matrix Composites Reinforced with Nano-Sized SiC_p Fabricated by Semisolid Stirring Process
Reprinted from: *Metals* **2017**, *7*, 49, doi: 10.3390/met7020049 . . . . **69**

Amit Kumar, Khin Sandar Tun, Amit Devendra Kohadkar and Manoj Gupta
Improved Compressive, Damping and Coefficient of Thermal Expansion Response of Mg–3Al–2.5La Alloy Using Y_2O_3 Nano Reinforcement
Reprinted from: *Metals* **2017**, *7*, 104, doi: 10.3390/met7030104 . . . . **77**

Nguyen Thi Hoang Oanh, Nguyen Hoang Viet, Ji-Soon Kim and Alberto Moreira Jorge Junior
Characterization of In-Situ Cu–TiH2–C and Cu–Ti–C Nanocomposites Produced by Mechanical Milling and Spark Plasma Sintering
Reprinted from: *Metals* **2017**, *7*, 117, doi: 10.3390/met7040117 . . . . **88**

Ahmed Nassef, Waleed H. El-Garaihy and Medhat El-Hadek
Characteristics of Cold and Hot Pressed Iron Aluminum Powder Metallurgical Alloys
Reprinted from: *Metals* **2017**, *7*, 170, doi: 10.3390/met7050170 . . . . **100**

Ahmed Nassef, Waleed H. El-Garaihy and Medhat El-Hadek
Mechanical and Corrosion Behavior of Al-Zn-Cr Family Alloys
Reprinted from: *Metals* **2017**, *7*, 171, doi: 10.3390/met7050171 . . . . **112**

Hyun Min Nam, Duck Min Seo, Hyung Duk Yun, Gurunathan Thangavel, Lee Soon Park and Su Yong Nam
Transparent Conducting Film Fabricated by Metal Mesh Method with Ag and Cu@Ag Mixture Nanoparticle Pastes
Reprinted from: *Metals* **2017**, *7*, 176, doi: 10.3390/met7050176 . . . . **124**

Linhui Zhang, Yan Jiang, Qianfeng Fang, Rui Liu, Zhuoming Xie, Tao Zhang, Xianping Wang and Changsong Liu
Comparative Investigation of Tungsten Fibre Nets Reinforced Tungsten Composite Fabricated by Three Different Methods
Reprinted from: *Metals* **2017**, *7*, 249, doi: 10.3390/met7070249 . . . . **132**

Sónia Simões, Filomena Viana, Marcos A. L. Reis and Manuel F. Vieira
Aluminum and Nickel Matrix Composites Reinforced by CNTs: Dispersion/Mixture by Ultrasonication
Reprinted from: *Metals* **2017**, *7*, 279, doi: 10.3390/met7070279 . . . . **144**

Milli Suchita Kujur, Ashis Mallick, Vyasaraj Manakari, Gururaj Parande, Khin Sandar Tun and Manoj Gupta
Significantly Enhancing the Ignition/Compression/Damping Response of Monolithic Magnesium by Addition of Sm_2O_3 Nanoparticles
Reprinted from: *Metals* **2017**, *7*, 357, doi: 10.3390/met7090357 . . . . **155**

Hajo Dieringa, Lydia Katsarou, Ricardo Buzolin, Gábor Szakács, Manfred Horstmann, Martin Wolff, Chamini Mendis, Sergey Vorozhtsov and David StJohn
Ultrasound Assisted Casting of an AM60 Based Metal Matrix Nanocomposite, Its Properties, and Recyclability
Reprinted from: *Metals* **2017**, *7*, 388, doi: 10.3390/met7100388 . . . . **172**

Cristina Arévalo, Isabel Montealegre-Melendez, Eva M. Pérez-Soriano, Enrique Ariza, Michael Kitzmantel and Erich Neubauer
Study of the Influence of TiB Content and Temperature in the Properties of In Situ Titanium Matrix Composites
Reprinted from: *Metals* **2017**, *7*, 457, doi: 10.3390/met7110457 . . . . **185**

Youhong Sun, Chi Zhang, Baochang Liu, Qingnan Meng, Shaoming Ma and Wenhao Dai
Reduced Graphene Oxide Reinforced 7075 Al Matrix Composites: Powder Synthesis and Mechanical Properties
Reprinted from: *Metals* **2017**, *7*, 499, doi: 10.3390/met7110499 . . . . **198**

Abdollah Saboori, Seyed Kiomars Moheimani, Matteo Pavese, Claudio Badini and Paolo Fino
New Nanocomposite Materials with Improved Mechanical Strength and Tailored Coefficient of Thermal Expansion for Electro-Packaging Applications
Reprinted from: *Metals* **2017**, *7*, 536, doi: 10.3390/met7120536 . . . . **212**

Suqing Zhang, Tijun Chen, Jixue Zhou, Dapeng Xiu, Tao Li and Kaiming Cheng
Mechanical Properties of Thixoforged In Situ Mg_2Si_p/AM60B Composite at Elevated Temperatures
Reprinted from: *Metals* **2018**, *8*, 106, doi: 10.3390/met8020106 . . . . **226**

Josef Zapletal, Zuzanka Trojanová, Pavel Dolevolumezal, Stanislava Fintová and Michal Knapek
Elastic and Plastic Behavior of the QE22 Magnesium Alloy Reinforced with Short Saffil Fibers and SiC Particles
Reprinted from: *Metals* **2018**, *8*, 133, doi: 10.3390/met8020133 . . . . **239**

Sravya Tekumalla, Najib Farhan, Tirumalai S. Srivatsan and Manoj Gupta
Nano-ZnO Particles' Effect in Improving the Mechanical Response of Mg-3Al-0.4Ce Alloy
Reprinted from: *Metals* **2016**, *6*, 276, doi: 10.3390/met6110276 . . . . **252**

Gururaj Parande, Vyasaraj Manakari, Harshit Gupta and Manoj Gupta
Magnesium-β-Tricalcium Phosphate Composites as a Potential Orthopedic Implant:
A Mechanical/Damping/Immersion Perspective
Reprinted from: *Metals* **2018**, *8*, 343, doi: 10.3390/met8050343 . . . . **263**

About the Special Issue Editor

Manoj Gupta was a former Head of the Materials Division of the Mechanical Engineering Department and Director designate of the Materials Science and Engineering Initiative at NUS, Singapore. He obtained his Ph.D. from the University of California, Irvine, USA (1992) and conducted postdoctoral research at the University of Alberta, Canada (1992). In August 2017, he was highlighted among the top 1% scientists by The Universal Scientific Education and Research Network and among the top 2.5% scientists by ResearchGate. He developed the Disintegrated Melt Deposition technique and the Hybrid Microwave Sintering technique, an energy-efficient, solid-state processing method to synthesize alloys/micro/nano-composites. He has published over 490 peer-reviewed journal papers and owns two US patents. His current h-index is 56 and his RG index is 46; his publications have received more than 13,000 citations. He has also co-authored six books, published by John Wiley, Springer, and MRF-USA. He is Editor-in-Chief and Editor of 12 international peer-reviewed journals. In 2018, he was announced World Academy Championship Winner in the area of Biomedical Sciences by the International Agency for Standards and Ratings. A multiple award winner, he actively collaborates with various scientist and visits Japan, France, Saudi Arabia, Qatar, China, USA, and India.

metals

MDPI

Editorial

Metal Matrix Composites

Manoj Gupta

Materials Group, Department of Mechanical Engineering, National University of Singapore, 9 Engineering Drive 1, Singapore 117576, Singapore; mpegm@nus.edu.sg

Received: 21 May 2018; Accepted: 23 May 2018; Published: 24 May 2018

Metal Matrix Composites (MMCs) are a unique class of materials capable of providing design freedom to material scientists, allowing the creation of materials that can be targeted to a wide spectrum of applications [1,2]. The key factors that affect the design of MMCs include the following [1–3]:

(a)   Choice of matrix material.
(b)   Type, size (length scale) and amount of reinforcement.
(c)   Type of processing (primary and secondary), as this controls the microstructure including matrix-reinforcement integrity.
(d)   Heat treatment procedure.

The choice of matrix material and reinforcement is primarily influenced by the end application. For example, nickel-based materials are commonly chosen as matrices for high temperature applications, while titanium- and magnesium-based materials are considered for both light weighting of engineering structures and biomedical applications [1–5]. Similarly, different processing methods (primary and secondary) are chosen depending on many factors, such as cost of the end product, type of microstructure desired and volume of production. As an example, conventional casting is always preferred when high volume production is required. The proper utilization of heat treatment also plays a crucial role in selectively enhancing the properties of composites. In view of the dynamic scientific and application potential of metal matrix composites, the present thematic issue was launched and was highly successful, with 20 papers contributed by researchers from all over the world and accepted after rigorous peer review. Key materials that were investigated include the following:

(a)   Aluminum-based materials.
(b)   Magnesium-based materials.
(c)   Titanium-based materials.
(d)   Copper-based materials.
(e)   Silver-based materials.
(f)   Tungsten-based materials.

Reinforcement investigated in the above matrices included at least one of the following:

(a)   ZnO in nanolength scale.
(b)   SiC in nanolength scale.
(c) Y_2O_3 in nanolength scale.
(d) Sm_2O_3 in nanolength scale.
(e)   Carbon-based reinforcements such as CNTs and graphene oxide.
(f)   Metal-based reinforcement such as complex alloy reinforcement in micron length scale.
(g)   Tungsten fiber nets.
(h)   TiB.
(i)   Intermetallic reinforcement such as Mg_2Si.

(j)   Saffil fibers.

(k)   β-Tricalcium phosphate.

The types of processing that were employed to synthesize composites included the following:

(a)   Liquid based methods such as the disintegrated melt deposition method including ultrasonication.

(b)   Semi-solid or two-phase methods.

(c)   In-situ method.

(d)   Solid state method including cold and hot pressing, mechanical milling and spark plasma sintering.

(e)   Thixoforging.

The reported properties of the metal matrix composites in this special issue included:

(a)   Mechanical properties such as tensile and compression response.

(b)   Ignition response.

(c)   Damping response.

(d)   Coefficients of thermal expansion.

(e)   Elastic and plastic deformation behavior.

(f)   Corrosion response.

The choice of matrix, reinforcement, processing method and characterization results reported by researchers in the twenty papers included in this special issue clearly indicate the prevalence of scientific curiosity and the optimism of researchers worldwide to create new metal-based composite materials for a wide range of applications, spanning from the biological sector to multiple engineering sectors.

This special issue certainly provides a succinct description of research activities conducted across the world and thus will be very useful for students and researchers in both academia and industry.

Finally, I would like to thank all the authors for their excellent contributions to this issue, to the reviewers for making useful comments, and to the Metals editorial staff for publishing these articles promptly.

Conflicts of Interest: The authors declare no conflict of interest.

References

1.   Ibrahim, I.A.; Mohamed, F.A.; Lavernia, E.J. Particulate reinforced metal matrix composites—A review. *J. Mater. Sci.* **1991**, *26*, 1137–1156. [CrossRef]

2.   Lloyd, D.J. Particle reinforced aluminium and magnesium matrix composites. *Int. Mater. Rev.* **1994**, *39*, 1–23. [CrossRef]

3.   Ceschini, L.; Dahle, A.; Gupta, M.; Jarfors, A.E.W.; Jayalakshmi, S.; Morri, A.; Rotundo, F.; Toschi, S.; Singh, R.A. *Aluminum and Magnesium Metal Matrix Nanocomposites*; Springer: Singapore, 2016; ISBN 978-981-10-2680-5 (Print); 978-981-10-2681-2 (Online).

4.   Gupta, M.; Seetharaman, S. Magnesium Based Nanocomposites for Cleaner Transport. In *Nanotechnology for Energy Sustainability*; Raj, B., Van de Voorde, M., Mahajan, Y., Eds.; Wiley-VCH: Weinheim, Germany, 2017.

5.   *Insight into Designing Biocompatible Magnesium Alloys and Composites*; Gupta, M., Meenashisundaram, G.K., Eds.; Springer: Singapore, 2015.

metals

MDPI

Review

Metallurgical Challenges in Carbon Nanotube-Reinforced Metal Matrix Nanocomposites

Abolfazl Azarniya [1], Mir Saman Safavi [2], Saeed Sovizi [1], Amir Azarniya [1], Biao Chen [3], Hamid Reza Madaah Hosseini [1,*] and Seeram Ramakrishna [4,*]

[1] Department of Materials Science and Engineering, Sharif University of Technology, Azadi Avenue, Tehran 11155-9466, Iran; azarniya_abolfazl@mehr.sharif.edu (Ab.A.); sovizi71@gmail.com (S.S.); azarniya.amir@gmail.com (Am.A.)

[2] Department of Materials Science and Engineering, University of Tabriz, Tabriz 51666-16471, Iran; m.s.safavi94@ms.tabrizu.ac.ir

[3] Joining and Welding Research Institute (JWRI), Osaka University, Osaka 567-0047, Japan; chen-b@jwri.osaka-u.ac.jp

[4] Department of Mechanical Engineering, National University of Singapore, 9 Engineering Drive 1, Singapore 117576, Singapore

* Correspondence: madaah@sharif.edu (H.R.M.H.); seeram@nus.edu.sg (S.R.); Tel.: +98-922-451-4107 (H.R.M.H.)

Received: 14 July 2017; Accepted: 11 September 2017; Published: 22 September 2017

Abstract: The inclusion of carbon nanotubes (CNTs) into metallic systems has been the main focus of recent literature. The aim behind this approach has been the development of a new property or improvement of an inferior one in CNT-dispersed metal matrix nanocomposites. Although it has opened up new possibilities for promising engineering applications, some practical challenges have restricted the full exploitation of CNTs' unique characteristics. Non-uniform dispersion of CNTs in the metallic matrix, poor interfacial adhesion at the CNT/metal interface, the unfavorable chemical reaction of CNTs with the matrix, and low compactability are the most significant challenges, requiring more examination. The present paper provides a broad overview of the mentioned challenges, the way they occur, and their adverse influences on the physicomechanical properties of CNT-reinforced metal matrix nanocomposites. The suggested solutions to these issues are fully addressed.

Keywords: carbon nanotube; nanocomposite; dispersion; interfacial adhesion; phase transformation; physicomechanical properties

---

1. Introduction

Carbon nanotubes (CNTs) as one of carbon allotropes have been discovered in 1991 by Iijima [1]. There are two main types of CNTs including single-walled (SWNTs) and multi-walled nanotubes (MWNTs). CNTs are the monolithic cylinders of one or more layers of the graphene, so that if only one graphene layer is rolled, it is called SWNTs. MWNTs can be also understood by rolling several parallel graphene layers. Both SWNTs and MWNTs are formed with open or closed ends [2]. The morphology of MWNTs highly depends on their fabrication process, where Russian doll tubes are formed using perfectly concentric cylinders, and scroll tubes can be formed when a single graphene sheet is rolled as scroll [3]. All carbon atoms in the perfect structure of CNTs (except those present on the edges) are bonded in a hexagonal lattice. The presence of impurities drastically degrades the final properties of the tubes. MWNTs have larger diameters than SWNTs, where the diameter of SWNTs ranges from 0.8 to 2 nm and the diameter of MWNTs varies between 5 and 20 nm [2]. Furthermore, the size and surface area of CNTs are of prime significance, so that they can play an important role in the antibacterial applications, where a decrement in CNTs size may significantly improve their antibacterial efficiency through the increase of the specific surface area [4]. Also, the variation of geometrical dimensions in CNTs can significantly affect their final properties due to a change in the length to diameter ratio. However, the final properties of SWNTs and MWNTs do not change in the same manner with the aspect ratio variation. For instance, the Young's modulus of SWNTs deeply depends on the size and chirality, while the elastic modulus of MWNTs slightly changes with the diameter variation [5]. Although CNTs are among the strongest materials in nature, their mechanical properties vary in different directions. The results confirm the higher Young's modulus and tensile strength in the axial direction compared to the radial one [6].

Owing to their appropriate mechanical, electrical, and thermal properties, CNTs have been extensively used in a wide range of the engineering applications such as high-strength and conductive composites, energy storage devices, and hydrogen storage [7]. Moreover, due to a favorable combination of strength and weight, CNTs are commonly used in the aerospace and automobile industries [8]. In these cases, CNTs are used whether in pure state or as a reinforcement in the metal, ceramic, and polymer matrix nanocomposites [9–11]. As to CNT-reinforced nanocomposites, the aim is the development of a new property or improvement of present properties.

In general, the incorporation of CNTs into metallic matrices improves the mechanical properties of such binary systems (e.g., microhardness and fracture toughness) or electrical properties. However, they may adversely affect the properties [12]. Despite the mentioned advantages, this approach mainly faces four major challenges as follow:

(i) Favorable dispersion of CNTs throughout the matrix. A composite with improved properties will be obtained when the reinforcements are uniformly distributed through the matrix. Otherwise, the micro-pores as well as agglomerated particles may form all over the microstructure [13,14]. To overcome this challenge, a broad spectrum of dispersion methods is developed. Among these techniques, the mechanical methods [15], surface treatment [16], and chemical methods [17] are the most conventional. Each of these techniques has their own advantages and disadvantages, being discussed in the next sections.

(ii) Unfavorable chemical reaction of CNTs with matrix at high pressures, elevated temperatures, and induced strains. It is usually accompanied by the formation of defects. On the other words, to thermally decompose CNTs in exposure to a metallic matrix, the presence or formation of defects is required. It is shown that the thermal decomposition of CNTs can bilaterally affect the final properties of the nanocomposites. In other words, the final properties strongly depend on the chemical composition of the formed intermetallics [18,19].

(iii) Poor interfacial adhesion between CNTs and the matrix due to the hydrophobic nature of the CNTs. This shortcoming deteriorates the load bearing between the matrix and CNTs. Moreover, intensive phonon scattering arisen from the insufficient adhesion can significantly enhance the electrical resistivity [20]. It is noteworthy that the interfacial adhesion can be improved whenever a controlled superficial chemical reaction between CNTs and metallic matrix occurs. On the other words, the poor interfacial adhesion can be considered as a sub-challenge under the title of "chemical reaction of CNTs with metallic matrix". This review paper has adopted such a policy.

(iv) Low compactability of metallic powders. The incorporation of CNTs into metallic powders can decline the relative density of final CNT-metal compacts, if the agglomeration of CNTs is heavy and their volume fraction is exceedingly high.

As a general conclusion, the aforementioned challenges should be considered simultaneously in order to produce a CNT-reinforced metal matrix composite with superior properties. The present paper provides a broad overview of these challenges and their potential effects on the properties of CNT-dispersed metal matrix nanocomposites in some details. Also, the practical solutions to these challenges are introduced and their positive influences on the physicomechanical properties of the nanocomposites are studied. It is noteworthy that the full review of the potential effects of CNT addition on the physicomechanical properties of metallic systems is not the main purpose of the present review paper. However, one can find such information in [21,22].

2. Metallurgical Challenges in CNT-Metal Matrix Nanocomposites

2.1. Dispersion of CNTs (Carbon Nanotubes)

To date, a wide variety of research works are conducted with the aim to produce CNT-reinforced metal matrix nanocomposites with superior physicomechanical features for a broad range of functional and structural applications. However, the incorporation of CNTs into the metallic matrices is severely restricted due to both outstanding challenges: (i) Non-uniform dispersion of CNTs throughout the metallic matrices, and (ii) weak interfacial adhesion between CNTs and metallic matrix. In comparison to carbon particles and fibers, CNTs are more likely to be non-uniformly distributed inside the metallic systems due to their comparatively high aspect ratio and extraordinary specific surface area [23].

2.1.1. Water Solubilization of CNTs

A majority of electronic, thermal, and optical applications in which CNTs are used for the fabrication of novel devices or development of new emerging platforms need the large-scale production of stable CNT suspensions with a uniform dispersion and no agglomeration [24]. It is while the van der Waals attraction between CNTs results in their agglomeration and the formation of bundles (Figure 1) due to high surface area [25]. To overcome this practical challenge, some solutions are suggested. Better dispersion of CNTs in water as an important medium for biomedical and biochemistry applications is recognized by various methods such as using surfactants, polymers or chemical functionalization [24]. As to the chemical functionalization, a covalent bond is established between the chemical species and carbon atoms located at side walls or end caps of CNTs. However, this method suffers from two main problems [23,26–28]:

(i) CNTs may shorten, get severe damages during the functionalization process, and lose their superior mechanical properties and distribution of π-electrons. As a general rule, the electron transport strongly depends on the surface structure of CNTs, because the structural damages can serve as phonon scattering sites and interrupt the transport properties of nanotubes.

(ii) The strong acid solutions normally used for functionalization are hazardous to environment and hard to handle.

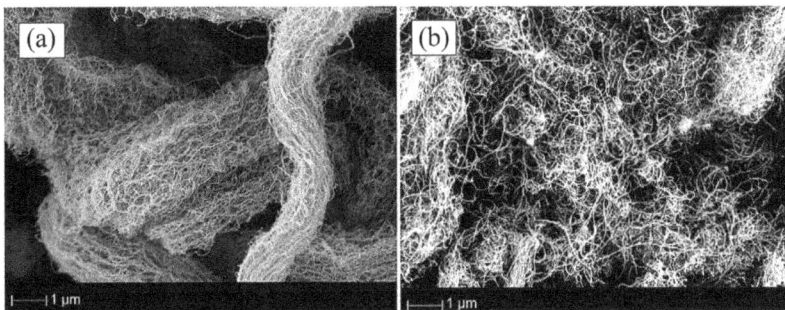

Figure 1. SEM (scanning electron microscope) images of (**a**) multi-walled carbon nanotube (MWCNT) bundles and (**b**) agglomerations [29] (Reproduced with permission from [29], Elsevier, 2012).

The aforementioned disadvantages have persuaded the researchers to develop other dispersing methods based on non-covalent modifications to uniformly distribute CNTs in water. These techniques benefit from two advantages over covalent functionalization methods:

(i) They induce no damages to CNTs, so that the electron transportation along the nanotubes remains intact; and

(ii) CNTs are capable of forming an ordered network through the supermolecules acting as non-covalent modifiers. Among these modifiers, polymers and surfactants are the most conventional. These agents can thread themselves onto CNTs (Figure 2a) or wrap themselves around them (Figure 2b) [28,30,31]. Surfactants are also employed for dispersing CNTs in metal matrix nanocomposites.

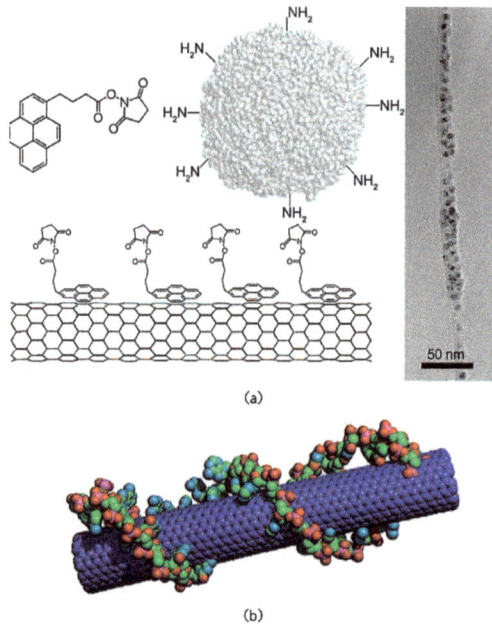

(a)

(b)

Figure 2. Non-covalent functionalization of carbon nanotubes (CNTs): (**a**) With a surfactant and (**b**) with a polymeric agent [32] (Adapted with permission from [32], Elsevier, 2016).

2.1.2. Dispersion of CNTs into Metallic Systems

The practical results show that the dispersion process of CNTs inside the metallic matrices can be completely different and more difficult than water media. This is because an intrinsically different phase (i.e., metal) is added to the system which makes the process more complex. Among these complexities, the CNT/metal interfaces and the poor wettability of CNTs by metal matrices are of prime significance. As generally accepted, an increment in CNT content can result in further agglomeration due to increased possibility of contacts between CNTs and decreased wettability. Therefore, dispersing individual CNTs in metallic systems with the reasonable spacing is a critical challenge for engineers and researchers [10,25].

The agglomeration of CNTs decreases the superior strengthening effects of nanotubes in CNT-metal matrix nanocomposites and deteriorates their physicomechanical properties. Moreover, the inhomogeneous dispersion of CNTs gives rise to non-homogeneity and anisotropy in these binary systems. On the other hand, it may impede the full sintering and reduces the relative density, wettability of CNTs, and resultant mechanical properties [16,33–35].

The above-mentioned challenges about the agglomeration and poor wettability of CNTs in metallic matrices have persuaded the researchers to discover more-efficient methods to uniformly distribute the nanotubes in these material systems. Their huge explorations have resulted in a wide variety of practical dispersion methods, although they are not classified into generally accepted standard categories. In fact, some literature has suggested different arbitrary classifications of the newly developed dispersion methods. For instance, they can be classified into four main groups: (i) Pre-treatments such as functionalizing, ultrasonication, and surface treatments; (ii) Mixing CNT with a metal precursor such as electrodeposition; (iii) Mixing CNT with a metal powders through milling prior to the consolidation routes; and (iv) Post-treatments such as extrusion and rolling.

There are some different approaches for categorization of CNT dispersion methods. From the phenomenological perspective, they may be classified into three main groups: (i) Colloidal mixing in which physical reactions occur between CNTs and other species present in aqueous and inorganic media; (ii) Chemical mixing through which CNTs react chemically with other components; and (iii) Mechanical mixing which uses a mechanical force to detach CNTs from each other. However, all the categorized groups have no distinct boundaries, so that one dispersing method can belong to two or three groups in the same time. For instance, the ultrasonication method, which usually belongs to colloidal mixing, can be categorized in the chemical mixing group, if a chemical reaction occurs during the sonication of CNTs in the aqueous medium. As another example, the metallization is considered as a chemical mixing method, but it needs a complementary mixing method such as milling. As a result, the researchers have their arbitrary classification principles which are not general rules. Furthermore, CNT-metal dispersion methods are progressively updated and modified, so that the sharp boundaries between them are gradually faded. Table 1 summarizes some of the common dispersion methods among which the milling, ultrasonication, the application of surfactants, and metallization are explained in the next sections. Also, their possible challenges and suggested solutions are extensively discussed.

Ball Milling

In this procedure, a milling container with several hard balls is used for dispersing CNTs inside the metallic systems. During the milling, metal powders become fractured and welded again, so that CNTs are entrapped between them [15,36]. Since the primary powders as well as CNT bundles are entrapped between balls or balls and the container wall, the grinding motion of milling balls separates the nanotubes from each other, destroys the agglomerations, and disperses CNTs among the metallic powders. The milling time is one effective processing factor determining the degree of dispersion. As the mixing proceeds, the agglomeration is alleviated. Moreover, in the initial steps of milling, metallic powders have round shapes, and CNTs are dispersed at their surfaces. However, as the mixing continues, soft metallic powders flatters, and CNTs are embedded in the powders [15,36]. Generally, the mechanical milling is less able to inhibit CNT agglomerations, as seen in Figure 3 [37].

Figure 3. CNTs agglomeration in 5 wt % CNT-Al blends processed by mechanical milling for 30 min [38] (Adapted with permission from [38], Elsevier, 2011).

Table 1. A short description of CNTs (carbon nanotubes) dispersion methods in metal matrix nanocomposites, their advantages and disadvantages.

Dispersion Method	Principles	Advantages	Disadvantages	Studied Metallic Systems
Ball milling	In this method, CNTs and metallic powders are milled in a container with hard balls.	- Mass production - Commercial applications - Easy processing - Low cost - Possibility of in-situ functionalization during the mixing - The possibility of in-situ alloying	- Generation of defects in CNTs - Cutting or breaking of CNTs - Inclusion of contaminants into the system - Long-time procedure	Al [39], Ni [40], Mg [12], Cu [41], Fe [42], and their alloys
Ultrasonication	In this method, CNTs are dispersed in a solution composed of organic solvents or aqueous surfactant by using high frequency sound waves.	- Mass production - Commercial applications - Easy processing - The possibility of in-situ functionalization or SDS (sodium dodecyl sulfate) treatment during the mixing - Low cost	- Generation of defects in CNTs (lower than the mechanical milling) - Low CNT content is homogeneously dispersed	Al [14], Cu [43], Fe [44] and their alloys
Surfactants	In this method, non-covalent surfactants are embedded on CNTs surfaces with the aim to decrease the CNTs affinity to stick together.	- Mass production - Commercial applications - Easy processing - The possibility of in-situ functionalization or SDS treatment during the mixing - Low cost	- Low CNT content can be uniformly dispersed - Need for a mechanical method to disperse CNTs throughout the metallic powders	Al [45], Ti [18], Ni [46], and their alloys such as CuZr and CuZrAl [47]
Metallization (i.e., decoration with metals)	In this method, CNTs are coated with a metallic layer such as Cu, Ni, Co, Mo, and W before their introduction into the matrix.	- High efficiency (depending on the dispersion method used for metalized CNTs)	- Need for an additional mixing method such as mechanical mixing for producing dense parts	Al [16] and Cu [48]
Hetero-agglomeration principle	In this method, oppositely charged CNTs and metallic powder are co-deposited. CNTs and metallic powders are charged through a chemical step such as ultrasonication in acidic media.	- The possibility of uniformly dispersing CNTs up to 5 vol % - Mass production - Production of CNT-containing systems with no damage	- Relatively high cost	Al [17]
Molecular level dispersion	A chemical or physical reaction occurs between functionalized CNTs and metallic ions in a solution medium.	- The possibility of uniformly dispersing high CNTs content (up to 10–15 vol %) in metallic matrices - The possibility of embedding CNTs inside the metallic powders rather than dispersing them at particles surfaces	- Practical complexity - Limited production scale	Cu [49,50], Ag [51], Ni [52], AgPd [53], Al [54], and AlCu [55]
In-situ chemical synthesis	In this method, CNTs are directly synthesized from the vapor phase on the metallic powders.	- Mass production - The possibility of automatization - The possibility of uniformly dispersing a high CNTs content in metallic systems	- Need for expensive equipment	Mg [56], FeCr [57], Cu [58], AlCu [59], Al [60], and Fe [61]

Table 1. *Cont.*

Dispersion Method	Principles	Advantages	Disadvantages	Studied Metallic Systems
Electroless deposition	In this method, CNTs are immersed in an electrolyte solution containing metal ions. During the process, the nanotubes are covered by ions through developing potential. A substrate can be used on which CNTs and metallic ions are simultaneously deposited.	- Commercial applications - High efficiency - The possibility of combination with other mechanical mixing techniques such as ultrasonication for better CNTs dispersion - The possibility of uniformly dispersing other components	- Only used for some metals - High cost - Need for a supporting mixing method to obtain a full efficiency	Cu [62], Ni [63], Al [64], Au, and Pt [65]
Electrodeposition	In this method, CNTs are embedded in an electrochemically formed metallic film. This film is deposited as metallic ions transfer from cathode or electrolyte solution to the anode.	- Commercial applications - High efficiency - The possibility of combination with other mechanical mixing techniques such as ultrasonication for better CNTs dispersion - The possibility of uniformly dispersing other components	- Only used for certain metals - High cost - Not easy to handle - Need for a supporting mixing method to obtain a full efficiency	Ni [66], Cu [67], Sn-Bi [68], Cu-Ni [69], and Ni-Co [70]
Using hybrid reinforcements	In this method, CNTs are added to other components such as SiC, A_2O_3, and graphite for better distribution. The ball milling is the most conventional method for mixing the hybrid reinforcements.	- The possibility of obtaining higher mechanical properties due to strengthening effects of hybrid reinforcements - High efficiency - Cost-effective	- Need for an additional mixing process such as ball milling - The possibility of formation of unwanted intermetallic compounds	Al [71], Cu [72], Mg [73]
Preform infiltration	In this method, an initial preform of CNTs and other components is fabricated, and the molten metal is infiltrated into it.	- The possibility of selectively dispersing CNTs at a macro level - High efficiency to prevent the agglomeration	- Metallic component needs to be melted - Restrictions imposed by shape geometry - Need for a two-step procedure - Relatively high cost	Al [74,75]
FSP/SLM induced dispersion	After the fabrication of bulk CNT/metal nanocomposites, friction stir processing (FSP) or selective laser melting (SLM) is used for re-dispersing CNTs.	- High efficiency (depending on the dispersion method used before the fabrication)	- Imposition of additional cost - Only used for finished CNT/metal bulk nanocomposites	Al [76], Cu [77]
Induction melting-based dispersion	In this method, a solid metal with a CNT-containing flux is poured into an induction furnace. The presence of such a flux decreases the melting temperature of the solid metal. Finally, CNTs are dispersed by induction-formed fluctuations.	- Mass production - Easy to handle	- Additional cost due to melting of the used metal - Need for expensive instruments	Al [78], Ti [79]
Casting-based methods such as stir casting and squeeze casting	In these methods, CNTs are dispersed in a molten metal through stirring or squeezing.	- Mass production - Easy to handle - Comparatively low cost	- Additional cost due to the melting of the metals - Comparatively low efficiency	Al [80]

To improve the CNT dispersion in metallic matrices, a process controlling agent (PCA) such as polyacrylic acid (PAA) or methanol is usually used. These chemicals are utilized not only for better dispersion, but also for CNTs functionalization (Figure 4).

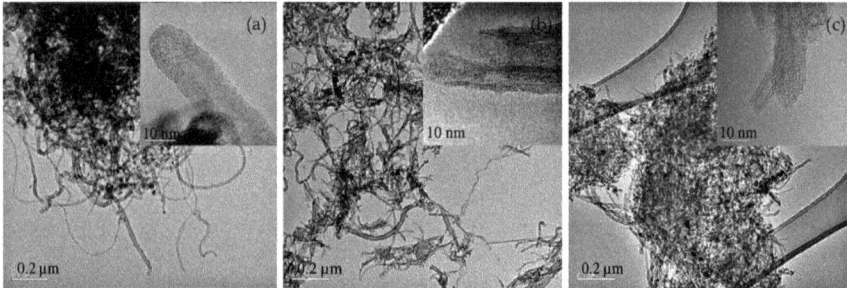

Figure 4. TEM (transmission electron microscopy) images of (**a**) as-synthesized CNTs, (**b**) CNTs ball milled with NH_4HCO_3 as a functionalizing agent and process control agent (PCA), and (**c**) CNTs milled with no PCA. The presence of chemicals is shown to result in improved CNTs dispersion [81] (Reproduced with permission from [81], Elsevier, 2008).

Despite the ability to functionalize CNTs, the mechanical milling is capable of in-situ alloying as a subsidiary mechanism for strengthening CNT-metal matrix nanocomposites. Moreover, it can produce the metallic nanostructures whose mechanical strength is higher than that of conventional nanocomposites. Such grain structures arise from the formation of large numbers of dislocations inside the metallic powders due to high plastic deformation.

As another advantage, a close control over the particle size and particle size distribution in ball milled powder systems is possible. These characteristics are controlled by two competing mechanisms during the milling: (i) Cold working, which results in decreased ductility, particle fracturing, and reduced particle size; and (ii) Cold welding, which results in increased particle size [40,82,83].

Besides the aforementioned advantages, the mechanical milling may induce structural damages to CNTs such as breaking or shortening, especially when the dwell time is extended. These defects confine the processing time [84]. They may be also formed when the milling is conducted under inappropriate practical conditions. The non-controlled milling results in formation of carbides through a chemical reaction between the damaged nanotubes and reactive metallic powders such as Al. Such reactions may occur even at comparatively low milling temperatures.

Among the various milling apparatuses, the planetary milling imposes less energy on metallic powders and CNTs, especially in comparison with high-energy mechanical milling. The higher the energy imposed on the powder blend, the higher the volume fraction of defective nanotubes will be [82]. In contrast, high energy milling can evenly embed CNTs in the metallic powders, while the planetary milling is usually incapable of uniformly dispersing CNTs. It may persuade the CNT agglomerations and locate them at the external surface of metallic powders rather the diffusion into them. Furthermore, a useful combination of practical conditions is required to obtain high-quality nanocomposites [85]. Figure 5 shows the way the nanotubes may be mechanically dispersed throughout Al matrix.

It seems that the utilization of a chemical modification or a processing control agent as well as the low energy milling technique can successfully reduce the structural damages of CNTs [82].

Another practical method to prevent the chemical reactions during the milling is the suppression of heat-induced changes in metallic powders through cryogenic techniques. Cryogenic milling is referred to as the milling procedure at very low temperatures (usually below 100 K). Such temperatures give rise to the increase in brittleness of metallic powder, reduction in grain growth, resultant grain refinement, and decrement in grain size. In this case, the hardness of CNT-metal matrix nanocomposites increases by the Hall-Petch mechanism. Therefore, such methods are able to reduce the processing time and bring forth finer grains with the lowest possibility of chemical reactions. Interestingly, if the cryogenic milling is conducted with the presence of carbon-containing PCA such as stearic acid, their carbons may react with metallic powders and form carbides during the milling or subsequent sintering processes [82].

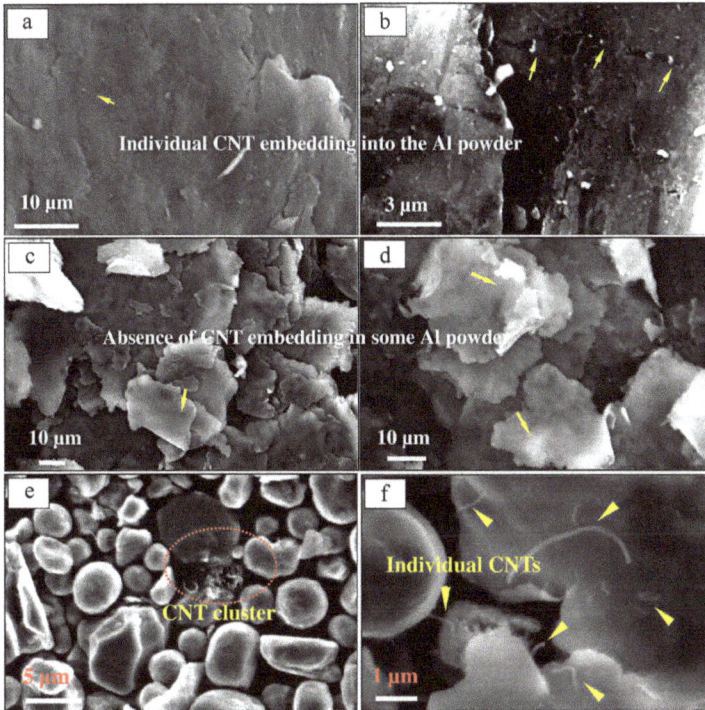

Figure 5. SEM images of a CNT-Al blend after milling by high energy milling (**a–d**) and low energy milling (**e,f**) [85]. The CNTs are uniformly dispersed and embedded in Al powders during high energy ball milling, as shown by yellow arrows (Reproduced with permission from [85], Elsevier, 2011).

The CNTs are uniformly dispersed and embedded in Al powders during high energy ball milling. However, their dispersion has not a uniform pattern, so that there are some Al powders with no CNT. Conversely, all Al powders may obtain CNTs during the milling by low energy milling, but CNT agglomerations still remain, and CNTs are located at the surface of Al powders [85].

In summary, if the best equipment as well as better control over processing conditions is utilized to disperse CNTs among the metallic powders, each method has its own intrinsically certain ability, so that if the CNT loading exceeds the optimal value of each technique (4–5 wt % CNT at the best conditions), the nanotubes tend to form agglomeration. In this case, the pores and poor interfacial bonding can degrade the mechanical properties as well as thermal and electrical characteristics of manufactured CNT-containing metal matrix nanocomposites [86,87]. Consequently, many researchers believe that such a dispersion method and arbitrary practical conditions have low potential to uniformly disperse CNTs in metallic matrix. Some pros and cons of mechanical milling are listed in Table 1.

Ultrasonication

In this method, sound waves with high frequency are used for dispersing CNTs in a solution composed of organic solvents or aqueous surfactants. In order to increase the solubilization of CNTs in water, the hydrophobic CNT surfaces can be modified by a wide range of surfactants and polymer adsorbates. The mechanism by which the ultrasonication debunds CNTs in a solution containing an aqueous surfactant is composed of four step: (i) A high local shear stress at the end of CNT bundles is induced by the collapse of micro-bubbles formed during the ultrasonication; (ii) Surfactants diffuse into and absorb on CNTs as soon as a gas bubble or gap is formed; (iii) The unzipping process takes place along the longitudinal axis of a particular CNT; (iv) The individual surfactant-coated CNT is released inside the solution [88].

It is worth mentioning that unzipping the individual CNTs from the bundles will not be completed if there is no polymer absorbate or surfactant. This is because CNTs have van der waals attraction to each other which may again [88].

However, the formation of hotspots during the ultrasonication which simultaneously increases the local pressure and temperature may induce some defects in CNTs, form carbon-dangling bonds at their sidewalls, and cut or chop the nanotubes. Such drawbacks decrease the effectiveness of the ultrasonication as a useful method for dispersing CNTs in CNT-containing metal matrix nanocomposites. In order to reduce these disadvantages, organic molecules, for example, poly (methyl methacrylate) (PMMA) or monochlorobenzene (MCB) can be added to form reactive species during the sonication. These species which are generated from decomposition of added organic molecules at the hotspots can chemically react with dangling bonds of CNTs and are easily wiped through burning in oxygen gas. This event may leave holes on the sidewalls [89].

As a holistic conclusion, the ultrasonication has a finite ability to disperse CNTs in metal matrix nanocomposites. According to the literature, the acid-treated CNTs can be uniformly dispersed up to 1 wt % in metallic matrices (such as Al) by this technique. Further addition of CNTs to the matrix results in agglomeration and generation of micropores during the subsequent consolidation stage. These defects may significantly decrease the mechanical, electrical and tribological properties of these binary systems [13,14].

The Application of Surfactants

Active surface agents such as surfactants can significantly improve the surface conditions of CNTs and prevent from their agglomeration and re-bonding. In general, these surface agents consist of two parts: (i) A hydrophilic part with a polar head and (ii) A hydrophobic part coupled with a hydrocarbon chain at the tail of the surfactant molecule. These modifiers enhance the metastability of the colloidal suspensions by providing an electrostatic and/or steric repulsive force between CNTs and reducing their surface energy. Depending on their head group charges, the surfactants are usually classified into three different groups: cationic, anionic, and nonionic or zwitterionic [90]. One of the most conventional anionic surfactants for CNT-reinforced metal matrix nanocomposites is SDS (soduim dodecyl sulfate), wherein the sulfates and hydrocarbon chains act as hydrophilic and hydrophobic parts, respectively. In fact, CNTs can interact with the hydrophobic part of SDS, so that an electrostatic repulsion force is formed between the negatively charged CNTs and the hydrophilic sulfate groups of SDS. Thus, it can remarkably affect the stabilization of CNTs in aqueous media [28]. There are quite a few studies evaluating the application of SDS as a surfactant in metallic matrices. The surveys confirm that there is a defined limit for inclusion of CNTs into the matrices. This is because if the critical level is exceeded, the uniform dispersion of CNTs throughout the metallic matrices is not possible and further incorporation leads to the formation of agglomerated particles [45]. The agglomerated particles can impressively degrade the mechanical properties and functionality of the nanocomposites. However, the introduction of SDS into the metallic systems can improve the CNTs dispersion.

Zwitterionic surfactants are a new class of widely used surface agents for dispersing CNTs in metal matrices. Whereas they are polar and soluble in water, their solubility in organic media is limited. They chiefly consist of cationic and anionic parts attached to the same molecule. The self-assembled monolayers (SAMs) of these surfactants can be also formed on CNT bundles (Figure 6). In this case, the strong dipole/dipole attraction between SAMs and CNTs overcomes the present van der Waals forces between the individual nanotubes and disassembles the ubiquitous bundles [91,92].

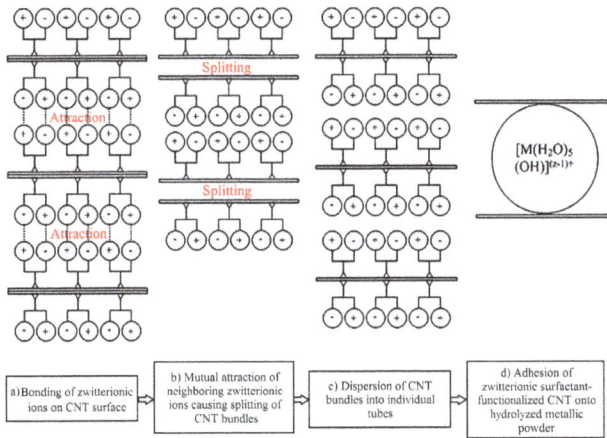

Figure 6. A schematic illustration of the mechanism by which zwitterionic surfactants disassemble CNT bundles [47] (Reproduced with permission from [47], Elsevier, 2014).

The experimentations show that if the CNTs content exceeds 1 wt %, the agglomerated particles will form all over the microstructure of the nanocomposites. Moreover, the processing parameters deeply affect the way the nanotubes are dispersed throughout the matrix, where the mechanical properties can be degraded because of employing an unsuitable process. It is ascribed to the formed pores and degraded interfacial adhesion due to evaporation of the surfactant at high temperatures. Post treatments (e.g., heat treatment) is often employed to remove such pores [92].

Another disadvantage is the lack of control on the reactions between CNTs and the metal matrices, where the intermetallics can be generated as a result of unwanted reactions during the sintering process [92].

It seems that there is an urgent need for other surface treatments such as coating to uniformly disperse the high volume fraction of CNTs throughout the metal matrices.

Metallization

The outermost surface of CNTs can be coated by several metallic elements such as Cu, Ni, Co, Mo, and W, if enhanced interfacial adhesion between CNTs and the matrix is greatly required. Such coatings may contribute to the prevention of CNTs agglomeration all over the matrix [93,94]. It is demonstrated that W-coated CNTs by metal-organic chemical vapor deposition (MOCVD) method can significantly inhabit the CNTs agglomeration. This fact is shown in Figure 7. As clearly seen, the pores or air bags at CNT/metal interface is removed as the main reason for improved interfacial adhesion [48]. The enhanced thermal conductivity of the composites is ascribed to the formation of desirable interface which can remarkably reduce the thermal lost.

Similarly, it is shown that there is no intermetallic formation in the microstructure of (Mo-coated CNT)-Al nanocomposites [16]. It is attributed to the presence of Mo layer, acting as an appropriate barrier to the interfacial reaction between CNTs and the matrix. Additionally, the applied coating is able to impressively increase the volume fraction of the uniformly dispersed CNTs. The mechanical properties of these composites are proved to improve due to desirable dispersion of CNTs throughout the matrix. Despite the improved mechanical behavior, the exploitation of such coatings may slow down the electron movement all over the nanocomposites and deteriorate the electrical conductivity.

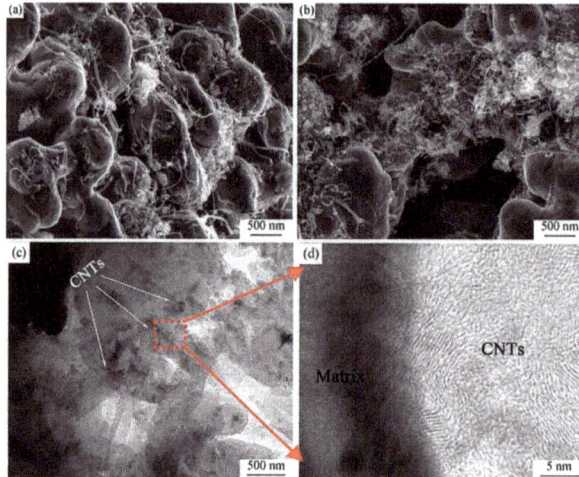

Figure 7. SEM micrographs of (W-coated CNT)-Al nanocomposites fabricated by ultrasonication method: (**a**) 5 vol % CNTs, (**b**) 10 vol % CNTs, (**c**) TEM micrograph of (W-coated 5 vol % CNT)-Al nanocomposites, and (**d**) higher magnification of the selected area in (**c**) [48] (Reproduced with permission from [48], Springer, 2012).

Effects of Dispersion Methods on Physicomechanical Properties

In addition to the consolidation and sintering techniques, the dispersion methods may greatly influence the properties of CNT-containing metal matrix nanocomposites. A review on the literature confirms that whenever the used dispersion method is inappropriately selected, so that it keeps or forms CNT aggregates, the mechanical and functional properties of produced nanocomposites may be degraded due to the formation of pores and poor interfacial boundaries between nanotubes and metallic matrix. Figure 8 shows the hardness of some spark plasma sintered (SPSed) CNT-Al nanocomposites as a function of CNT content for different dispersion methods. The aim behind the selection of SPS as the sintering method and Al as the metal matrix is to eliminate the potential effects of sintering process and chemical composition of metal matrix on the properties of these systems as much as possible. As clearly seen, the highest hardness values can be achieved by the mechanical milling due to the formation of severe plastic deformation-induced dislocations and defects in Al matrix. However, if the dispersing process is not suitably controlled, its intrinsic potential will be obtained. For instance, some dispersion techniques result in decreased properties when the CNT content exceeds 1–1.5 wt %. As another example, if the metallization is not done in a suitable manner, or the coating layer is not truly selected for the metal matrix, it cannot enhance the mechanical properties. Such a justification is also applicable to ultrasonication. If the ultrasonication is carried out under non-controlled conditions, the CNTs cannot play their positive role and reduce the mechanical properties after adding a low CNT content.

The tensile strength of MMNCs-reinforced metal matrix nanocomposites has similar trends. The incorporation of CNTs increases the tensile strength of metallic matrices, if they are homogeneously dispersed and make a proper interfacial bonding with the matrix. However, by adding CNTs more than the critical content, a heavy agglomeration occurs and the tensile strength is degraded (Figure 9).

Similar trends are also seen in the thermal conductivity of SPSed CNT-Cu or Al matrix nanocomposites (Figure 10). Generally, the addition of CNTs to the metallic matrices decreases the thermal conductivity due to the grain boundary-induced electron scattering. However, the metallization can increase such a property through enhanced interfacial boundaries between nanotubes and metallic matrix. Besides the metallization, the polymer wrapping and acid treatment can decrease the degradation of thermal conductivity of Cu matrix by adding CNT due to the formation of enhanced interfacial boundaries. Such methods also have a limited potential to disperse CNTs. If the CNT content exceeds an optimum limit, the agglomeration of CNTs will decrease the thermal conductivity of these binary systems.

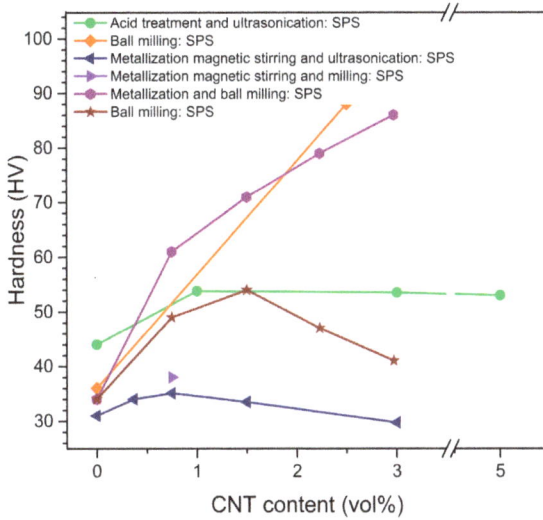

Figure 8. Hardness of some spark plasma sintered (SPSed) CNT-Al matrix nanocomposites fabricated by various dispersion methods [13,16,36,95].

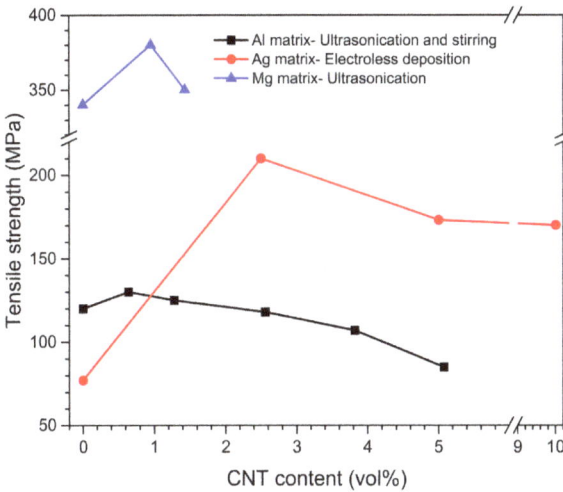

Figure 9. The tensile strength of some SPSed CNT-metal matrix nanocomposites fabricated by various dispersion methods [14,96,97].

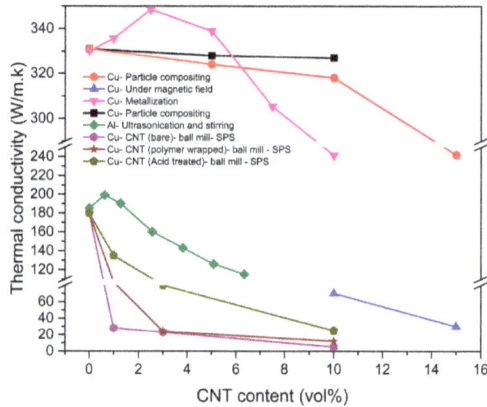

Figure 10. Thermal conductivity of some SPSed CNT-Cu and CNT-Al matrix nanocomposites fabricated by various dispersion methods [48,98–103].

2.2. Consolidation Challenge

The consolidation of pure metallic powders and their alloys is not a critical challenge, and pure metallic parts with the full density can be obtained by various sintering methods such as hot pressing, hot isostatic pressing, and spark plasma sintering. However, the addition of CNTs to the metallic matrix may greatly affect the consolidation rate and relative density of the binary system. In fact, the poor wettability between CNTs and metallic matrices and the formation of CNT agglomeration-induced pores impede the densification and decrease the relative density of the nanocomposites; no matter what the sintering method is. As a result, producing a fully dense CNT-reinforced metal matrix nanocomposite is a critical challenge which highly depends on the dispersion of CNTs in the metallic matrix. Actually, the selection of a proper dispersion method capable of homogeneously dispersing CNTs can approach the relative density of the CNT-metal bulk systems to that of pure metals.

Figure 11 shows the relative density of different CNT-reinforced metal matrix nanocomposites as a function of CNT content and dispersion method. As seen, the CNT ddition usually decreases the relative density and sinterabillity of the metallic matrices. Moreover, an increment in CNT content generally brings about more CNTs agglomeration and pores, decreasing the relative density. However, if a proper dispersion method such as the metallization for Cu and electroless deposition for Ag is utilized, CNTs will be homogeneously dispersed in metallic matrices with no declined density. It is noteworthy that each dispersion method has a certain ability to homogeneously disperse CNTs in a specific metallic matrix. If the CNT content exceeds the maximum ability of a mixing method, the agglomeration will occur. For instance, as seen in Figure 11, CNTs up to 2.5 vol % are homogeneously dispersed in Ag matrix through the electrodeposition method. Further an increase in CNT content may degrade the consolidation by inducing the agglomerates and pores.

As a conclusion, the consolidation of CNT-containing nanocomposites highly depends on the uniformity degree of dispersed CNTs. It implies the importance of the dispersion method to achieve a fully dense part. The employment of a suitable dispersion method and addition of CNTs below the critical content may assure the manufacturing of fully dense CNT-reinforced metal matrix nanocomposites.

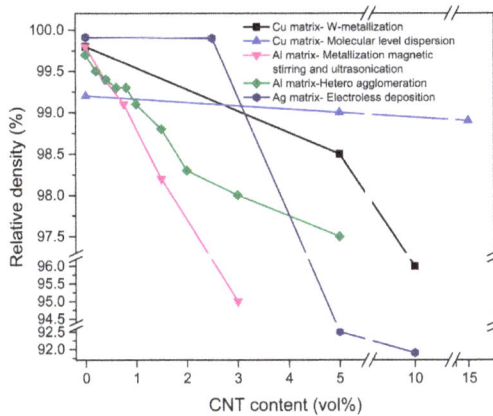

Figure 11. The relative density of some SPSed CNT-metal matrix nanocomposites fabricated by different dispersion methods [16,48,49,96,104].

2.3. Chemical Reaction with Metallic Matrix

2.3.1. Thermal Decomposition of Pure CNTs

A review on the literature confirms the effective function of CNTs on the structural and functional characteristics of CNT-containing bulk materials. To exploit the full potential of CNTs in nanocomposites, one should deeply understand the structural phenomena and microstructural interactions likely occurring during the synthesis or post-treatments of these new emerging materials. The first key to this issue is the profound consideration of structural transitions and phase transformations of CNTs in pure state and in exposure to a metallic matrix, because they can adversely affect the intrinsically superior properties of pure CNTs [105–109]. Table 2 summarizes the recently reported structural transitions of pure CNTs in a variety of densification methods and their influences on the properties. As seen, less empirical results are reported in the literature addressing these structural changes. In general, any allotropic transition in CNTs is highly undesirable, because it can limit the intrinsically unique features of nanotubes in electronic aspects and load bearing performances [110–112].

An overwhelming majority of reports about the conversion of CNTs to other allotropes is associated with spark plasma sintering, where high pressure and elevated sintering temperatures persuade any thermally activated phase transformation [105,106,113,114]. However, other consolidation techniques such as hydrogen plasma [115], high-pressure/high-temperature annealing heat treatment [116–118] and chemical vapor deposition (CVD) [107,108,119] can thermally decompose the structural geometry of nanotubes. As a practical conclusion, the engineers often strive to prevent from the thermal degradation of CNTs in both pure state and composite form, since it can vanish the final properties [120]. Hence, one can determine two temperature regimes for CNT allotropic phase transformations depending on whether CNTs are preserved: (i) low-power regime and (ii) high-power regime. In the first case, known as safe regime, the processing pressure and temperature are sufficiently low, so that the thermal power cannot destroy the integrity and tubular morphology of nanotubes and thermally activate the mechanisms by which CNTs can structurally convert to other allotropes. In the second case, the elevated temperature or high applied pressure can persuade the thermal decomposition of nanotubes. There is not any sharp temperature boundary between these regimes, depending on the kind of consolidation procedures, type of matrix exposed to CNTs, and activated sintering mechanisms [121].

Table 2. Overview over a variety of conventional manufacturing techniques, allowable processing windows and structural changes of pure CNTs and their effects on the physicomechanical properties of nanotubes.

Fabrication Process	Processing Temperature (°C)	Applied Pressure (MPa)	Processing Duration (min)	Final Phase Structure	Reported Mechanisms for Structural Changes	Reported Changes in Properties Due to CNTs Structural Changes	Ref.
Spark plasma sintering	1500	80	20	diamond	Not reported	Not reported	[105]
Spark plasma sintering	1500	80	20	diamond + graphite	Not reported	Not reported	[113]
Spark plasma sintering + polishing	1600	60	1	graphene	Shear stress induced by polishing which leads to peeling graphene away from CNTs	A significant reduction in friction coefficient due to the much lower friction between formed graphene layers	[106]
Spark plasma sintering	1000, 1500, 2000	50	5	graphene	Not reported	Enhanced thermal stability due to the formation of graphene at a high sintering temperature	[114]
Annealing heat treatment	(i) Below 2000 (ii) 2100–2400 (iii) 2500–2800	-	30	- No structural changes - the formation of large-sized double-walled carbon nanotubes (DWCNTs) - the graphitization of multi-walled nanotubes (MWNTs) and other kinds of crystalline sp² hybridized carbon structures -	Not reported	Not reported	[116]
Spark plasma sintering	1500	80	20	- Partial conversion of CNTs into diamond - The formation of n-diamond	High localized temperatures arisen from the presence of plasma during the sintering as a prerequisite for the formation of diamond	Not reported	[122]
Spark plasma sintering	1700, 1800, 1900, 2000	100	5	The presence of graphene in all targeted temperatures	The formation of graphene nanosheets due to the evaporation of carbon atoms during the heating stage of SPS and their agglomeration during the cooling stage	An increment in electrical conductivity due to the formation of graphene nanosheets	[123]
Spark plasma sintering + Fe85Ni catalysts	1200	70	20	diamond + graphite	The formation of diamond through a layer-by-layer mechanism	Not reported	[124]
Chemical vapor deposition	700	-	30	graphite	Not reported	Not reported	[107]

Table 2. *Cont.*

Fabrication Process	Processing Temperature (°C)	Applied Pressure (MPa)	Processing Duration (min)	Final Phase Structure	Reported Mechanisms for Structural Changes	Reported Changes in Properties Due to CNTs Structural Changes	Ref.
Chemical vapor deposition	700	-	56	amorphous carbon	Not reported	Not reported	[108]
Compaction + laser beam heating	2226	17,000	-	diamond	Direct conversion of CNTs to nano-sized diamonds with no an intermediate step such as melting or dissolution	Not reported	[125]
Annealing at high pressure	1300	4500	12	diamond	The conversion of CNTs to quasi-spherical onion-like structures followed by the formation of diamond crystals	Not reported	[117]
Chemical vapor deposition	500	-	2160	highly ordered graphite	CNT → carbon nanofiber → highly ordered graphite	Not reported	[119]
Electrical breakdown method	-	-	-	graphene nanoribbon	Fast unwrapping of MWNTs with no an intermediate step	Not reported	[126]
Hydrogen plasma	726	-	-	diamond	The formation and growth of diamond particles due to the formation of amorphous carbon clusters with sp^3 bonds	Not reported	[115]
Solution-based oxidative process	55–70	-	60	graphene nanoribbons + graphite oxide	The formation of graphene nanoribbons due to complete unravelling of CNT side walls	Not reported	[109]
High temperature annealing heat treatment	1100–2000	-	240 min heating + 480 min cooling process	graphite nanoribbons	The thermal activation of three different mechanisms for different heat treatment temperatures: − The coalescence of SWNTs at 1400 °C followed by rearrangement of the present bonds at 1600 °C to form MWNT − The collapse of MWNTs into graphite nanoribbons at temperatures higher than 1800 °C − Complete conversion of MWNTs to graphite nanoribbons at temperatures higher than 2000 °C	Not reported	[118]
High pressure and high temperature heat treatment	1600–2000	≥15,000	10–10,000 s (Exact dwelling time has not been reported)	diamond	The formation of diamond phase as a result of atomic diffusion with no graphitization or formation of intermediate phases	Not reported	[127]

In summary, CNTs may transform into other thermodynamically stable allotropes during the heat treatment process. Among these allotropes, cubic diamond [105], n-diamond [122], mono/multi-layer graphene [106], and graphite sheets [107] are the most extensively reported. However, the way these allotropes can affect the final properties are highly contradictory. Whereas some literatures refer to the deterioration of physicomechanical properties of CNT-based bulk materials, others put emphasis on the drastic improvement of some properties [106,123]. On the other hand, there is no consensus among researchers regarding what temperature range guarantees the conversion of CNTs to other carbon allotropes and which forms of new phases can be formed. For instance, Zhang et al. [123] evaluated the microstructure, mechanical and electrical properties of spark plasma sintered CNTs in pure state and demonstrated the in-situ formation of graphene nanosheets at 1700–2000 °C. It was shown that the formed graphene nanosheets are responsible for improved electrical conductivity in these conditions. Figure 12 indicates the TEM image of this formed graphene nanosheet at 2000 °C.

Figure 12. TEM image of graphene nanosheet formed during spark plasma sintering at 2000 °C [123] (Adapted with permission from [123], Elsevier, 2013).

2.3.2. Chemical Reactions

The chemical reactions between CNTs and metallic matrices can be categorized into two different aspects: (i) The partial chemical reaction at CNT/metal interface and (ii) the complete chemical reaction of CNTs with metallic matrix. If controlled, the former is favorable for the load bearing applications, but the latter can bilaterally affect the physicomechanical properties of CNT-dispersed metal matrix nanocomposites. In the next sections, the practical and scientific aspects of these chemical reactions will be addressed in some details.

Complete Chemical Reactions

(a) CNT-Reinforced Al Matrix Nanocomposites

A review on the literature confirms that Al_4C_3 is the most common in situ formed intermetallic in CNT-reinforced pure Al matrix nanocomposites. However, other intermetallics such as SiC can be formed during the reaction between the alloy matrix and CNTs. In fact, the chemical composition of the formed intermetallics strongly depends on the fabrication temperature and the chemical composition of the alloy matrix. The kinetics of the carbide formation at temperatures lower than Al melting point is negligible. Moreover, the presence of some alloying elements can significantly control the final properties of the nanocomposites. Two different sources are suggested for the formation of Al_4C_3: (i) Used process control agents (PCA) during the milling process such as stearic acid and (ii) incorporated CNTs. Also, the processing parameters such as temperature, post treatment variables and the size of initial CNTs play an important role in the formation of carbides and controlling their volume fraction, morphology, and geometrical dimensions. It is found that the morphology of Al_4C_3 can be easily changed by applying the post heat treatment or the presence of initial defects in CNT structures. Four types of morphologies including whisker-like, dumbbell-like, needle-like, and tube-like ones can be observed for Al_4C_3 particles, depending on the post treatment conditions and the purity of initial CNTs. Similarly, the volume fraction of the formed carbides deeply depends on the geometrical size of initial CNTs. Generally speaking, the presence of carbides at CNT/Al interfaces improves the mechanical properties of the nanocomposites. It is attributed to the enhanced load transfer between the matrix and reinforcements as a result of carbide formation at the interface. The strengthening effect of the formed carbides depends on their thickness, so that the formed carbides will dissolve back into the melt if the critical thickness is less than a critical limit.

The in situ formation of Al_4C_3 at the Al/CNT interface has been subjected to the extensive research studies. Some research works have considered the precipitation of Al_4C_3 as one of the predominant strengthening mechanisms in CNT-dispersed Al matrix nanocomposites. This compound forms during a chemical reaction between Al matrix and CNTs as:

$$4Al + 3C = Al_4C_3$$
$$\Delta G = -289{,}512 + 60T, \ T < 660 \ ^\circ C \tag{1}$$

The free energy (ΔG) of this reaction is negative by the melting point (660 °C) of Al. It means that Al_4C_3 is thermodynamically stable at temperatures lower than 660 °C. Coincidently, the formation of this carbide at temperatures lower than 660 °C is also reported [128].

The mentioned reaction can be explored for ball milled CNT-Al powder blend. In this case, Al_4C_3 can be formed when the energy state of the mixed powders reaches to a sufficiently high level. It is believed that the long-term milling process can seriously damage the CNTs structure and lead to the in situ formation of Al_4C_3 [87,129]. For instance, Ostovan et al. [87] evaluated the microstructure of the milled powders for 8 and 12 h. It was found that the needle-like Al_4C_3 is just formed after the milling for 12 h. It was attributed to the long-term milling-induced damage in CNTs. As a basic conclusion, the possible CNT amorphization during the sintering process as well as the formation of crystallographic defects in CNTs can significantly facilitate the formation of Al_4C_3. It is ascribed to the synergetic effects of instability of milled powders after the long-term milling, and generated driving force as a result of thermal processing such as sintering [87].

The literature have suggested two different sources for Al_4C_3 formation: (i) Used process control agents (PCA) during the milling process, (e.g., stearic acid) and (ii) incorporated CNTs. Since the stearic acid is composed of oxygen, carbon, and hydrogen atoms, it can be decomposed into its constituent elements during the long-term milling. On the other hand, the milling process can significantly damage the CNTs structure and enhance their amorphization [130]. The produced atomic carbon from the mentioned sources can easily react with the matrix Al to form Al_4C_3. More importantly, the formation of Al_4C_3 may consume the CNT walls or the entire CNT [129].

If the alloy matrix is used, other intermetallics can be formed beside Al_4C_3. The chemical composition of the formed intermetallics strongly depends on the alloying elements. For instance, Al_4C_3 and SiC can be formed as a result of the reaction between Al-Si alloy matrix and CNTs. Bakshi and coworkers [131] showed that the formation of each intermetallic corresponds to the Si weight percent at the fabrication temperature, as shown in Figure 13. As seen, the more the Si content in the alloy, the more SiC particles will form. It is ascribed to increased amount of Si in the liquid melt due to the reaction between Al and C to form Al_4C_3. It can increase the activity of Si to form SiC. Both reactions will proceed in the same manner until all of carbon content is completely exhausted.

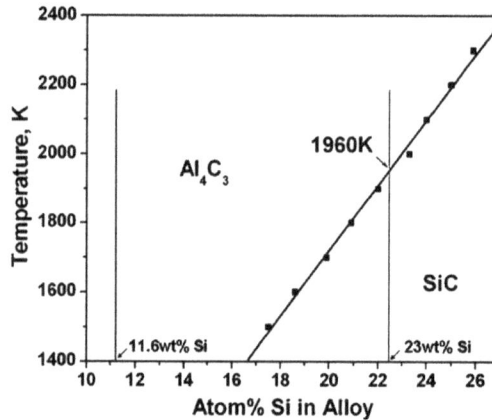

Figure 13. The chemical composition dependence of formed intermetallics on Si weight percent and fabrication temperature [131] (Reproduced with permission from [131], Elsevier, 2009).

In CNT-(Al-Si) nanocomposites, the poor wettability arises from a major difference between the surface tension values of the alloy matrix (about 800 mN·m^{-1}) and CNTs (about 45 mN·m^{-1}) [33]. However, the literature have reported improved wettability with the generation of some intermetallics. It is attributed to reduce contact angle between the alloy matrix and CNTs through the infiltration of Al-Si melt into CNT clusters [131].

The formation kinetics of Al_4C_3 at temperatures lower than Al melting point is slow. For the binary composites in which CNTs react with Al matrix chemically, the presence of alloying elements can significantly control the final properties, because the chemical composition may affect the thermodynamic aspects of the Al_4C_3-forming reaction [132].

The processing parameters such as temperature, post treatment variables and the size of initial CNTs play an important role in the formation, volume fraction, morphology, and size of the formed carbides in CNT-Al matrix nanocomposites. These compounds can be precipitated at different locations. For instance, the nano-sized Al_4C_3 particles often form on the external surface of CNTs as a result of a reaction between Al and CNT-derived amorphous carbon. In the case of partially graphitized CNTs, the carbides are evolved at the end of damaged CNTs. It is shown that only a little amount of MWNTs can react with the matrix to generate carbides and contribute to enhanced properties of the composites [133].

There are two main factors affecting the morphology of the formed compounds during the reaction between the matrix and CNTs: (i) Post treatment, and (ii) internal defects of CNTs. The morphology of Al_4C_3 precipitates strongly depends on post heat treatment temperature. It has been reported that the carbides grow in the form of nanosized whiskers at 950 °C, while they are more likely to disappear at comparatively lower temperatures [133]. Additionally, one can suppress the formation of these carbides through the close control over the sintering conditions, i.e., the sintering temperature should be kept below Al melting point [134]. Figure 14 exhibits the elongated Al_4C_3 precipitates formed as a result of a high-temperature heat treatment. As obviously seen, the longer the heat treatment temperature, the more elongated the carbides are.

In addition to the processing parameters, the internal defects of nanotubes can also change the way the precipitates are evolved. Two fascinating types of morphologies are reported for Al_4C_3 particles, i.e., dumbbell-like and tube-like. Whereas the first morphology arises from the CNTs tips, the second case originates from defective CNTs [135]. Figure 15 indicates SEM images of grown nano-whiskers during annealing treatment at 950 °C.

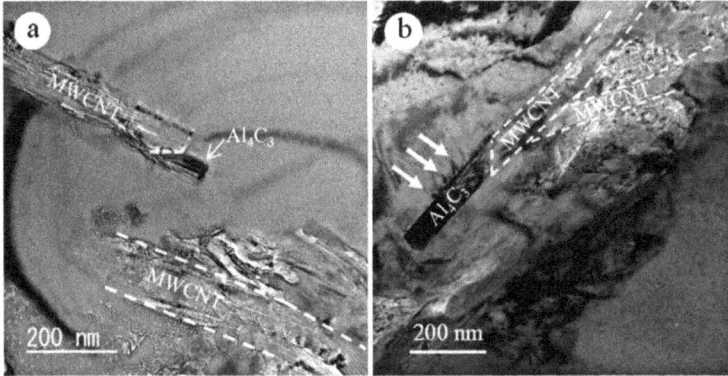

Figure 14. HRTEM (High resolution transmission electron microscope) images of MWNT (multi-walled nanotubes)-reinforced Al matrix nanocomposites heat treated at: (**a**) 873 K for 6 min and (**b**) 883 K for 60 min [132] (Adapted with permission from [132], Elsevier, 2016).

Figure 15. SEM images of (**a**) grown nano-whiskers during annealing treatment at 950 °C, and (**b**) the formation of nano-whiskers on the Al grains, as shown by the white arrows [133] (Adapted with permission from [133], Elsevier, 2006).

The geometrical size of initial CNTs is one of the affecting parameters controlling the volume fraction of the formed carbides. The empirical results show that Al_4C_3 particles can easily arise from MWNTs with 40 nm in diameter, while the amount of Al_4C_3 seems to decrease with an increase in diameter by 140 nm examined from the XRD results [38].

Albeit the development of Al_4C_3 carbides in CNT-Al matrix nanocomposites is of prime significance, the way they can affect the microstructure-related and mechanical properties is more important. As generally believed, the presence of carbides at CNT/Al interfaces improves the mechanical behavior of these nanocomposites. This is because the carbides can enhance the matrix capability to efficiently transfer the external load to nanotubes. This approach is in good agreement with the results reported by Tjong et al. [136], Esawi et al. [137], Kwon et al. [130,135] and Chen et al. [138]. Table 3 gives some information about the presence of Al_4C_3 in some CNT/Al composites with high strengthening effect. The results indicate that the formation of Al_4C_3 maybe doesn't have an adverse effect on the composite strength. The recent study by Chen et al. [138] reveals that nano-sized interfacial Al_4C_3 can noticeably enhance the load transfer efficiency, resulting in improved composite strength.

The way the intermetallics affect the final properties of the nanocomposites strongly depends on their thickness. The critical thickness is estimated as [131]:

$$t_{cri} = -V_M \times (\Delta\gamma/\Delta G_f) \tag{2}$$

where V_M is the molar volume of the formed carbide, $\Delta\gamma$ is the increase in total surface energy due to newly formed interfaces, and ΔG_f is the free energy of formation per mole of carbide. Further growth of the formed carbides can be achieved when the thickness reaches to a critical value. It can remarkably improve the wettability through diminishing the contact angle [131].

Table 3. The variation of tensile strength in CNT-dispersed Al matrix nanocomposites with the volume fraction of Al_4C_3 precipitates (SPS: Spark plasma sintering).

Amount of Added CNTs	Fabrication Method	Increase in Tensile Strength Value (%)	Ref.
2 wt % MWNTs	Sintering → hot extrusion	57.5	[139]
5 vol % MWNTs	SPS → hot extrusion	128	[120]
5 wt % MWNTs	Sintering → pressing → annealing	184	[128]

The in situ carbides evolved in CNT-(Al-Si) nanocomposites grow through two different pathways: (i) At the Al_4C_3/Al-Si interface and (ii) at the Al_4C_3/CNT interface. In accordance with the suggested model, the diffusion of carbon atoms present in CNTs structures act as the predominant mechanism for vertical growth of Al_4C_3 [131]. Figure 16 gives a schematic illustration of vertical and parallel growth of the carbides on CNTs.

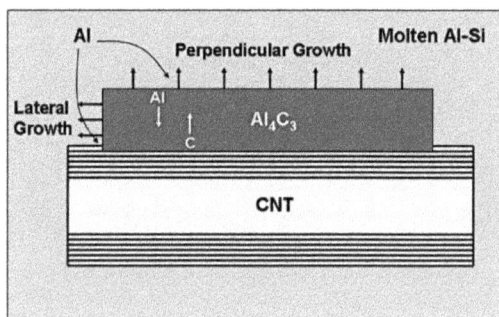

Figure 16. A schematic view of growth mechanisms of Al_4C_3 dispersoids on CNTs. It indicate the vertical migration of Al and C atoms for the formation of Al_4C_3 carbides [131] (Reproduced with permission from [131], Elsevier, 2009).

As a summary, the formation, volume fraction, morphology, and size of the formed carbides in CNT-Al matrix nanocomposites depend on the processing parameters such as temperature, post treatment and purity of the initial CNTs. The in situ formed intermetallics can improve the mechanical properties of CNT-reinforced Al matrix nanocomposites through enhanced load transfer between the matrix and the reinforcements.

(b) CNT-Reinforced Ti Matrix Nanocomposites

A review on the literature confirms that TiC is the most common in situ formed intermetallic in CNT-reinforced Ti matrix nanocomposites. This compound often forms at the boundaries of the sintered powders, if the initial CNTs are wrapped by initial Ti powders. The minimum reaction temperature required for the generation of TiC precipitates is 800 °C. The evaluation of the kinetic aspects of TiC formation in CNT-Ti binary systems has proved the faster formation kinetics of TiC compared to CNTs. The volume fraction of the formed TiC dispersoids differs with the chemical composition of the incorporated carbon allotrope into Ti matrix, e.g., CNT and graphite. TiC particles may form in two different morphologies including spherical and elongated ones. The applied dispersion method for mixing the initial powders, processing parameters such as sintering temperature and the initial volume fraction of CNT, as well as the application of post heat treatments can remarkably affect the mechanical properties of the composites. Generally, the empirical results have demonstrated the positive effects of the formed intermetallics on the physicomechanical properties of the nanocomposites.

There are several intermetallic compounds which can be formed during the reaction between Ti and carbon allotropes. Among these compounds, TiC is the most common [18,140]. It can be formed during a solid-state reaction between Ti and CNTs [18,140,141]. The standard free energy of the reaction may be obtained using the following equation:

$$\Delta G = -184{,}571.8 + 41.382T - 5.042T \times \ln T + 2.425 \times 10^{-3}T^2 - 9.79 \times 105/T \ (T < 1939 \ \text{K}) \quad (3)$$

The minimum reaction temperature required for the generation of TiC precipitates in CNT-Ti binary systems is 800 °C. These particles often locate at the boundaries of the sintered powders, provided that the initial CNTs are wrapped by initial Ti powder. It is found that reaction between the nanotubes and Ti matrix is promoted by an increment in the sintering temperature [141]. As to CNTs, the mentioned reaction is fast and produces more TiC particles. Thus, TiC dispersoids are the dominant phase in the microstructure.

The volume fraction of the formed TiC precipitates strongly depends on which carbon allotrope is incorporated into Ti matrix. Figure 17 shows the microstructure of Ti matrix reinforced with 0.4 wt % CNTs or 0.4 wt % graphite. As clearly seen, higher amounts of TiC dispersoids are present throughout Ti matrix in the MWNTs-reinforced composite (as indicated by white arrows) [142].

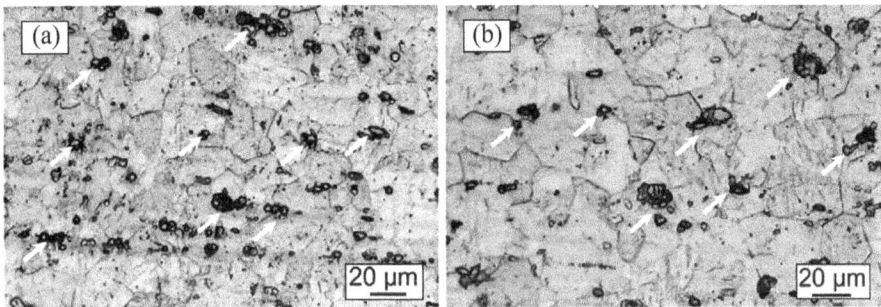

Figure 17. Optical images of Ti matrix reinforced with: (**a**) CNTs and (**b**) graphite [142] (Adapted with permission from [142], Elsevier, 2013).

TiC particles can form in different morphologies. For instance, Kondoh et al. [140] have shown the formation of two different shapes of TiC particles, namely spherical particles and elongated ones. Spherical particles are evolved as a result of incomplete disassembling of MWNTs, while extrusion-induced severe plastic deformation leads to the generation of elongated ones. Thus, the post treatment can affect the morphology of the formed intermetallics. Figure 18 shows the microstructure of the extruded composites, wherein Figure 18a indicates the uniformly distributed reinforcements throughout Ti matrix and Figure 18b confirms the desirable bonding between Ti matrix and TiC particles.

The experimental measurements confirm the positive influence of TiC precipitation at CNT/Ti interfaces on the physicomechanical properties of these nanocomposites [18,35,50,140]. The higher the volume fraction of incorporated CNTs, the superior the mechanical properties of the composites are [18,140]. However, the exact strengthening mechanism of intact CNTs has not been clarified [18]. As practical variables, the used method for mixing the initial powders, processing parameters such as sintering temperature, and initial volume fraction of CNTs, as well as the post heat treatment can remarkably affect the mechanical properties of the composites [18]. In the case of the mixing method, Munir et al. [143] have demonstrated that unlike the solution ball milling (SBM) process, TiC can be formed when the initial powders (Ti powder + 0.5 wt % MWNTs) are mixed by high energy ball milling. However, the in situ TiC particles can be formed by using both methods whenever 1.0 wt % MWNTs are incorporated into Ti matrix. As to the potential effects of the processing parameters on the mechanical behavior of CNT-Ti matrix nanocomposites, Xue et al. [141] have demonstrated that the compressive yield strength of the nanocomposites decreases with an increase in the sintering temperature before a slight rise. This is originated from the combined effect of increased relative density and structural damage of CNTs. Finally, the relative density of in situ formed TiC dispersoids increases with an increment in the volume fraction of initial CNTs. Thus, the higher the initial CNTs volume fraction, the better the mechanical properties will be [142]. In the case of the way a post heat treatment can affect the mechanical behavior, it is shown that the properties can be degraded upon increasing the post annealing temperature. This decrease is "gradual", because the presence of TiC dispersoids throughout Ti matrix may drastically prevent Ti particles from the extreme coarsening through the particle-pinning mechanism [18].

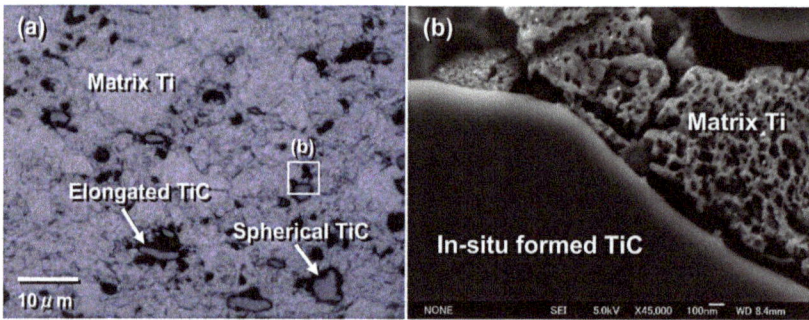

Figure 18. Extruded CNT-dispersed metal matrix nanocomposites: (**a**) Optical image of the binary systems and (**b**) SEM image of TiC/Ti interface [140] (Reproduced with permission from [140], Elsevier, 2009).

As a summary, the incorporation of CNTs into the Ti matrix can drastically enhance the tensile strength, yield strength, and microhardness values of the matrix. It may be ascribed to the dispersion effect of un-bundled CNTs and in situ formed TiC particles, as shown in Figure 19.

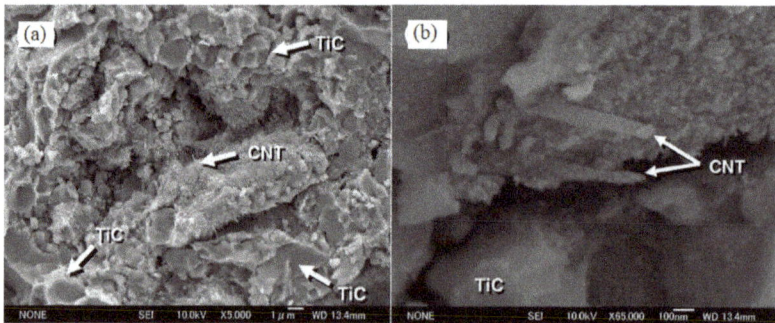

Figure 19. SEM image of fracture surface of CNT-Ti nanocomposites in different magnifications [140]: (**a**) ×5000; (**b**) ×65,000 (Adapted with permission from [140], Elsevier, 2009).

(c) CNT-Reinforced Mg Matrix Nanocomposites

Mg and its alloys are not able to sufficiently wet CNTs. To solve this practical challenge, two general approaches have been suggested: (i) Coating of CNTs by a metallic thin film and (ii) alloying Mg matrix with appropriate elecments. The first solution can efficiently enhance the interfacial adhesion through the prevention of CNTs from clustering. Clustering of CNTs adversely alter the interfacial bonding, resulting in deteriorated mechanical properties. It is found that an intermetallic compound (Mg_2Ni) can be generated through wrapping CNTs by a Ni-based coating and alleviate decreased properties in Mg matrix composites. It is ascribed to an improved bonding at the interface, enhancing the mechanical properties of the nanocomposites with no significant decrement in ductility [12]. From this perspective, the utilization of Mg-Al alloys (AZ-type) such as AZ31, AZ81, Mg-3Al-1Zn, and Mg-6 wt % Al instead of pure Mg is strongly recommended because of the easy formation of ternary Al_2MgC_2 compounds.

Depending on the chemical composition of used Mg alloy matrix and CNTs coatings, the several intermetallic phases of different stoichiometries may form among which Al_2MgC_2, Al_3Mg_2, $Al_{12}Mg_{17}$, Al_4C_3, and Mg_2Ni are of prime significance [144]. Figure 20 shows the needle-like morphology of Al_2MgC_2 ternary carbides.

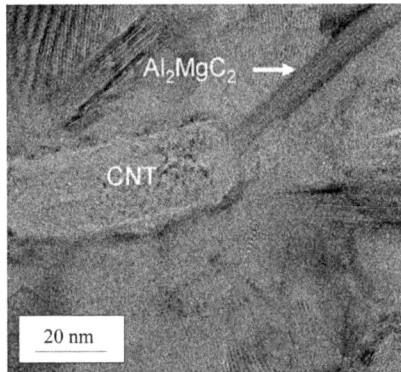

Figure 20. TEM image of (Mg-6 wt % Al)/CNT interface, indicating the in situ formation of needle-like ternary carbides (Al_2MgC_2) [144] (Reproduced with permission from [144], Elsevier, 2011).

The precipitation of the interfacial intermetallics has an optimum limit, so that the mechanical properties of the nanocomposites may be weakened whenever it is exceeded. Therefore, a close control over the interfacial reactions between CNTs and Mg is highly required. These reactions proceed by simultaneous diffusion of the matrix atoms and carbon atoms present on the superficial regions of CNTs toward the preferred reaction sites. To confine Mg-CNTs reactions along the interfaces, employing a short-term method such as microwave sintering has been suggested. The long-time contact between the matrix and reinforcements may drastically degrade the mechanical properties of the nanocomposites due to the formation of excessive content of the intermetallics [12].

As previously discussed, if the precipitation rate of the intermetallic compounds are closely controlled, enhanced properties can be obtained. This enhancement can be justified by three different mechanisms: (i) Grain refinement, (ii) improvement in load transfer from the matrix to reinforcements, and (iii) appropriate wettability. For instance, it is shown that the Al_4C_3 phase often acts as a suitable grain refiner and boosts the tensile strength of CNT-AZ81 nanocomposites [19]. As another example, the formation of SiC during the sintering of MWNTs-AZ91 powder blend facilitates the load transfer from the matrix to MWNTs and enhances the mechanical properties of produced composites [145]. In addition, the mechanically improved behavior of (Ni-coated MWNTs)-Mg composites is ascribed to the presence of an adherent and void-free MWNTs /Mg interface due to Mg_2Ni precipitation. Figure 21 shows Mg_2Ni dispersoids at the MWNTs/Mg interfaces [12].

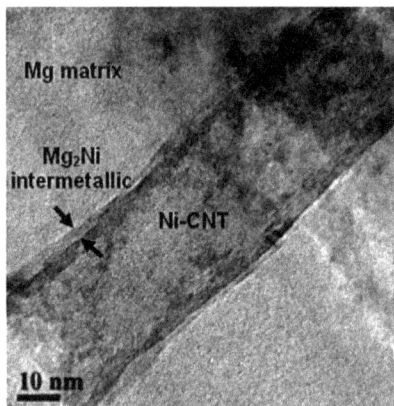

Figure 21. TEM image of (Ni-coated MWNT)-Mg composites, showing the presence of Mg_2Ni intermetallic at the interface [12] (Adapted with permission from [12], Elsevier, 2014).

Similar to CNT-Al and CNT-Ti binary systems, the precipitation of the mentioned intermetallics may bilaterally affect the mechanical properties of CNT-dispersed Mg matrix nanocomposites. The poor failure strain of the composites is attributed to the ubiquitous presence of coarse intermetallic particles. The initial volume fraction of incorporated CNTs, formation of intermetallics, and quality of the formed interface between the Mg and nanotubes are the important factors in determining the mechanical properties of the nanocomposites. It is shown that an increase in the initial amount of incorporated CNTs may promote the refinement of coarse grains and enhance the failure strain [146,147]. Figure 22 shows the evolved microstructure of CNT-AZ31 nanocomposites produced using the disintegrated melt deposition technique followed by hot extrusion. As seen, the formed $Mg_{17}Al_{12}$ particles are distributed at or along α phase boundaries and give rise to the grain refinement of Mg matrix [148].

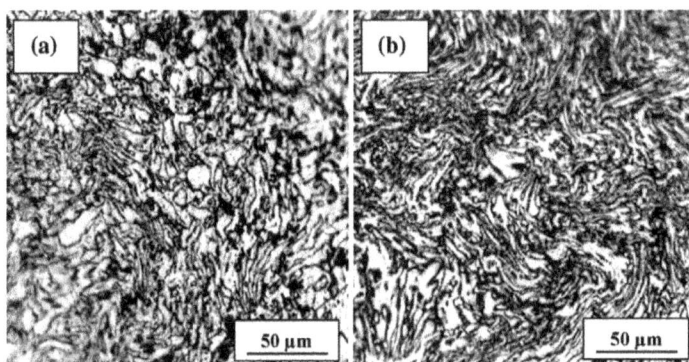

Figure 22. Optical image of (**a**) AZ31 and (**b**) 1 vol % CNT-AZ31 nanocomposites fabricated through the disintegrated melt deposition technique followed by hot extrusion at 350 °C. In these systems, $Mg_{17}Al_{12}$ intermetallics are abundantly formed. These compounds are manifested by black regions, while white regions exhibit α-Mg [148] (Reproduced with permission from [148], Springer, 2012).

The formation of the mentioned intermetallics enhances the mechanical properties of CNT-Ti matrix nanocomposites through guaranteeing the suitable wetting between the matrix and nanotubes. It can significantly prevent the microscale cavities which form at the Ti/CNT interfaces and contribute to improved mechanical properties [12]. Additionally, the formed intermetallics can improve the mechanical properties of the nanocomposites through the restriction of mechanical deformation [148]. The formation of a clean interface with no impurity or defect is shown to improve the mechanical properties [144]. A recent study by Rashad et al. [146] showed that the formed intermetallics in MWNTs-dispersed Mg-3Al-1Zn alloys are refined with the CNTs inclusion. Figure 23 indicates the schematic demonstration of $Mg_{17}Al_{12}$ generation at the interface of Mg-3Al-1Zn/MWNTs composite, where the coarse particles of $Mg_{17}Al_{12}$ can be refined through the incorporation of MWNTs into the matrix.

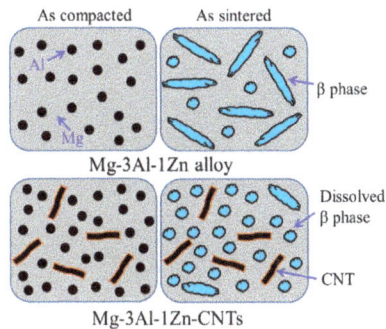

Figure 23. A schematic view of $Mg_{17}Al_{12}$ particles formed in Mg-3Al-1Zn/MWNTs nanocomposites prepared by powder metallurgy method. The incorporation of MWNTs can drastically refine the particles (β phase corresponds to $Mg_{17}Al_{12}$) [146] (Reproduced with permission from [146], Elsevier, 2015).

A review on the literature confirms that recent research works have strived to attribute the enhanced properties of CNT-Mg composites to the formation of different intermetallics. Fukuda et al. [144] have resorted to Al_2MgC_2 compound. However, an opposite trend has been observed by Pei et al. [110], where Al_2MgC_2 strengthens the carbon nanofiber/Mg interface, but simultaneously leads to decreased mechanical properties. It is ascribed to the difference in the extent of graphitization. In other words, produced Al_2MgC_2 intermetallic in CNT-reinforced composites insignificantly affects the CNTs strength. Table 4 provides the readers with an overview of the fabrication method, chemical composition of the possible intermetallic compounds, and variations in mechanical properties of the composites.

Table 4. An overview of fabrication method, formed intermetallic compounds, and mechanical properties of CNT-reinforced Mg matrix nanocomposites.

Fabrication Method	CNTs Volume Fraction	Type of Used CNTs	Chemical Composition of the Formed Intermetallics	Variations of Mechanical Properties	Suggested Reasons for Enhanced/Deteriorated Properties	Ref.
Powder metallurgy based wet processing	0.71–1.56 vol % MWNTs	Functionalized; using a zwitterionic surfactant	Al_2MgC_2	An increase in yield strength and elongation at failure by 28.5 MPa and 12.2%, respectively	- Presence of clean interface - CNTs-induced strengthening - Formation of Al_2MgC_2	[144]
Powder metallurgy technique	0.3 wt % MWNTs	Ni-coated	Mg_2Ni	An increase in microhardness, yield strength, and ultimate strength by 16 HV, 80 MPa, and 66 MPa, respectively. A decrease in ductility by 1.5%	- Grain refinement - CNTs-induced strengthening - Improved interfacial interactions	[12]
DMD (disintegrated melt deposition) technique followed by hot extrusion	1.5 vol % CNT nanoparticles	Unfunctionalized	Al_4C_3, $Al_{12}Mg_{17}$, Al_3Mg_2	An increase in yield strength, ultimate strength, and failure strain by 107 MPa, 114 MPa, and 7%, respectively	- Grain refinement due to the presence of Al_4C_3 - Crack formation-based tensile toughening mechanism	[19]
Squeeze infiltration method	1–5 wt % MWNTs	Acid treated	SiC	An increase in microhardness, yield strength, and ultimate strength by 80 HV, 170 MPa and 90 MPa, respectively	- Grain refinement - Improved load transfer owing to SiC formation at the interface	[145]
DMD method followed by hot extrusion	1.0 vol %	Acid treated	$Mg_{17}Al_{12}$	An increase inmicrohardness, yield strength, ultimate strength, and elongation at failure by 29 HV, 34 MPa, 8 MPa, and 3.6%, respectively	- Grain refinement - Strengthening by stable intermetallic phases - CNTs-induced strengthening	[148]
Powder metallurgy method	0.25–1.0 wt % MWNTs	Unfunctionalized	$Mg_{17}Al_{12}$	An increase in ultimate strength and failure strain by 10 MPa and 9%, respectively. Yield strength remains unchanged	- CNTs-induced strengthening - Refinement of coarse $Mg_{17}Al_{12}$ particles through the incorporation of CNTs into the matrix, which improves the ductility	[146]
Accumulative roll bonding	3 vol % MWNTs	Unfunctionalized	Al_3Mg_2, $Mg_{17}Al_{12}$	-	- CNTs-induced strengthening - Refinement of coarse $Mg_{17}Al_{12}$ particles through incorporation of CNTs into the matrix, which improves the ductility	[147]

As another interesting result, a slight increment in relative density of CNT-Mg nanocomposites has been observed owing to the formation of in situ intermetallics. For instance, it is found that the incorporation of 0.3 wt % Ni-coated CNTs into Mg matrix increases its density by 0.01 g/cm^3 [12].

(d) CNT-Reinforced Cu Matrix Nanocomposites

Similar to Mg and its alloys, Cu is not able to sufficiently wet CNTs. This is because there is a significant difference between the surface tension of Cu and CNTs. Furthermore, it is not expected that Cu reacts with CNTs in conformity with Cu-C phase diagram. Instead, some fissures may form at Cu/CNTs interfaces [149]. Figure 24a shows such an obvious crevice at the interface. The addition of Ti to Cu matrix eliminates these crevices through the formation of a thin TiC layer at the interface, as shown in Figure 24b. Such crevices can significantly degrade the final properties of the nanocomposites, e.g., thermal conductivity, which will be discussed in the next section.

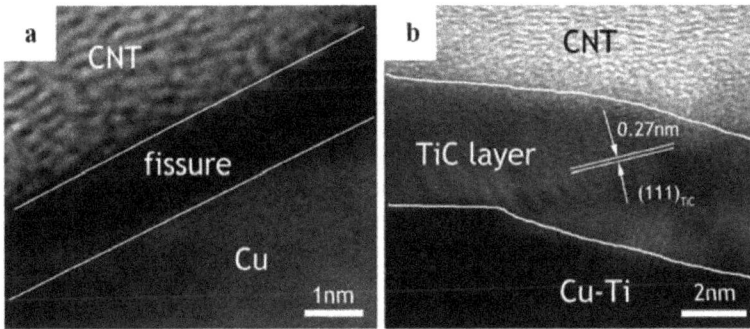

Figure 24. TEM images of CNT/Cu interfaces: (**a**) The formation of a pronounced crevice due to the lack of wettability and (**b**) the precipitation of in situ TiC layer as a result of Ti addition to pure Cu matrix [149] (Reproduced with permission from [149], Springer, 2013).

The improvement of the interfacial bonding between Cu matrix and CNTs has been the focus of many research works. One of the suggested solutions is alloying of Cu matrix using carbide forming elements such as Cr and Ti. It can lead to the precipitation of Cr_3C_2, TiC, and Cr_7C_3 carbides. The open ends of CNTs and carbon atoms present on the wall defects can serve as the carbon source reacting with Cr [150]. These carbides are generated at the appropriate sites including broken surfaces and reactive edges of the tubes through various bonding mechanisms such as rope anchors, chain-link structures and bridges [151]. The standard free energy (ΔG_T^{θ}) of the reactions may be obtained using the following equations [152]:

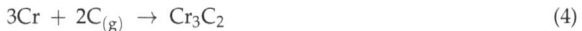

$$3Cr + 2C_{(g)} \rightarrow Cr_3C_2 \tag{4}$$

$$\Delta G_T^{\theta} = -72.333T \times \ln T + 0.062T^2 + 1,345,000T^{-1} - 0.00001T^3 + 744.180T - 110,762.634$$

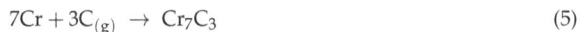

$$7Cr + 3C_{(g)} \rightarrow Cr_7C_3 \tag{5}$$

$$\Delta G_T^{\theta} = -114.064T \times \ln T + 0.108T^2 + 1,772,000T^{-1} - 0.0000192T^3 + 706.278T - 218,451.845$$

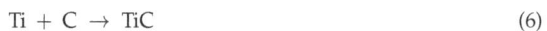

$$Ti + C \rightarrow TiC \tag{6}$$

$$\Delta G_T^{\theta} = -27.75T \times \ln T + 0.0241T^2 + 665,300T^{-1} - 0.0000003T^3 + 191.888T - 194,857.227$$

In general, Cu-Ti compounds precipitate during SPS, if the Cu-Ti alloy is used as matrix. It is while the Cu-Ti compounds may be decomposed into TiC during the sintering process. The formation of TiC proceeds through the reaction between CNTs and Ti diffused from the primary particle boundaries [20].

An island-shaped Cr_7C_3 can form as a result of a chemical reaction between the open tip of MWNTs and diffused Cr. Other carbides can be evolved due to a reaction between defective sidewalls of MWNTs and Cr atoms [49]. It is shown that the formed carbides at the tip of MWNTs can provide superior load transfer, because they are in contact with more amounts of graphene layers [153].

The studies show that the controlled formation of C-Cu intermetallic compounds at Cu/CNTs interfaces can noticeably improve the physicomechanical properties, thermal conductivity, pitting corrosion resistance, and electrical conductivity of these binary composites. The major strengthening mechanism responsible for this enhancement is the improved load transfer from the matrix to CNTs as a result of carbides precipitation. From mechanical point of view, the Cr_7C_3 particles are reported to effectively transfer the applied tensile loads to MWNTs in CNT-(Cu-Cr) nanocomposites [153]. Also, the yield strength of Cu-Ti alloy increases by 88% through the inclusion of CNTs. It is ascribed to the synergic combination of plastic deformation and stronger interfacial bonding induced by the formation of a thin TiC layer [154].

The literature confirms the degraded thermal conductivity of MWNTs-reinforced pure Cu matrix composites due to the poor interfacial bonding at the CNT/Cu interface. Alloying of Cu matrix may modify superficial regions of MWNTs and drastically enhance the thermal conductivity of the nanocomposites due to increased heat transfer. Improved interfacial bonding is attributed to the carbides formation. Another suggested approach to improving the interfacial bonding is the coating of MWNTs by a metallic layer. Nonetheless, the thermal conductivity of these composites may be considerably degraded due to the formation of excessive impurities during the coating process. In contrast, the generation of metallurgical bonds between Cu-Ti alloy matrix and MWNTs improves the thermal conductivity of the composites due to the formation of TiC dispersoids. This compound can facilitate the electron-phonon coupling and prevent their scattering [149]. As another typical example, the superior thermal conductivity of CNT-(Cu-Cr) nanocomposites can be achieved through the generation of Cr_3C_2 at the interface. These precipitates are perfectly capable of improving the heat transfer efficiency by strengthened interfacial bonding. Figure 25 indicates the precipitation of Cr_3C_2 at the CNT-(Cu-Cr) interface. As a general rule, if the intermetallic compounds form at the interface of these composites, the thermal conductivity will be enhanced. Furthermore, the volume fraction of initial CNTs strongly affects the thermal conductivity of the nanocomposites, wherein the thermal conductivity increases with an increment in the initial volume fraction of CNTs, as shown in Figure 26.

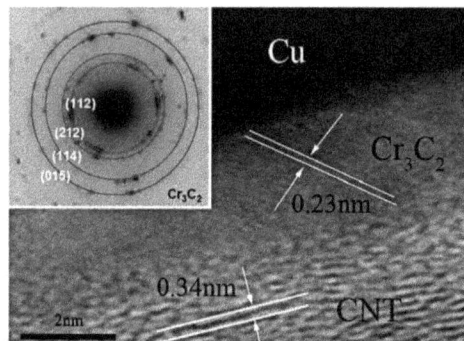

Figure 25. TEM image of Cr_3C_2 precipitates at the CNT-(Cu-Cr) interface [150] (Adapted with permission from [150], Springer, 2013).

As to the electrochemical activity, the presence of the mentioned intermetallics may increase the pitting corrosion resistance of MWNTs-dispersed Cu matrix nanocomposites. This effect is related to enhanced passivation originated from the formation of the carbides. This is the case for the electrical conductivity. The enhancement in electrical properties of Cu-Ti solid solutions reinforced with MWNTs can be obtained upon the formation of ubiquitous intermetallic compounds. These carbides diminish the concentration of Ti atoms in the matrix solid solution and statistically reduce the scattering sites of charge carriers [20]. For the pure Cu matrix, a similar improvement is observed. It is ascribed to enhanced electron transport as a result of strong interfacial bonding originated from TiC formation at MWNTs-Cu interfaces [154].

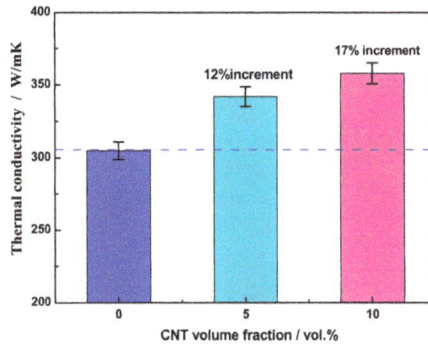

Figure 26. Thermal conductivity of CNT-(Cu-Cr) nanocomposites as a function of initial volume fraction of CNTs [150] (Reproduced with permission from [150], Springer, 2013).

In summary, the mechanical, corrosion-related, and electrical properties of the nanocomposites can be enhanced through the in situ formation of some intermetallics as a consequence of improved load transfer, enhanced passivation and suitable electron transport.

(e) Other Composites Systems

Intermetallic compounds can be formed at the interface of pure Ni matrix and CNTs. In most cases, the formation of intermetallic compounds such as a metastable hexagonal Ni_3C at the defect sites of tubes can significantly improve the mechanical properties of CNT-dispersed Ni matrix nanocomposites due to the formation of strong interfacial bonds [52]. Such a scenario is observed in Fe alloys. In these alloys, the iron carbides form as a result of a chemical reaction between the iron matrix and included CNTs. Microstructure-related and mechanical properties of the composites are drastically affected by the carbides formation. Lin et al. [155] proved the deterioration of mechanical properties in these composites with carbides precipitation; albeit the reason has not been discussed.

Partial Chemical Reaction

The strength of CNT/metal interfacial bonding is one of the key factors controlling the physicomechanical features of CNT-reinforced metal matrix nanocomposites as well as the uniform distribution of CNTs and high relative density [49]. As a general rule, the weak wettability and poor interfacial compatibility between CNTs and metal matrix degrades the fracture toughness due to the limited interfacial load transfer [156]. Among the widely used methods for the fabrication of metal matrix nanocomposites (MMNCs), powder metallurgy routes provide a better interfacial bonding as compared to casting processes [157]. A model is developed by Coleman et al. [158] describing the tensile strength of CNT-containing composites (σ_C) based on the shear strength of the interface (σ_{shear}). Equation (7) describes this model:

$$\sigma_C = \left(1 + \frac{b}{R}\right)\left[\frac{l_{CNT}}{2R}\sigma_{shear} - \left(1 + \frac{b}{R}\right)\sigma_M\right]V_{CNT} + \sigma_M \qquad (7)$$

where b is thickness of the interface layer, σ_M is the strength of the matrix and R, l_{CNT} and V_{CNT} are the radius, length and volume fraction of CNTs, respectively.

Zhou et al. [159] used an in-situ pull-out technique to quantitatively evaluate interfacial shear strength (IFSS) in CNT-Al nanocomposites. They bonded the end of the protruding CNTs directly to the tip of an atomic force microscopy (AFM) cantilever and pulled them out from Al matrix. They reported that the estimated tensile strength of the nanocomposite based on the shear lag model using the obtained IFSS values is consistent with the experimental ones. To date, the direct microscopic method for determination of IFSS values have also been employed for CNT-filled ceramic [160] and polymer [161] matrix nanocomposites. It is inferred that the tensile strength of the nanocomposites can be maximized by thickening the interfacial region through the formation of a crystalline coating around CNTs or their functionalization or in situ formation of metal oxide/carbide interlayer at boundary zones. For instance, shear strength of CNT-Al nanocomposites is enhanced by the formation of interfacial Al_4C_3 carbides during a chemical reaction between defective CNTs and molten Al in the interboundary areas [120,162]. Figure 27 shows the interfacial regions in CNT-Al matrix nanocomposites. As seen, several in situ phases are embedded in the grain boundary layer of Al-CNT sintered compacts including Al_2O_3, amorphous carbon black, graphite and Al_4C_3 phases [120]. The formation of carbides on the surface of CNTs at Al/CNT interfaces diminishes the contact area between CNTs and Al matrix with poor wettability and substitutes the Al/Al carbide interfaces with a lower wetting angle than that of CNT/Al boundaries (~55° vs. 130–140°) [162,163]. The stronger the physicochemical interactions between these two phases, the more adhesive the interface will be. To simply describe metal/ceramic physical interaction, the work of adhesion (W_{ad}) is defined as: (i) The amount of energy released whenever two free surfaces are brought into contact or (ii) the work per unit area of interface which is reversibly performed to separate the two phases. W_{ad} can be given by the following equation [128,164]:

$$W_{ad} = \sigma_{LV}(1 + \cos\theta) \tag{8}$$

where σ_{LV} is the surface tension and θ is the contact angle between two phases. Therefore, W_{ad} as a measure of the interfacial adhesion could be enhanced by improving the wettability through extensive interfacial chemical reactions. From this point of view, Al_4C_3 formation at the CNT/Al interface increases the interfacial adhesion from 200 mJ/m² for Al-C system to 1156 mJ/m² for Al-Al_4C_3 (calculated at 1100 °C) and consequently counteracts delamination and improves the load transfer efficiency [128].

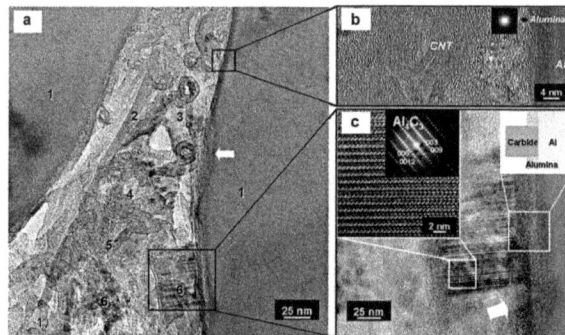

Figure 27. TEM images of the grain boundary regions in CNT-dispersed Al matrix nanocomposites: (**a**) Micrograph of the whole grain boundary including (1) Al, (2) alumina, (3) CNT, (4) amorphous carbon black, (5) graphite, and (6) Al_4C_3 phases; (**b**) interfacial region between CNT and Al matrix; and (**c**) HRTEM image and SAD (Selected area diffraction) pattern of Al_4C_3 phase. The region with the broken alumina phase is indicated with the white arrow [120] (Reproduced with permission from [120], Elsevier, 2009).

On the contrary, some research works have reported the adverse effects of the carbide formation on the CNT/Al interfacial bonding and degradation of mechanical properties in CNT-Al nanocomposites due to a weak interfacial adhesion [165–167]. However, some researchers have demonstrated directly or indirectly the desired influences of CNT/metal matrix interfacial reactions on the mechanical features of the nanocomposites regarding the fact that the CNT/metal interfacial load transfer is a direct function of the interfacial bonding. Zhou et al. [168] produced the Al_4C_3 nanostructure at the end of CNTs incorporated into Al matrix through an appropriate heat treatment and investigated the thermal expansion behavior of the nanocomposite as a criterion of CNT/Al matrix interfacial bonding. They observed stress contrast around carbides as an evidence of a friction trace, leading to the enhancement in their anchor effect from Al matrix. This carbide-induced anchor effect minimizes the local interfacial slippage and constrains the matrix deformation. Consequently, the thermal expansion exhibits a linear and reversible behavior under cyclic thermal load, indicating a reduced interfacial slippage between CNTs and Al matrix at the presence of Al_4C_3 precipitates.

It is proven that CNT/metal interfacial adhesion strongly depends on the applied method for dispersion of CNTs inside the matrix. Among the various discussed methods, molecular-level process and metallization of CNTs are the most effective. Kim et al. [49] used molecular-level process prior to spark plasma sintering to fabricate CNT-filled Cu nanocomposites. They reported enhanced hardness and wear resistance in spark plasma sintered CNT-Cu nanocomposites than pure Cu with good CNT/Cu interfacial bonding as well as uniform CNTs distribution in the matrix and high relative density. Metallization of CNTs using chemical vapor deposition (CVD) [16] and electroless coating [95] is one of the effective approaches to achieve a strong metal/CNT interfacial bonding as well as a homogeneous dispersion of CNTs inside the matrix. It can improve the load transfer from matrix to CNTs through the metallic coating. In another study, He et al. [128] fabricated CNT-reinforced Al nanocomposites with enhanced mechanical properties via the in situ chemical vapor deposition process and ascribed the strong interfacial bonding to the formation of transition thin layer of Al_4C_3 between CNTs and Al matrix.

Clean oxide-free surfaces of metallic powders are a prerequisite for obtaining a good CNT/metal interfacial adhesion. To meet this challenge, some additives in CNT-metal systems are utilized to reduce the oxide content. Owing to its negligible solid solubility, good adhesion with Cu, and high thermodynamic stability, 0.5 wt % ruthenium (Ru) as an air-stable transition metal was exploited in a recent survey to reduce non-protective oxide films on Cu surfaces. Furthermore, Ru contributes to increased thermal conductivity of CNT-Cu binary systems as a result of enhanced CNT/Cu interfacial adhesion and inhibition of the heat carrying phonons scattering at the CNT/Cu interfaces and weak interfacial bonding-induced pores. However, if Ru content exceeds 2.5 wt %, a reduction in hardness is observed due to the restricted densification of the powder system. It arises from high melting point of Ru, serving as a diffusion barrier during the consolidation process. Also, using 1 wt % Ru decreases the electrical conductivity of pristine copper, because the interfacial resistance occurs due to high electrical resistivity of Ru than Cu (7.1×10^{-6} vs. 1.68×10^{-6} Ω·cm) [169].

Interestingly, metal oxide formation on the surface of the particles may strengthen the interfacial adhesion in some CNT-reinforced metallic systems. For instance, Kondoh et al. [97] fabricated CNT-Mg nanocomposites by spark plasma sintering of CNT-coated Mg particles and reported in-situ formation of MgO dispersoids during the consolidation process as well as a thin MgO film on the surface of Mg particles due to the atmospheric corrosion. They confirmed that the diffusion of carbon atoms into the MgO structure provides an enhanced interfacial adhesion with CNTs. This phenomenon is mechanistically similar to the epitaxial growth of chromium carbide nanostructures mostly Cr_7C_3 at the CNT/Cu-Cr interface as a result of the substitutional diffusion of Cr atoms into radially unzipped defects of CNTs. It is schematically shown in Figure 28. The results show that the formation of Cr carbide nanostructures may potentially increase the efficiency of interfacial load transfer while preserving CNTs structure [170]. It is further surveyed by Kathrein et al. [171].

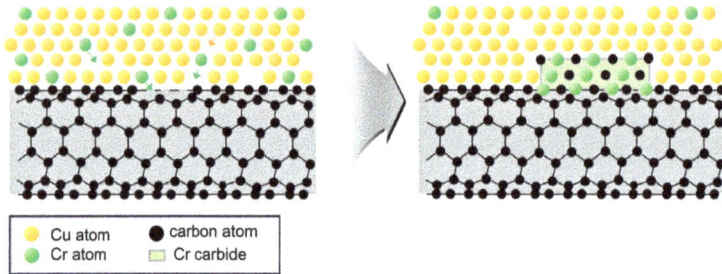

Figure 28. A schematic view of epitaxial growth of Cr carbide at the CNT/Cu-Cr interface via the substitutional diffusion of Cr atoms into the nanotubes during sintering [170] (Reproduced with permission from [170], Elsevier, 2013).

The nanocomposite processing parameters such as applied temperature/pressure and post thermal/mechanical treatments are among factors affecting CNT/matrix interfacial adhesion. Guo et al. [172] fabricated CNT-Al nanocomposites through spark plasma sintering followed by hot rolling process and reported that higher sintering temperature (630 °C vs. 590 °C) provides a stronger cohesion force for Al-CNTs and Al-Al particles due to higher degree of densification.

3. CNT Pinning-Induced Grain Refinement

In addition to the load bearing capacity, CNTs can affect the grain size of the metal matrix nanocomposites. In fact, the incorporation of CNTs into a metallic matrix results in lower grain size due to the CNT-induced pinning effect. Such an effect arises from the low dimensions of CNTs which can pin the grains of the metal matrix similar to nanosized round particles. The more the CNT content, the higher the effectiveness of boundary pinning and the lower the grain size are [42,101,173,174]. Figure 29 shows the STEM (Scanning transmission electron microscope) micrographs of Cu and CNT-Cu nanocomposites after the consolidation by high pressure torsion (HPT). As obviously seen, an increase in CNT content may feasibly reduce the grain size and narrow its distribution thank to the grain pinning effect [42].

The mean grain size (D) of reinforcement-containing metal matrices can be estimated by the Zener pinning relation [174]:

$$D = \frac{k \times r}{f^n} \tag{9}$$

where k is a proportional dimensionless constant, f is the volume fraction of the secondary phase, and r is the mean reinforcement radius, respectively. The secondary phase particle (i.e., CNTs) acts as a frictional force against the grain boundary migration and hinders the grain growth. As seen in Equation (7), the strengthening effect of particles depends on their radius (diameter). The smaller the diameter of CNTs, the smaller the mean grain size of the metallic matrix will be. Moreover, the average distance between CNTs is another factor which greatly affects the pinning effect. An increase in CNT content gives rise to lower distance between CNTs, increased interfacial energy absorption and decreased grain size. Additionally, for a constant CNT content, the average distance between short CNTs are lower than that between long ones. As a result, short CNTs have stronger pinning effect and result in lower grain size [174].

Such a grain refinement affects the mechanical properties of CNT-dispersed metal matrix nanocomposites. In general, CNTs can increase the mechanical strength of metal matrices through a variety of mechanisms: (i) Load transfer from the metallic matrix to CNTs, (ii) reduction in the matrix grain size through the pinning effect (Hall-Petch relation), (iii) CNT-induced Orowan looping mechanism, (iv) solid solution strengthening due to the diffusion of carbon atoms from CNTs to the matrix, (v) secondary phase (particle) strengthening due to the in-situ formation of carbide particles by a chemical reaction between CNTs and the metal matrix, (vi) work hardening of the matrix due to the formation and accumulation of dislocations at CNT/metal interfaces caused by thermal mismatch between CNTs and metal matrix, and (vii) the strengthening effect induced by impurities originating from the mixing methods [175]. Depending on the type of matrix and the fabrication method, one or some of these mechanisms may be stimulated. Another factor affecting the strengthening mechanisms is the physical and size-dependent properties of CNTs. For instance, if the thermal expansion coefficient of CNTs is considerably different with that of the matrix, the dislocations will build up at CNT/metal interfaces and enhance the mechanical strength of the nanocomposite. Moreover, the diameter and length (i.e., aspect ratio) of CNTs may influence the strengthening mechanisms in metal matrices. Among the aforementioned strengthening mechanisms, the load transfer mechanism is the most influential. Typically, the higher length or lower diameter of CNTs (i.e., higher aspect ratio) results in better load transfer and higher mechanical properties such as improved yield strength and elastic modulus. However, when the aspect ratio of CNTs is lower than a critical value, the Orowan mechanism is dominant [175,176].

Beside the positive effect of long CNTs on enhanced mechanical properties than short ones, long nanotubes are usually more sensitive to agglomeration. As a consequence, the dispersion of shorter CNTs is an easier task than that of longer ones [34]. Using a high content of long CNTs may induce the agglomeration and degrade the mechanical properties of final nanocomposite.

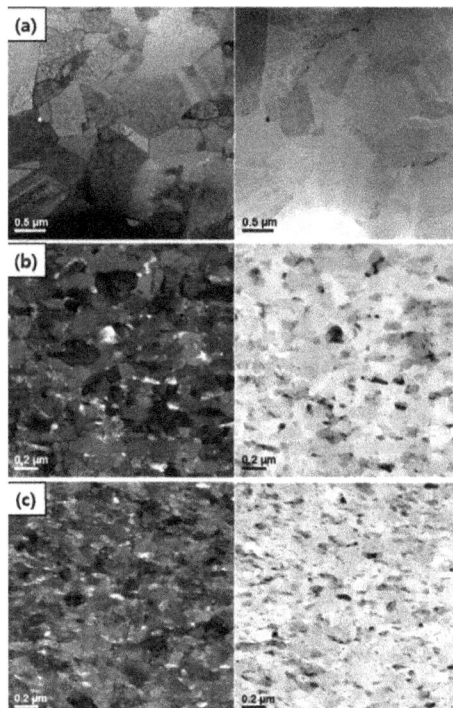

Figure 29. STEM (scanning transmission electron microscope) images of Cu and CNT-Cu nanocomposites fabricated by high pressure torsion (HPT): (**a**) Pure Cu, (**b**) 5 vol % CNT-Cu, and (**c**) 10 vol % CNT-Cu nanocomposites. The left images are provided by bright-field mode, and the right ones by high-angle annular dark-field mode [41] (Reproduced with permission from [41], Springer, 2013).

4. Conclusions and Outlook

The present review paper has provided a broad overview of practical challenges in the fabrication of CNT-dispersed metal matrix nanocomposites. These challenges have been categorized into four main groups: (i) Non-uniform dispersion of CNTs throughout the metallic matrix, (ii) thermal decomposition of CNTs and chemical reaction with the metallic matrix, (iii) poor interfacial adhesion, and (iv) low compactability. One can obtain a CNT-metal matrix nanocomposite with superior properties if all of four challenges are appropriately overcome. A large number of research works have focused on these issues and strived to suggest feasible solutions to them. Their strategy is often based on the better dispersion of CNTs through the effective prevention of their agglomeration and close control over the processing parameters for suppressing the unwanted phase transformations and preventing from unfavorable microstructural features. Although these solutions remarkably enhance the physicomechanical properties, each of them has its own limitations. It seems that the future research works should find their way toward development of new solutions for uniform dispersion of CNTs in metallic matrices, so that the electrical and thermal properties are improved as well as mechanical ones. It seems that the future research works should focus on the development of more efficient ways to uniformly disperse CNTs in the metallic systems or introduction of new methods for efficient consolidation of metallic powders mixed with CNTs.

Acknowledgments: The present work has been conducted by some active members of the "Advanced Materials Research Group (AMRG)" founded by Abolfazl Azarniya in 2016. The authors would like to acknowledge all team members including Anne Jung (from Saarland University, Saarbrücken, Germany), Mohammad Mirzaali (from Delft University of Technology, Delft, The Netherlands), Dariusz Garbiec (from Metal Forming Institute, Poznań, Poland), Ridvan Yamanoglou (from Kocaeli University, Kocaeli, Turkey), Temel Varol (from Karadeniz Technical University, Trabzon, Turkey), Mohammad Abedi and Dmitry Moskovskikh (from National University of Science and Technology MISIS, Moscow, Russia), Flávio Bartolomeu and Georgina Miranda (from University of Minho, Braga, Portugal), Xabier Garmendia (from University of Liverpool, Liverpool, UK), Joseph Ahn (from Imperial College London, London, UK), Yaser Shanjani (from Stanford University, Stanford, CA, USA), Chor Yen Yap, Sing Swee Leong, and Wai Yee Yeong (from Nanyang Technological University, Singapore), Marek Weglowski (from Institute of Welding, Gliwice, Poland), Wessel Wits (from University of Twente, Enschede, The Netherlands) and some others for their best efforts. These scientific partners are a part of AMRG's forthcoming research publications.

Author Contributions: Abolfazl Azarniya, Mir Saman Safavi, Saeed Sovizi, Amir Azarniya, and Biao Chen outlined the review, performed the literature search and wrote the manuscript. Hamid Reza Madaah Hosseini and Seeram Ramakrishna helped in compiling the wide information related to the topic and data presentation and performed technical editing for all corrections. Hamid Reza Madaah Hosseini and Seeram Ramakrishna also helped in finalizing the manuscript and developed the idea of the topic and were responsible for the correspondence.

Conflicts of Interest: The authors declare no conflict of interest.

References

1. Iijima, S. Helical microtubules of graphitic carbon. *Nature* **1991**, *354*, 56. [CrossRef]
2. De Volder, M.F.; Tawfick, S.H.; Baughman, R.H.; Hart, A.J. Carbon nanotubes: Present and future commercial applications. *Science* **2013**, *339*, 535–539. [CrossRef] [PubMed]
3. Duclaux, L. Review of the doping of carbon nanotubes (multiwalled and single-walled). *Carbon* **2002**, *40*, 1751–1764. [CrossRef]
4. Kang, S.; Herzberg, M.; Rodrigues, D.F.; Elimelech, M. Antibacterial effects of carbon nanotubes: Size does matter! *Langmuir* **2008**, *24*, 6409–6413. [CrossRef]
5. Kaushik, B.K.; Majumder, M.K. Carbon Nanotube: Properties and Applications. In *Carbon Nanotube Based VLSI Interconnects*; Springer: Berlin/Heidelberg, Germany, 2015; pp. 17–37.
6. Ibrahim, K.S. Carbon nanotubes? properties and applications: A review. *Carbon Lett.* **2013**, *14*, 131–144. [CrossRef]
7. Baughman, R.H.; Zakhidov, A.A.; de Heer, W.A. Carbon nanotubes—The route toward applications. *Science* **2002**, *297*, 787–792. [CrossRef]

8. Bakshi, S.; Lahiri, D.; Agarwal, A. Carbon nanotube reinforced metal matrix composites—A review. *Int. Mater. Rev.* **2010**, *55*, 41–64. [CrossRef]

9. Chen, B.; Li, S.; Imai, H.; Jia, L.; Umeda, J.; Takahashi, M.; Kondoh, K. Load transfer strengthening in carbon nanotubes reinforced metal matrix composites via in-situ tensile tests. *Compos. Sci. Technol.* **2015**, *113*, 1–8. [CrossRef]

10. Zapata-Solvas, E.; Gómez-García, D.; Domínguez-Rodríguez, A. Towards physical properties tailoring of carbon nanotubes-reinforced ceramic matrix composites. *J. Eur. Ceram. Soc.* **2012**, *32*, 3001–3020. [CrossRef]

11. Haque, A.; Ramasetty, A. Theoretical study of stress transfer in carbon nanotube reinforced polymer matrix composites. *Compos. Struct.* **2005**, *71*, 68–77. [CrossRef]

12. Nai, M.H.; Wei, J.; Gupta, M. Interface tailoring to enhance mechanical properties of carbon nanotube reinforced magnesium composites. *Mater. Des.* **2014**, *60*, 490–495. [CrossRef]

13. Kim, I.Y.; Lee, J.H.; Lee, G.S.; Baik, S.H.; Kim, Y.J.; Lee, Y.Z. Friction and wear characteristics of the carbon nanotube–aluminum composites with different manufacturing conditions. *Wear* **2009**, *267*, 593–598. [CrossRef]

14. Wu, J.; Zhang, H.; Zhang, Y.; Wang, X. Mechanical and thermal properties of carbon nanotube/aluminum composites consolidated by spark plasma sintering. *Mater. Des.* **2012**, *41*, 344–348. [CrossRef]

15. Morsi, K.; Esawi, A.; Borah, P.; Lanka, S.; Sayed, A. Characterization and spark plasma sintering of mechanically milled aluminum-carbon nanotube (CNT) composite powders. *J. Compos. Mater.* **2010**, *44*, 1991–2003. [CrossRef]

16. Nie, J.H.; Jia, C.C.; Shi, N.; Zhang, Y.F.; Li, Y.; Jia, X. Aluminum matrix composites reinforced by molybdenum-coated carbon nanotubes. *Int. J. Miner. Metall. Mater.* **2011**, *18*, 695–702. [CrossRef]

17. Kurita, H.; Kwon, H.; Estili, M.; Kawasaki, A. Multi-walled carbon nanotube-aluminum matrix composites prepared by combination of hetero-agglomeration method, spark plasma sintering and hot extrusion. *Mater. Trans.* **2011**, *52*, 1960–1965. [CrossRef]

18. Kondoh, K.; Threrujirapapong, T.; Umeda, J.; Fugetsu, B. High-temperature properties of extruded titanium composites fabricated from carbon nanotubes coated titanium powder by spark plasma sintering and hot extrusion. *Compos. Sci. Technol.* **2012**, *72*, 1291–1297. [CrossRef]

19. Paramsothy, M.; Tan, X.; Chan, J.; Kwok, R.; Gupta, M. Carbon nanotube addition to concentrated magnesium alloy AZ81: Enhanced ductility with occasional significant increase in strength. *Mater. Des.* **2013**, *45*, 15–23. [CrossRef]

20. Imai, H.; Kondoh, K.; Li, S.; Umeda, J.; Fugetsu, B.; Takahashi, M. Microstructural and Electrical Properties of Copper–Titanium Alloy Dispersed with Carbon Nanotubes via Powder Metallurgy Process. *Mater. Trans.* **2014**, *55*, 522–527. [CrossRef]

21. Azarniya, A.; Sovizi, S.; Azarniya, A.; Varol, T.; Nithyadharsenia, P.; Hosseini, H.R.M.; Ramakrishna, S.; Reddy, M.V. Physicomechanical properties of spark plasma sintered carbon nanotube-containing ceramic matrix nanocomposites. *Nanoscale* **2017**, *9*, 12779–12820. [CrossRef] [PubMed]

22. Azarniya, A.; Azarniya, A.; Sovizi, S.; Hosseini, H.R.M.; Varol, T.; Kawasaki, A.; Ramakrishna, S. Physicomechanical properties of spark plasma sintered carbon nanotube-reinforced metal matrix nanocomposites. *Prog. Mater. Sci.* **2017**, *90*, 276–324. [CrossRef]

23. Ma, P.C.; Siddiqui, N.A.; Marom, G.; Kim, J.K. Dispersion and functionalization of carbon nanotubes for polymer-based nanocomposites: A review. *Compos. Part A Appl. Sci. Manuf.* **2010**, *41*, 1345–1367. [CrossRef]

24. Islam, M.; Rojas, E.; Bergey, D.; Johnson, A.; Yodh, A. High weight fraction surfactant solubilization of single-wall carbon nanotubes in water. *Nano Lett.* **2003**, *3*, 269–273. [CrossRef]

25. Dai, H. Carbon nanotubes: Opportunities and challenges. *Surf. Sci.* **2002**, *500*, 218–241. [CrossRef]

26. Bandyopadhyaya, R.; Nativ-Roth, E.; Regev, O.; Yerushalmi-Rozen, R. Stabilization of individual carbon nanotubes in aqueous solutions. *Nano Lett.* **2002**, *2*, 25–28. [CrossRef]

27. Wang, Y.; Iqbal, Z.; Mitra, S. Rapidly functionalized, water-dispersed carbon nanotubes at high concentration. *J. Am. Chem. Soc.* **2006**, *128*, 95–99. [CrossRef] [PubMed]

28. Vaisman, L.; Wagner, H.D.; Marom, G. The role of surfactants in dispersion of carbon nanotubes. *Adv. Colloid Interface Sci.* **2006**, *128*, 37–46. [CrossRef] [PubMed]

29. Alig, I.; Pötschke, P.; Lellinger, D.; Skipa, T.; Pegel, S.; Kasaliwal, G.R.; Villmow, T. Establishment, morphology and properties of carbon nanotube networks in polymer melts. *Polymer* **2012**, *53*, 4–28. [CrossRef]

30. Star, A.; Stoddart, J.F.; Steuerman, D.; Diehl, M.; Boukai, A.; Wong, E.W.; Yang, X.; Chung, S.W.; Choi, H.; Heath, J.R. Preparation and properties of polymer-wrapped single-walled carbon nanotubes. *Angew. Chem. Int. Ed.* **2001**, *40*, 1721–1725.

31. Hirsch, A. Functionalization of single-walled carbon nanotubes. *Angew. Chem. Int. Ed.* **2002**, *41*, 1853–1859. [CrossRef]

32. Grumezescu, A. *Engineering of Nanobiomaterials: Applications of Nanobiomaterials*; William Andrew: Norwich, NY, USA, 2016.

33. Kwon, H.; Kawasaki, A. *Effect of Spark Plasma Sintering in Fabricating Carbon Nanotube Reinforced Aluminum Matrix Composite Materials*; INTECH: Vienna, Austria, 2011.

34. Bakshi, S.R.; Musaramthota, V.; Lahiri, D.; Singh, V.; Seal, S.; Agarwal, A. Spark plasma sintered tantalum carbide: Effect of pressure and nano-boron carbide addition on microstructure and mechanical properties. *Mater. Sci. Eng. A* **2011**, *528*, 1287–1295. [CrossRef]

35. Daoush, W.M. Processing and characterization of CNT/Cu nanocomposites by powder technology. *Powder Metall. Met. Ceram.* **2008**, *47*, 531–537. [CrossRef]

36. Yadav, V.; Harimkar, S.P. Microstructure and properties of spark plasma sintered carbon nanotube reinforced aluminum matrix composites. *Adv. Eng. Mater.* **2011**, *13*, 1128–1134. [CrossRef]

37. Umma, A.; Maleque, M.A.; Iskandar, I.Y.; Ali, M.Y. Effect of Ball Milling Parameters on the Synthesization of Carbon Nanotube Aluminium Nano Composite. In *Advanced Materials Research*; Trans Tech Publ: Zurich, Switzerland, 2013.

38. Esawi, A.; Morsi, K.; Sayed, A.; Taher, M.; Lanka, S. The influence of carbon nanotube (CNT) morphology and diameter on the processing and properties of CNT-reinforced aluminium composites. *Compos. Part A Appl. Sci. Manuf.* **2011**, *42*, 234–243. [CrossRef]

39. Esawi, A.; Morsi, K. Dispersion of carbon nanotubes (CNTs) in aluminum powder. *Compos. Part A Appl. Sci. Manuf.* **2007**, *38*, 646–650. [CrossRef]

40. Rather, S.U.; Nahm, K.S. Hydrogen uptake of high-energy ball milled nickel-multiwalled carbon nanotube composites. *Mater. Res. Bull.* **2014**, *49*, 525–530. [CrossRef]

41. Yoon, E.Y.; Lee, D.J.; Park, B.; Akbarpour, M.; Farvizi, M.; Kim, H.S. Grain refinement and tensile strength of carbon nanotube-reinforced Cu matrix nanocomposites processed by high-pressure torsion. *Met. Mater. Int.* **2013**, *19*, 927. [CrossRef]

42. Vishlaghi, M.B.; Ataie, A. Investigation on solid solubility and physical properties of Cu–Fe/CNT nano-composite prepared via mechanical alloying route. *Powder Technol.* **2014**, *268*, 102–109.

43. Zhao, S.; Zheng, Z.; Huang, Z.; Dong, S.; Luo, P.; Zhang, Z.; Wang, Y. Cu matrix composites reinforced with aligned carbon nanotubes: Mechanical, electrical and thermal properties. *Mater. Des.* **2016**, *675*, 82–91. [CrossRef]

44. Xu, J.D.; Zhu, K.T.; Weng, X.F.; Weng, W.Z.; Huang, C.J.; Wan, H.L. Carbon nanotube-supported Fe–Mn nanoparticles: A model catalyst for direct conversion of syngas to lower olefins. *Catal. Today* **2013**, *215*, 86–94. [CrossRef]

45. Liao, J.Z.; Tan, M.J.; Sridhar, I. Spark plasma sintered multi-wall carbon nanotube reinforced aluminum matrix composites. *Mater. Des.* **2010**, *31*, S96–S100. [CrossRef]

46. Nguyen, J.; Holland, T.; Wen, H.; Fraga, M.; Mukherjee, A.; Lavernia, E. Mechanical behavior of ultrafine-grained Ni-carbon nanotube composite. *J. Mater. Sci.* **2014**, *49*, 2070–2077. [CrossRef]

47. Nguyen, J.; Wen, H.; Zhang, Z.; Yaghmaie, F.; Lavernia, E. Surfactant assisted dispersion and adhesion behavior of carbon nanotubes on Cu–Zr and Cu–Zr–Al amorphous powders. *J. Mater. Sci. Technol.* **2014**, *30*, 847–853. [CrossRef]

48. Nie, J.H.; Jia, C.C.; Jia, X.; Li, Y.; Zhang, Y.F.; Liang, X.B. Fabrication and thermal conductivity of copper matrix composites reinforced by tungsten-coated carbon nanotubes. *Int. J. Miner. Metall. Mater.* **2012**, *19*, 446–452. [CrossRef]

49. Kim, K.T.; Cha, S.I.; Hong, S.H. Hardness and wear resistance of carbon nanotube reinforced Cu matrix nanocomposites. *Mater. Sci. Eng. A* **2007**, *449*, 46–50. [CrossRef]

50. Xue, Z.; Wang, L.; Zhao, P.; Xu, S.; Qi, J.; Fei, W. Microstructures and tensile behavior of carbon nanotubes reinforced Cu matrix composites with molecular-level dispersion. *Mater. Des.* **2012**, *34*, 298–301. [CrossRef]

51. Pal, H.; Sharma, V.; Sharma, M. Thermal expansion behavior of CNT/Ag nanocomposite. *Int. J. Mater. Res.* **2014**, *105*, 566–570. [CrossRef]

52. Hwang, J.; Lim, B.; Tiley, J.; Banerjee, R.; Hong, S. Interface analysis of ultra-high strength carbon nanotube/nickel composites processed by molecular level mixing. *Carbon* **2013**, *57*, 282–287. [CrossRef]

53. Sharma, H.P.V. Mechanical, electrical, and thermal expansion properties of carbon nanotube-based silver and silver-palladium alloy composites. *Int. J. Miner. Metall. Mater.* **2014**, *21*, 1132–1140.

54. Maqbool, A.; Khalid, F.A.; Hussain, M.A.; Bakhsh, N. Synthesis of Copper Coated Carbon Nanotubes for Aluminium Matrix Composites. In *IOP Conference Series: Materials Science and Engineering*; IOP Publishing: Bristol, UK, 2014.

55. Nam, D.H.; Cha, S.I.; Lim, B.K.; Park, H.M.; Han, D.S.; Hong, S.H. Synergistic strengthening by load transfer mechanism and grain refinement of CNT/Al–Cu composites. *Carbon* **2012**, *50*, 2417–2423. [CrossRef]

56. Sun, F.; Shi, C.; Rhee, K.Y.; Zhao, N. In situ synthesis of CNTs in Mg powder at low temperature for fabricating reinforced Mg composites. *J. Alloys Compd.* **2013**, *551*, 496–501.

57. Moonngam, S.; Tunjina, P.; Deesom, D.; Banjongprasert, C. Fe-Cr/CNTs nanocomposite feedstock powders produced by chemical vapor deposition for thermal spray coatings. *Surf. Coat. Technol.* **2016**, *306*, 323–327. [CrossRef]

58. Wang, S.; Liu, P.; Chen, X.; Liu, X.; Li, W.; Ma, F.C.; He, D.H. Effects of Growth Parameters on the Morphology of CNTs/Cu Composite Powder Prepared Using Cr/Cu Catalyst by Chemical Vapor Deposition. *Rare Met. Mater. Eng.* **2015**, *44*, 1832–1837.

59. Meng, X.; Liu, T.; Shi, C.; Liu, E.; He, C.; Zhao, N. Synergistic effect of CNTs reinforcement and precipitation hardening in in-situ CNTs/Al–Cu composites. *Mater. Sci. Eng. A* **2015**, *633*, 103–111. [CrossRef]

60. Yang, X.; Zou, T.; Shi, C.; Liu, E.; He, C.; Zhao, N. Effect of carbon nanotube (CNT) content on the properties of in-situ synthesis CNT reinforced Al composites. *Mater. Sci. Eng. A* **2016**, *660*, 11–18. [CrossRef]

61. Kozhuharova-Koseva, R.; Hofmann, M.; Leonhardt, A.; Mönch, I.; Mühl, T.; Ritschel, M.; Büchner, B. Relation between Growth Parameters and Morphology of Vertically Aligned Fe-filled Carbon Nanotubes. *Fuller. Nanotub. Carbon Nanostruct.* **2007**, *15*, 135–143. [CrossRef]

62. Silvain, J.F.; Vincent, C.; Heintz, J.M.; Chandra, N. Novel processing and characterization of Cu/CNF nanocomposite for high thermal conductivity applications. *Compos. Sci. Technol.* **2009**, *69*, 2474–2484. [CrossRef]

63. Sung-Kyu, K.; Tae-Sung, O. Electrodeposition behavior and characteristics of Ni-carbon nanotube composite coatings. *Trans. Nonferrous Met. Soc. China* **2011**, *21*, s68–s72.

64. Ko, S.Y.; Kim, B.Y.; Kim, Y.I.; Kim, T.Y.; Kim, K.T.; McKay, B.J.; Shin, J.S. Manufacture of CNTs-Al Powder Precursors for Casting of CNTs-Al Matrix Composites. In *Materials Science Forum*; Trans Tech Publ: Zurich, Switzerland, 2013.

65. Dai, L.; Qu, L. Substrate-Enhanced Electroless Deposition (SEED) of Metal Nanoparticles on Carbon Nanotubes. U.S. Patent 7,538,062, 26 May 2009.

66. Jeon, Y.; Byun, J.; Oh, T. Electrodeposition and mechanical properties of Ni-carbon nanotube nanocomposite coatings. *J. Phys. Chem. Solids* **2008**, *69*, 1391–1394. [CrossRef]

67. Arai, S.; Saito, T.; Endo, M. Cu–MWCNT composite films fabricated by electrodeposition. *J. Electrochem. Soc.* **2010**, *157*, D147–D153. [CrossRef]

68. Zhang, S.; Chen, Q. Fabrication of MWCNT incorporated Sn–Bi composite. *Compos. Part B Eng.* **2014**, *58*, 275–278. [CrossRef]

69. Ji, K.; Zhao, H.; Zhang, J.; Chen, J.; Dai, Z. Fabrication and electromagnetic interference shielding performance of open-cell foam of a Cu–Ni alloy integrated with CNTs. *Appl. Surf. Sci.* **2014**, *311*, 351–356. [CrossRef]

70. Karslioglu, R.; Akbulut, H. Comparison microstructure and sliding wear properties of nickel–cobalt/CNT composite coatings by DC, PC and PRC current electrodeposition. *Appl. Surf. Sci.* **2015**, *353*, 615–627. [CrossRef]

71. Kwon, H.; Leparoux, M.; Kawasaki, A. Functionally graded dual-nanoparticulate-reinforced aluminium matrix bulk materials fabricated by spark plasma sintering. *J. Mater. Sci. Technol.* **2014**, *30*, 736–742. [CrossRef]

72. Akbarpour, M.; Salahi, E.; Hesari, F.A.; Simchi, A.; Kim, H. Fabrication, characterization and mechanical properties of hybrid composites of copper using the nanoparticulates of SiC and carbon nanotubes. *Mater. Sci. Eng. A* **2013**, *572*, 83–90. [CrossRef]

73. Thakur, S.K.; Kwee, G.T.; Gupta, M. Development and characterization of magnesium composites containing nano-sized silicon carbide and carbon nanotubes as hybrid reinforcements. *J. Mater. Sci.* **2007**, *42*, 10040–10046. [CrossRef]

74. Kim, H.; Babu, J.; Kang, C. Fabrication of A356 aluminum alloy matrix composite with CNTs/Al$_2$O$_3$ hybrid reinforcements. *Mater. Sci. Eng. A* **2013**, *573*, 92–99. [CrossRef]

75. Babu, J.; Srinivasan, A.; Kang, C. Nano and macromechanical properties of aluminium (A356) based hybrid composites reinforced with multiwall carbon nanotubes/alumina fiber. *J. Compos. Mater.* **2017**, *51*, 1631–1642.

76. Du, Z.; Tan, M.J.; Guo, J.F.; Wei, J. Aluminium-carbon nanotubes composites produced from friction stir processing and selective laser melting. *Materialwiss. Werkst.* **2016**, *47*, 539–548. [CrossRef]

77. Jafari, J.; Givi, M.K.B.; Barmouz, M. Mechanical and microstructural characterization of Cu/CNT nanocomposite layers fabricated via friction stir processing. *Int. J. Adv. Manuf. Technol.* **2015**, *78*, 199–209. [CrossRef]

78. Mansoor, M.; Shahid, M. Carbon nanotube-reinforced aluminum composite produced by induction melting. *J. Appl. Res. Technol.* **2016**, *14*, 215–224. [CrossRef]

79. Wilson, K.; Barrera, E.; Bayazitoglu, Y. Processing of titanium single-walled carbon nanotube metal-matrix composites by the induction melting method. *J. Compos. Mater.* **2010**, *44*, 1037–1048. [CrossRef]

80. Elshalakany, A.B.; Osman, T.; Khattab, A.; Azzam, B.; Zaki, M. Microstructure and mechanical properties of MWCNTs reinforced A356 aluminum alloys cast nanocomposites fabricated by using a combination of rheocasting and squeeze casting techniques. *J. Nanomater.* **2014**, *2014*, 386370. [CrossRef]

81. Ma, P.C.; Tang, B.Z.; Kim, J.K. Conversion of semiconducting behavior of carbon nanotubes using ball milling. *Chem. Phys. Lett.* **2008**, *458*, 166–169. [CrossRef]

82. Woo, D.J.; Hooper, J.P.; Osswald, S.; Bottolfson, B.A.; Brewer, L.N. Low temperature synthesis of carbon nanotube-reinforced aluminum metal composite powders using cryogenic milling. *J. Mater. Res.* **2014**, *29*, 2644–2656. [CrossRef]

83. Yoo, S.; Han, S.; Kim, W. A combination of ball milling and high-ratio differential speed rolling for synthesizing carbon nanotube/copper composites. *Carbon* **2013**, *61*, 487–500. [CrossRef]

84. Atif, R.; Inam, F. Reasons and remedies for the agglomeration of multilayered graphene and carbon nanotubes in polymers. *Beilstein J. Nanotechnol.* **2016**, *7*, 1174–1196. [CrossRef] [PubMed]

85. Liao, J.; Tan, M.J. Mixing of carbon nanotubes (CNTs) and aluminum powder for powder metallurgy use. *Powder Technol.* **2011**, *208*, 42–48. [CrossRef]

86. Zhou, S.; Wu, C.; Zhang, T.; Zhang, Z. Carbon nanotube-and Fe$_p$-reinforced copper–matrix composites by laser induction hybrid rapid cladding. *Scr. Mater.* **2014**, *76*, 25–28. [CrossRef]

87. Ostovan, F.; Matori, K.A.; Toozandehjani, M.; Oskoueian, A.; Yusoff, H.M.; Yunus, R.; Ariff, A.H.M.; Quah, H.J.; Lim, W.F. Effects of CNTs content and milling time on mechanical behavior of MWCNT-reinforced aluminum nanocomposites. *Mater. Chem. Phys.* **2015**, *166*, 160–166. [CrossRef]

88. Strano, M.S.; Moore, V.C.; Miller, M.K.; Allen, M.J.; Haroz, E.H.; Kittrell, C.; Hauge, R.H.; Smalley, R. The role of surfactant adsorption during ultrasonication in the dispersion of single-walled carbon nanotubes. *J. Nanosci. Nanotechnol.* **2003**, *3*, 81–86. [CrossRef] [PubMed]

89. Koshio, A.; Yudasaka, M.; Zhang, M.; Iijima, S. A simple way to chemically react single-wall carbon nanotubes with organic materials using ultrasonication. *Nano Lett.* **2001**, *1*, 361–363. [CrossRef]

90. Bonard, J.M.; Stora, T.; Salvetat, J.P.; Maier, F.; Stöckli, T.; Duschl, C.; Forró, L.; Heer, W.A.; Châtelain, A. Purification and size-selection of carbon nanotubes. *Adv. Mater.* **2010**, *9*, 827–831. [CrossRef]

91. Fugetsu, B.; Han, W.; Endo, N.; Kamiya, Y.; Okuhara, T. Disassembling single-walled carbon nanotube bundles by dipole/dipole electrostatic interactions. *Chem. Lett.* **2005**, *34*, 1218–1219. [CrossRef]

92. Kondoh, K.; Fukuda, H.; Umeda, J.; Imai, H.; Fugetsu, B. Microstructural and mechanical behavior of multi-walled carbon nanotubes reinforced Al–Mg–Si alloy composites in aging treatment. *Carbon* **2014**, *72*, 15–21. [CrossRef]

93. Assovskiy, I.G.; Berlin, A.A. Metallized Carbon Nanotubes. *Int. J. Energ. Mater. Chem. Propuls.* **2009**, *8*, 281–289. [CrossRef]

94. Cross, R.; Cola, B.A.; Fisher, T.; Xu, X.; Gall, K.; Graham, S. A metallization and bonding approach for high performance carbon nanotube thermal interface materials. *Nanotechnology* **2010**, *21*, 445705. [CrossRef] [PubMed]

95. Jagannatham, M.; Sankaran, S.; Haridoss, P. Microstructure and mechanical behavior of copper coated multiwall carbon nanotubes reinforced aluminum composites. *Mater. Sci. Eng. A* **2015**, *638*, 197–207. [CrossRef]

96. Daoush, W.M.; Lim, B.K.; Nam, D.H.; Hong, S.H. Microstructure and mechanical properties of CNT/Ag nanocomposites fabricated by spark plasma sintering. *J. Exp. Nanosci.* **2014**, *9*, 588–596. [CrossRef]

97. Kondoh, K.; Fukuda, H.; Umeda, J.; Imai, H.; Fugetsu, B.; Endo, M. Microstructural and mechanical analysis of carbon nanotube reinforced magnesium alloy powder composites. *Mater. Sci. Eng. A* **2010**, *527*, 4103–4108. [CrossRef]

98. Chu, K.; Wu, Q.; Jia, C.; Liang, X.; Nie, J.; Tian, W.; Gai, G.; Guo, H. Fabrication and effective thermal conductivity of multi-walled carbon nanotubes reinforced Cu matrix composites for heat sink applications. *Compos. Sci. Technol.* **2010**, *70*, 298–304. [CrossRef]

99. Khaleghi, E.; Torikachvili, M.; Meyers, M.A.; Olevsky, E.A. Magnetic enhancement of thermal conductivity in copper–carbon nanotube composites produced by electroless plating, freeze drying, and spark plasma sintering. *Mater. Lett.* **2012**, *79*, 256–258. [CrossRef]

100. Thostenson, E.T.; Karandikar, P.G.; Chou, T.W. Fabrication and characterization of reaction bonded silicon carbide/carbon nanotube composites. *J. Phys. D Appl. Phys.* **2005**, *38*, 3962. [CrossRef]

101. George, R.; Kashyap, K.; Rahul, R.; Yamdagni, S. Strengthening in carbon nanotube/aluminium (CNT/Al) composites. *Scr. Mater.* **2005**, *53*, 1159–1163. [CrossRef]

102. Varo, T.; Canakci, A. Effect of the CNT Content on Microstructure, Physical and Mechanical Properties of Cu-Based Electrical Contact Materials Produced by Flake Powder Metallurgy. *Arab. J. Sci. Eng.* **2015**, *40*, 2711–2720. [CrossRef]

103. Yang, C.; Chen, Q. Electric resistance of carbon nanotube with a Cu chain: A first-principle calculation. *Proc. Inst. Mech. Eng. Part N J. Nanomater. Nanoeng. Nanosyst.* **2013**, *227*, 115–119. [CrossRef]

104. Kurita, H.; Estili, M.; Kwon, H.; Miyazaki, T.; Zhou, W.; Silvain, J.F.; Kawasaki, A. Load-bearing contribution of multi-walled carbon nanotubes on tensile response of aluminum. *Compos. Part A Appl. Sci. Manuf.* **2015**, *68*, 133–139. [CrossRef]

105. Zhang, F.; Shen, J.; Sun, J.; Zhu, Y.Q.; Wang, G.; McCartney, G. Conversion of carbon nanotubes to diamond by spark plasma sintering. *Carbon* **2005**, *43*, 1254–1258. [CrossRef]

106. Li, J.; Wang, L.; He, T.; Jiang, W. Surface graphitization and mechanical properties of hot-pressed bulk carbon nanotubes compacted by spark plasma sintering. *Carbon* **2007**, *45*, 2636–2642. [CrossRef]

107. Zanganeh, N.; Rajabi, A.; Torabi, M.; Allahkarami, M.; Moghaddas, A.; Sadrnezhaad, S. Growth and microstructural investigation of multiwall carbon nanotubes fabricated using electrodeposited nickel nanodeposits and chemical vapor deposition method. *J. Mol. Struct.* **2014**, *1074*, 250–254. [CrossRef]

108. Fu, X.; Cui, X.; Wei, X.; Ma, J. Investigation of low and mild temperature for synthesis of high quality carbon nanotubes by chemical vapor deposition. *Appl. Surf. Sci.* **2014**, *292*, 645–649. [CrossRef]

109. Kosynkin, D.V.; Higginbotham, A.L.; Sinitskii, A.; Lomeda, J.R.; Dimiev, A.; Price, B.K.; Tour, J.M. Longitudinal unzipping of carbon nanotubes to form graphene nanoribbons. *Nature* **2009**, *458*, 872–876. [CrossRef] [PubMed]

110. Pei, Z.; Li, K.; Gong, J.; Shi, N.; Elangovan, E.; Sun, C. Micro-structural and tensile strength analyses on the magnesium matrix composites reinforced with coated carbon fiber. *J. Mater. Sci.* **2009**, *44*, 4124–4131. [CrossRef]

111. Guo, C.; Zhan, Z.; Zhang, D. Influence of Different Preparation Processes on the Mechanical Properties of Carbon Nanotube-Reinforced Copper Matrix Composites. *Strength Mater.* **2015**, *47*, 143–149. [CrossRef]

112. Tatami, J.; Katashima, T.; Komeya, K.; Meguro, T.; Wakihara, T. Electrically Conductive CNT-Dispersed Silicon Nitride Ceramics. *J. Am. Ceram. Soc.* **2005**, *88*, 2889–2893. [CrossRef]

113. Zhang, F.; Shen, J.; Sun, J.; McCartney, D. Direct synthesis of diamond from low purity carbon nanotubes. *Carbon* **2006**, *44*, 3136–3138. [CrossRef]

114. Hojati-Talemi, P.; Kannan, A.G.; Simon, G.P. Fusion of carbon nanotubes for fabrication of field emission cathodes. *Carbon* **2012**, *50*, 356–361. [CrossRef]

115. Sun, L.; Gong, J.; Zhu, Z.; Zhu, D.; He, S.; Wang, Z.; Chen, Y.; Hu, G. Nanocrystalline diamond from carbon nanotubes. *Appl. Phys. Lett.* **2004**, *84*, 2901–2903. [CrossRef]

116. Kim, Y.; Muramatsu, H.; Hayashi, T.; Endo, M.; Terrones, M.; Dresselhaus, M. Thermal stability and structural changes of double-walled carbon nanotubes by heat treatment. *Chem. Phys. Lett.* **2004**, *398*, 87–92. [CrossRef]

117. Wang, W.; Cao, L. Transformation of carbon nanotubes to diamond at high pressure and high temperature. *Russ. Phys. J.* **2001**, *44*, 178–182. [CrossRef]
118. Gutierrez, H.; Kim, U.; Kim, J.; Eklund, P. Thermal conversion of bundled carbon nanotubes into graphitic ribbons. *Nano Lett.* **2005**, *5*, 2195–2201. [CrossRef] [PubMed]
119. Che, G.; Lakshmi, B.; Martin, C.; Fisher, E.; Ruoff, R.S. Chemical vapor deposition based synthesis of carbon nanotubes and nanofibers using a template method. *Chem. Mater.* **1998**, *10*, 260–267. [CrossRef]
120. Kwon, H.; Estili, M.; Takagi, K.; Miyazaki, T.; Kawasaki, A. Combination of hot extrusion and spark plasma sintering for producing carbon nanotube reinforced aluminum matrix composites. *Carbon* **2009**, *47*, 570–577. [CrossRef]
121. Yang, K.; Hitchcock, D.; He, J.; Rao, A.M. Tuning Electrical Properties of Carbon Nanotubes via Spark Plasma Sintering. In *Encyclopedia of Nanotechnology*; Springer: Berlin/Heidelberg, Germany, 2012; pp. 2780–2788.
122. Zhang, F.; Mihoc, C.; Ahmed, F.; Lathe, C.; Burkel, E. Thermal stability of carbon nanotubes, fullerene and graphite under spark plasma sintering. *Chem. Phys. Lett.* **2011**, *510*, 109–114. [CrossRef]
123. Zhang, Z.H.; Qi, L.; Shen, X.B.; Wang, F.C.; Lee, S.K. Microstructure and mechanical properties of bulk carbon nanotubes compacted by spark plasma sintering. *Mater. Sci. Eng. A* **2013**, *573*, 12–17. [CrossRef]
124. Zhang, F.; Adam, M.; Ahmed, F.; Otterstein, E.; Burkel, E. Pulsed electric field induced diamond synthesis from carbon nanotubes with solvent catalysts. *Diam. Relat. Mater.* **2011**, *20*, 853–858. [CrossRef]
125. Yusa, H. Nanocrystalline diamond directly transformed from carbon nanotubes under high pressure. *Diam. Relat. Mater.* **2002**, *11*, 87–91. [CrossRef]
126. Kim, K.; Sussman, A.; Zettl, A. Graphene nanoribbons obtained by electrically unwrapping carbon nanotubes. *ACS Nano* **2010**, *4*, 1362–1366. [CrossRef] [PubMed]
127. Sumiya, H.; Yusa, H.; Inoue, T.; Ofuji, H.; Irifune, T. Conditions and mechanism of formation of nano-polycrystalline diamonds on direct transformation from graphite and non-graphitic carbon at high pressure and temperature. *High Press. Res.* **2006**, *26*, 63–69. [CrossRef]
128. He, C.; Zhao, N.; Shi, C.; Song, S. Mechanical properties and microstructures of carbon nanotube-reinforced Al matrix composite fabricated by in situ chemical vapor deposition. *J. Alloys Compd.* **2009**, *487*, 258–262. [CrossRef]
129. Simões, S.; Viana, F.; Reis, M.A.; Vieira, M.F. Influence of dispersion/mixture time on mechanical properties of Al–CNTs nanocomposites. *Compos. Struct.* **2015**, *126*, 114–122. [CrossRef]
130. Kwon, H.; Leparoux, M. Hot extruded carbon nanotube reinforced aluminum matrix composite materials. *Nanotechnology* **2012**, *23*, 415701. [CrossRef] [PubMed]
131. Bakshi, S.R.; Keshri, A.K.; Singh, V.; Seal, S.; Agarwal, A. Interface in carbon nanotube reinforced aluminum silicon composites: Thermodynamic analysis and experimental verification. *J. Alloys Compd.* **2009**, *481*, 207–213. [CrossRef]
132. Zhou, W.; Bang, S.; Kurita, H.; Miyazaki, T.; Fan, Y.; Kawasaki, A. Interface and interfacial reactions in multi-walled carbon nanotube-reinforced aluminum matrix composites. *Carbon* **2016**, *96*, 919–928. [CrossRef]
133. Ci, L.; Ryu, Z.; Jin-Phillipp, N.Y.; Rühle, M. Investigation of the interfacial reaction between multi-walled carbon nanotubes and aluminum. *Acta Mater.* **2006**, *54*, 5367–5375. [CrossRef]
134. Sridhar, I.; Narayanan, K.R. Processing and characterization of MWCNT reinforced aluminum matrix composites. *J. Mater. Sci.* **2009**, *44*, 1750–1756. [CrossRef]
135. Kwon, H.; Park, D.H.; Silvain, J.F.; Kawasaki, A. Investigation of carbon nanotube reinforced aluminum matrix composite materials. *Compos. Sci. Technol.* **2010**, *70*, 546–550. [CrossRef]
136. Tjong, S.C. Recent progress in the development and properties of novel metal matrix nanocomposites reinforced with carbon nanotubes and graphene nanosheets. *Mater. Sci. Eng. R Rep.* **2013**, *74*, 281–350. [CrossRef]
137. Esawi, A.; Morsi, K.; Sayed, A.; Taher, M.; Lanka, S. Effect of carbon nanotube (CNT) content on the mechanical properties of CNT-reinforced aluminium composites. *Compos. Sci. Technol.* **2010**, *70*, 2237–2241. [CrossRef]
138. Chen, B.; Shen, J.; Ye, X.; Imai, H.; Umeda, J.; Takahashi, M.; Kondoh, K. Solid-state interfacial reaction and load transfer efficiency in carbon nanotubes (CNTs)-reinforced aluminum matrix composites. *Carbon* **2017**, *114*, 198–208. [CrossRef]

139. Pérez-Bustamante, R.; Gómez-Esparza, C.; Estrada-Guel, I.; Miki-Yoshida, M.; Licea-Jiménez, L.; Pérez-García, S.; Martínez-Sánchez, R. Microstructural and mechanical characterization of Al–MWCNT composites produced by mechanical milling. *Mater. Sci. Eng. A* **2009**, *502*, 159–163. [CrossRef]

140. Kondoh, K.; Threrujirapapong, T.; Imai, H.; Umeda, J.; Fugetsu, B. Characteristics of powder metallurgy pure titanium matrix composite reinforced with multi-wall carbon nanotubes. *Compos. Sci. Technol.* **2009**, *69*, 1077–1081. [CrossRef]

141. Xue, F.; Jiehe, S.; Yan, F.; Wei, C. Preparation and elevated temperature compressive properties of multi-walled carbon nanotube reinforced Ti composites. *Mater. Sci. Eng. A* **2010**, *527*, 1586–1589. [CrossRef]

142. Li, S.; Sun, B.; Imai, H.; Mimoto, T.; Kondoh, K. Powder metallurgy titanium metal matrix composites reinforced with carbon nanotubes and graphite. *Compos. Part A Appl. Sci. Manuf.* **2013**, *48*, 57–66. [CrossRef]

143. Munir, K.S.; Zheng, Y.; Zhang, D.; Lin, J.; Li, Y.; Wen, C. Improving the strengthening efficiency of carbon nanotubes in titanium metal matrix composites. *Mater. Sci. Eng. A* **2017**, *696*, 10–25. [CrossRef]

144. Fukuda, H.; Kondoh, K.; Umeda, J.; Fugetsu, B. Interfacial analysis between Mg matrix and carbon nanotubes in Mg–6 wt %Al alloy matrix composites reinforced with carbon nanotubes. *Compos. Sci. Technol.* **2011**, *71*, 705–709. [CrossRef]

145. Park, Y.; Cho, K.; Park, I.; Park, Y. Fabrication and mechanical properties of magnesium matrix composite reinforced with Si coated carbon nanotubes. *Procedia Eng.* **2011**, *10*, 1446–1450. [CrossRef]

146. Rashad, M.; Pan, F.; Asif, M.; Li, L. Enhanced ductility of Mg–3Al–1Zn alloy reinforced with short length multi-walled carbon nanotubes using a powder metallurgy method. *Prog. Natl. Sci. Mater. Int.* **2015**, *25*, 276–281. [CrossRef]

147. Yoo, S.; Han, S.; Kim, W. Magnesium matrix composites fabricated by using accumulative roll bonding of magnesium sheets coated with carbon-nanotube-containing aluminum powders. *Scr. Mater.* **2012**, *67*, 129–132. [CrossRef]

148. Srivatsan, T.; Godbole, C.; Paramsothy, M.; Gupta, M. Influence of nano-sized carbon nanotube reinforcements on tensile deformation, cyclic fatigue, and final fracture behavior of a magnesium alloy. *J. Mater. Sci.* **2012**, *47*, 3621–3638. [CrossRef]

149. Chu, K.; Jia, C.C.; Li, W.S. Thermal conductivity enhancement in carbon nanotube/Cu-Ti composites. *Appl. Phys. A Mater. Sci. Process.* **2013**, *110*, 269–273. [CrossRef]

150. Kong, J.; Zhang, C.Y.; Cheng, X. Novel Cu-Cr alloy matrix CNT composites with enhanced thermal conductivity. *Appl. Phys. A* **2013**, *112*, 631–636. [CrossRef]

151. Gill, P.; Munroe, N. Study of carbon nanotubes in Cu-Cr metal matrix composites. *J. Mater. Eng. Perform.* **2012**, *21*, 2467–2471. [CrossRef]

152. Li, W.S.; Zhang, J.; Dong, H.F.; Chu, K.; Wang, S.C.; Liu, Y.; Li, Y.M. Thermodynamic and kinetic study on interfacial reaction and diamond graphitization of Cu-Fe-based diamond composite. *Chin. Phys. B* **2013**, *22*, 524–530. [CrossRef]

153. Cho, S.; Kikuchi, K.; Kawasaki, A.; Kwon, H.; Kim, Y. Effective load transfer by a chromium carbide nanostructure in a multi-walled carbon nanotube/copper matrix composite. *Nanotechnology* **2012**, *23*, 315705. [CrossRef] [PubMed]

154. Chu, K.; Jia, C.; Li, W.; Wang, P. Mechanical and electrical properties of carbon-nanotube-reinforced Cu–Ti alloy matrix composites. *Phys. Status Solidi Appl. Res.* **2013**, *210*, 594–599. [CrossRef]

155. Lin, D.; Liu, C.R.; Cheng, G.J. Laser sintering of separated and uniformly distributed multiwall carbon nanotubes integrated iron nanocomposites. *J. Appl. Phys.* **2014**, *115*, 113513. [CrossRef]

156. Bakshi, S.R.; Agarwal, A. An analysis of the factors affecting strengthening in carbon nanotube reinforced aluminum composites. *Carbon* **2011**, *49*, 533–544. [CrossRef]

157. Abdizadeh, H.; Ebrahimifard, R.; Baghchesara, M.A. Investigation of microstructure and mechanical properties of nano MgO reinforced Al composites manufactured by stir casting and powder metallurgy methods: A comparative study. *Compos. Part B Eng.* **2014**, *56*, 217–221. [CrossRef]

158. Coleman, J.N.; Cadek, M.; Blake, R.; Nicolosi, V.; Ryan, K.P.; Belton, C.; Fonseca, A.; Nagy, J.B.; Gun'ko, Y.K.; Blau, W.J. High performance nanotube-reinforced plastics: Understanding the mechanism of strength increase. *Adv. Funct. Mater.* **2004**, *14*, 791–798. [CrossRef]

159. Zhou, W.; Yamamoto, G.; Fan, Y.; Kwon, H.; Hashida, T.; Kawasaki, A. In-situ characterization of interfacial shear strength in multi-walled carbon nanotube reinforced aluminum matrix composites. *Carbon* **2016**, *106*, 37–47. [CrossRef]

160. Nozaka, Y.; Wang, W.; Shirasu, K.; Yamamoto, G.; Hashida, T. Inclined slit-based pullout method for determining interfacial strength of multi-walled carbon nanotube–alumina composites. *Carbon* **2014**, *78*, 439–445. [CrossRef]
161. Tsuda, T.; Ogasawara, T.; Deng, F.; Takeda, N. Direct measurements of interfacial shear strength of multi-walled carbon nanotube/PEEK composite using a nano-pullout method. *Compos. Sci. Technol.* **2011**, *71*, 1295–1300. [CrossRef]
162. Kwon, H.; Kawasaki, A. Extrusion of spark plasma sintered aluminum-carbon nanotube composites at various sintering temperatures. *J. Nanosci. Nanotechnol.* **2009**, *9*, 6542–6548. [CrossRef] [PubMed]
163. Kwon, H.; Takamichi, M.; Kawasaki, A.; Leparoux, M. Investigation of the interfacial phases formed between carbon nanotubes and aluminum in a bulk material. *Mater. Chem. Phys.* **2013**, *138*, 787–793. [CrossRef]
164. Ruehle, M.; Evans, A.G. Structure and chemistry of metal/ceramic interfaces. *Mater. Sci. Eng. A* **1989**, *107*, 187–197. [CrossRef]
165. Deng, C.; Wang, D.; Zhang, X.; Li, A. Processing and properties of carbon nanotubes reinforced aluminum composites. *Mater. Sci. Eng. A* **2007**, *444*, 138–145. [CrossRef]
166. Li, H.; Kang, J.; He, C.; Zhao, N.; Liang, C.; Li, B. Mechanical properties and interfacial analysis of aluminum matrix composites reinforced by carbon nanotubes with diverse structures. *Mater. Sci. Eng. A* **2013**, *577*, 120–124. [CrossRef]
167. Liu, Z.; Xiao, B.; Wang, W.; Ma, Z. Developing high-performance aluminum matrix composites with directionally aligned carbon nanotubes by combining friction stir processing and subsequent rolling. *Carbon* **2013**, *62*, 35–42. [CrossRef]
168. Zhou, W.; Yamaguchi, T.; Kikuchi, K.; Nomura, N.; Kawasaki, A. Effectively enhanced load transfer by interfacial reactions in multi-walled carbon nanotube reinforced Al matrix composites. *Acta Mater.* **2017**, *125*, 369–376. [CrossRef]
169. Sule, R.; Olubambi, P.; Sigalas, I.; Asante, J.; Garrett, J.; Roos, W. Spark plasma sintering of sub-micron copper reinforced with ruthenium–carbon nanotube composites for thermal management applications. *Synth. Met.* **2015**, *202*, 123–132. [CrossRef]
170. Cho, S.; Kikuchi, K.; Miyazaki, T.; Kawasaki, A.; Arami, Y.; Silvain, J.F. Epitaxial growth of chromium carbide nanostructures on multiwalled carbon nanotubes (MWCNTs) in MWCNT–copper composites. *Acta Mater.* **2013**, *61*, 708–716. [CrossRef]
171. Kathrein, H.; Gonska, H.; Freund, F. Subsurface segregation and diffusion of carbon in magnesium oxide. *Appl. Phys. A* **1983**, *30*, 33–41. [CrossRef]
172. Guo, B.; Ni, S.; Yi, J.; Shen, R.; Tang, Z.; Du, Y.; Song, M. Microstructures and mechanical properties of carbon nanotubes reinforced pure aluminum composites synthesized by spark plasma sintering and hot rolling. *Mater. Sci. Eng. A* **2017**, *698*, 292–298. [CrossRef]
173. Chen, B.; Li, S.; Imai, H.; Jia, L.; Umeda, J.; Takahashi, M.; Kondoh, K. Carbon nanotube induced microstructural characteristics in powder metallurgy Al matrix composites and their effects on mechanical and conductive properties. *J. Alloys Compd.* **2015**, *651*, 608–615. [CrossRef]
174. Suárez, S.; Ramos-Moore, E.; Lechthaler, B.; Mücklich, F. Grain growth analysis of multiwalled carbon nanotube-reinforced bulk Ni composites. *Carbon* **2014**, *70*, 173–178. [CrossRef]
175. Chen, B.; Shen, J.; Ye, X.; Jia, L.; Li, S.; Umeda, J.; Takahashi, M.; Kondoh, K. Length Effect of Carbon Nanotubes on the Strengthening Mechanisms in Metal Matrix Composites. *Acta Mater.* **2017**, *140*, 317–325. [CrossRef]
176. Hassanzadeh-Aghdam, M.; Mahmoodi, M. A comprehensive analysis of mechanical characteristics of carbon nanotube-metal matrix nanocomposites. *Mater. Sci. Eng. A* **2017**, *701*, 34–44. [CrossRef]

metals

MDPI

Article

Enhancing the Hardness and Compressive Response of Magnesium Using Complex Composition Alloy Reinforcement

Khin Sandar Tun [1], Yuming Zhang [2], Gururaj Parande [1], Vyasaraj Manakari [1] and Manoj Gupta [1,*]

1   Department of Mechanical Engineering, National University of Singapore, Singapore 117576, Singapore; mpekhst@nus.edu.sg (K.S.T.); gururaj.parande@u.nus.edu (G.P.); mbvyasaraj@u.nus.edu (V.M.)
2   Department of Mechanical & Industrial Engineering, University of Toronto St. George Campus, Toronto, ON M5S 1A1, Canada; yuming.zhang@mail.utoronto.ca
*   Correspondence: mpegm@nus.edu.sg; Tel.: +65-6516-6358

Received: 22 March 2018; Accepted: 14 April 2018; Published: 17 April 2018

Abstract: The present study reports the development of new magnesium composites containing complex composition alloy (CCA) particles. Materials were synthesized using a powder metallurgy route incorporating hybrid microwave sintering and hot extrusion. The presence and variation in the amount of ball-milled CCA particles (2.5 wt %, 5 wt %, and 7.5 wt %) in a magnesium matrix and their effect on the microstructure and mechanical properties of Mg-CCA composites were investigated. The use of CCA particle reinforcement effectively led to a significant matrix grain refinement. Uniformly distributed CCA particles were observed in the microstructure of the composites. The refined microstructure coupled with the intrinsically high hardness of CCA particles (406 HV) contributed to the superior mechanical properties of the Mg-CCA composites. A microhardness of 80 HV was achieved in a Mg-7.5HEA (high entropy alloy) composite, which is 1.7 times higher than that of pure Mg. A significant improvement in compressive yield strength (63%) and ultimate compressive strength (79%) in the Mg-7.5CCA composite was achieved when compared to that of pure Mg while maintaining the same ductility level. When compared to ball-milled amorphous particle-reinforced and ceramic-particle-reinforced Mg composites, higher yield and compressive strengths in Mg-CCA composites were achieved at a similar ductility level.

Keywords: magnesium; high entropy alloy; composite; hardness; compressive properties

1. Introduction

Magnesium (Mg) is the lightest of all structural metals and possesses the highest strength-to-density ratio. In addition, magnesium has other favorable advantages, including a high damping capacity, high dimensional stability, good machinability, good electromagnetic shielding characteristics, and recyclability. Accordingly, magnesium is the designers' choice for possible production of lightweight vehicles to meet the demand of reducing greenhouse emissions [1–3]. Magnesium is mostly used in the form of alloys in commercial applications [4,5]. With the advent of composite technology, research interest has been placed on the development of high-performance magnesium composites. By a careful selection of matrix and reinforcing phases, newly formed composite materials with significant improvements in elastic modulus, strength, ductility, and coefficient of thermal expansion can be fabricated. Composite materials are attractive because they offer the possibility for combining useful engineering properties of individual elements, which is otherwise not possible from monolithic materials. The attractive physical and mechanical properties obtained from metal matrix composites (MMCs) have made them potential candidates for aerospace, automotive, and other structural applications [6]. For the fabrication of magnesium composites, various types of ceramic reinforcements, such as alumina (Al_2O_3), yttria (Y_2O_3), zirconia (ZrO_2), silicon carbide (SiC), boron carbide (B_4C), and titanium carbide (TiC), have been used in Mg matrix [7–11]. In addition to ceramic reinforcements, research efforts have been made to synthesize Mg composites using metal particle reinforcements, such as copper, nickel, titanium, molybdenum, and aluminum [11,12]. Investigations have also been made on Mg composites containing hybrid reinforcements in the form of ceramic plus metal, ceramic plus ceramic, and ceramic plus carbon nanotube (CNT) besides single ceramic, metal, and CNT reinforcements [11,13–16]. In a recent development on magnesium composites, ball-milled amorphous particles were also used as a reinforcement in a pure magnesium matrix [17–19].

In the present study, an attempt is made to develop new magnesium composites using ball-milled complex composition alloy (CCA) particles. From the thermodynamic calculation, the current complex composition alloy has a high entropy value (ΔS_{mix}) of 12.97 J/K mol. Based on the concept of configurational entropy, the multicomponent alloys having a mixing or configurational entropy value which is equivalent to or greater than 12.471 J/K mol ($\Delta S_{conf} \geq 1.5R$ (12.471), where R is the gas constant), are regarded as high entropy alloys (HEAs). According to this concept, the CCA used in this study can be classified under the category of high entropy alloys. A careful examination of the published literature reveals that no attempt has been made to investigate the microstructure and mechanical properties of magnesium using ball-milled CCA particles. A powder metallurgy route incorporating microwave-assisted rapid sintering coupled with hot extrusion was used to synthesize Mg and Mg-CCA composites. The effect of the presence of CCA particles on microstructure and mechanical properties of magnesium was investigated. The interrelation between microstructure and mechanical properties of Mg-CCA composites was studied. The properties of Mg-CCA composites when varying the amount of reinforcement alloy particles are reported and the test results are benchmarked against pure magnesium.

2. Materials and Methods

2.1. Synthesis of Materials

In this study, magnesium powder of 98.5% purity and with a size range of 60–300 μm (Merck, Darmstadt, Germany) was used as the matrix material. Ball-milled CCA particles were used as the particulate reinforcements. Initially, CCA pieces from cast ingot with composition $Al_{35}Mg_{30}Si_{13}Zn_{10}Y_7Ca_5$ [20] were crushed into particles by ball milling the cast pieces at 200 rpm for 30 minutes with a ball to cast pieces weight ratio of 10:1 in a RETSCH PM-400 mechanical alloying machine (RETSCH, Haan, Germany). Monolithic magnesium and magnesium composites (Mg-2.5 wt % CCA, Mg-5 wt % CCA, and Mg-7.5 wt % CCA) were synthesized using a powder metallurgy technique. The synthesis process for Mg-CCA composites involved blending pure magnesium powder with CCA particles using the same machine at 200 rpm for 1 h without the use of grinding balls. The blended powder mixtures were then cold compacted using a 100-ton press to form billets that measured 35 mm in diameter and 45 mm in height. Monolithic magnesium was compacted using the same parameters without blending. The compacted billets were then sintered using a hybrid microwave sintering technique for 16 min to reach a temperature of 640 °C near the melting point of magnesium using a 900 W, 2.45 GHz SHARP microwave oven (Sharp Corporation, Osaka, Japan). The sintered billets were homogenized at 400 °C for 1 h and subsequently hot extruded at a temperature of 350 °C at an extrusion ratio of 20.25:1.

2.2. Characterization

Microstructural characterization studies were conducted on the extruded polished samples to determine the grain size, grain morphology, and presence and distribution of reinforcements. An Olympus metallographic optical microscope (Olympus Corporation, Shinjuku, Tokyo, Japan), MATLAB analysis software (R2013b, MathWorks, Natick, MA, USA), and a JEOL JSM-6010 scanning electron microscope (JEOL Ltd., Tokyo, Japan) were used for this purpose. X-ray diffraction analysis was conducted using an automated Shimadzu LAB-XRD-6000 (Shimadzu Corporation, Kyoto, Japan) (Cu Kα: λ = 1.54056 Å) spectrometer with a scan speed of 2 degrees per minute.

Microhardness measurements were performed on the polished samples using a Shimadzu-HMV automatic digital microhardness tester (Shimadzu Corporation, Kyoto, Japan) with a Vickers indenter. An indentation load of 245.5 mN and a dwell time of 15 s was used in accordance with the ASTM standard E384-08.

Room temperature compression tests were performed on cylindrical monolithic and composite samples according to ASTM E9-89a using an automated servo hydraulic testing machine MTS810 (MTS systems corporation, Eden Prairie, MN, USA). An extruded rod 8 mm in diameter was cut into 8 mm-long samples for compression tests to provide the aspect ratio (length/diameter) of unity. Samples were tested at a strain rate of 5×10^{-3} min^{-1} and the compression load was applied parallel to the extrusion direction.

Fracture surface characterization studies were carried out on the compressively fractured surfaces of Mg and Mg composites with the objective of establishing the failure mechanisms. Fractography was accomplished using a JEOL JSM-6010 scanning electron microscope (SEM).

3. Results and Discussion

3.1. Analysis on Reinforcement Particles

Figure 1a shows the size and morphology of the ball-milled CCA particles. The average particle size was calculated to be 2.7 ± 1.4 μm. The morphology of the particles can be seen as irregular shape although the smallest particles were almost spherical. The particle distribution can be seen in Figure 1b, and the particle sizes ranged from 1 μm to 7 μm with the dominant particle size in the range of 2–3 μm. The measured bulk microhardness of the cast materials was found to be very high at 406 ± 15 HV (4 ± 0.1 GPa) [20]. The XRD profiles of cast CCA and ball-milled CCA particles can be seen in Figure 2. The main phases found in the cast materials were maintained in the ball-milled version of the CCA material. The prominent change in the XRD pattern of ball-milled CCA particles is the broadening of the peaks, which may be attributed to the transformation of particles from the bulk material and an increase in the solid solubility of elements (Figure 2).

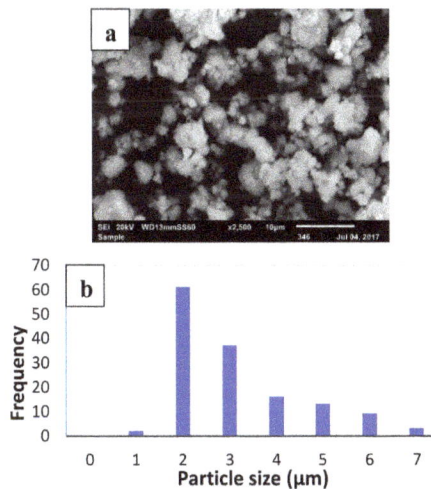

Figure 1. S scanning electron microscope (SEM) micrograph showing the morphology of ball-milled complex composition alloy (CCA) particles in (**a**) and the graph showing particle distribution in (**b**).

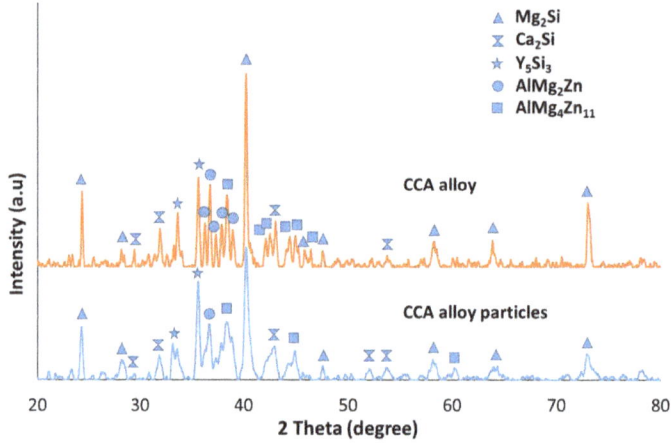

Figure 2. X-ray diffraction (XRD) pattern of the cast CCA alloy and ball-milled CCA alloy particles.

3.2. Microstructure

The distribution pattern of CCA reinforcement particles in the Mg matrix is shown in Figure 3b–d. For comparison purpose, the SEM micrograph of Mg is shown in Figure 3a. With the addition of 2.5 wt % CCA particles, the presence of uniformly distributed reinforcement particles can be seen in the micrograph (Figure 3b). In Mg-5CCA's composition, the reinforcement particles were reasonably distributed in the Mg matrix (Figure 3c). However, with the increased addition of CCA particles from 2.5 wt % to 5 wt %, the tendency for the formation of particle clustering can be seen in the micrograph and the particles were mostly decorated at the particle/grain boundaries (Figure 3c). In the case of Mg-7.5CCA's composition, in spite of having an increased amount of HEA particles, the HEA particles were uniformly distributed with limited evidence of particle clustering (Figure 3d) in the Mg matrix.

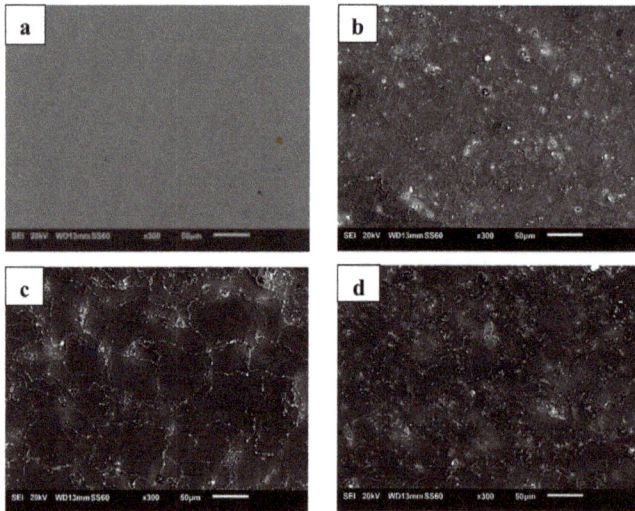

Figure 3. SEM micrographs of the: (**a**) Mg and CCA particle distribution in the (**b**) Mg-2.5CCA; (**c**) Mg-5CCA; and (**d**) Mg-7.5CCA composites.

The grain size and grain morphology of Mg and Mg-CCA composites are shown in Table 1 and Figure 5. The grain morphology was found to be nearly equiaxed and the grain size decreased with increasing presence of reinforcement particles up to 5 wt %. Further increase in the amount of particles from 5 wt % to 7.5 wt % had no effect on grain size reduction and the average grain size remained the same in both the Mg-5CCA and Mg-7.5CCA composites. However, from the grain size distribution analysis, the grain size distribution was more homogeneous in Mg-7.5CCA's composition (Figure 4d) when compared to Mg-5CCA's composition (Figure 4c). While having the same average grain size in both compositions, the presence of small-sized grains was found to be more frequent in Mg-5CCA's composition when compared to Mg-7.5CCA's composition. In fact, the distribution graph resembles a right-skewed distribution in Mg-5CCA's composition while a normal distribution pattern was observed in the Mg, Mg-2.5CCA, and Mg-7.5CCA composites.

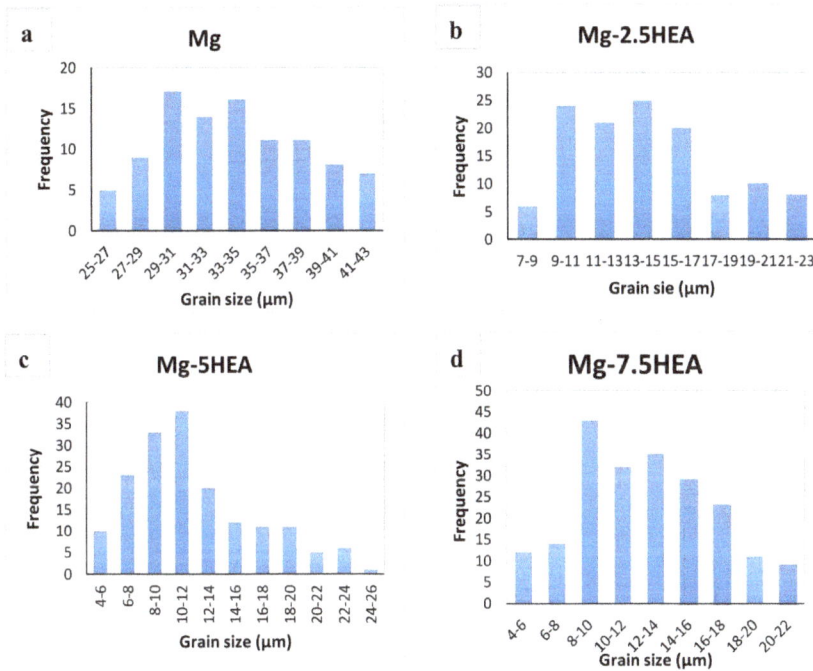

Figure 4. Grain size distribution graphs of the: (**a**) Mg and (**b**) Mg-2.5CCA; (**c**) Mg-5CCA; and (**d**) Mg-7.5CCA composites.

Table 1. Results of grain morphologyy and microhardness.

Materials	Grain Size (μm)	Aspect Ratio	Microhardness (HV)
Mg	34 ± 4	1.4 ± 0.3	47 ± 2
Mg-2.5 wt % CCA	14 ± 4	1.4 ± 0.3	56 ± 6
Mg-5.0 wt % CCA	12 ± 5	1.5 ± 0.3	70 ± 6
Mg-7.5 wt % CCA	12 ± 4	1.5 ± 0.3	80 ± 7

Since the reinforcement particle size used for the synthesis of Mg composites was at the micron length scale, particle-stimulated dynamic recrystallization during hot extrusion can be expected [21,22]. The more evenly distributed grain size (Figure 5) in the case of Mg-7.5CCA versus Mg-5HEA can be attributed to a more uniform distribution and a larger number of CCA particles in the former [23].

In extruded Mg composites, the recrystallized grains have common boundaries with the particles. With the more homogeneous distribution of particles throughout the matrix coupled with the increased presence of particles, it is possible that the migration of grain boundaries was prevented by the particles and hence the grain growth. This will provide more evenly distributed grains in the Mg-7.5CCA composite and hence while the largest grains observed in the Mg-5CCA composite measured in the range of 24–26 μm, it was only 20–22 μm in the case of the Mg-7.5CCA composite.

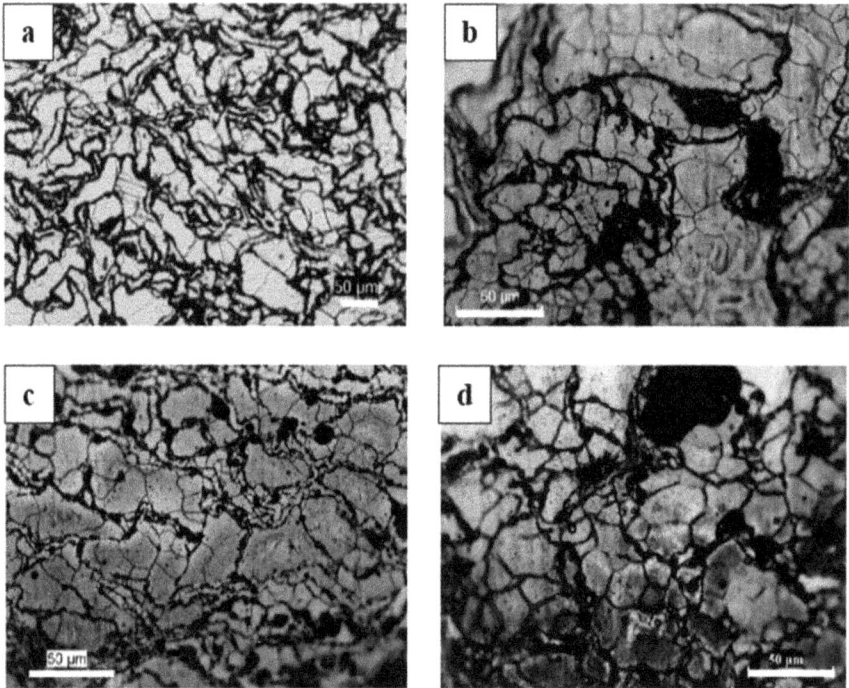

Figure 5. The grain size distribution pattern in the etched surfaces of the: (**a**) Mg; (**b**) Mg-2.5CCA; (**c**) Mg-5CCA; and (**d**) Mg-7.5CCA composites.

3.3. Mechanical Properties

The result of microhardness measurements on Mg and Mg-CCA composites is shown in Table 1. It was observed that the indentation resistance of Mg was significantly increased with the presence of CCA particles. When compared to pure Mg, an increment in hardness of 1.2 times, 1.5 times, and 1.7 times was observed in the Mg-2.5CCA, Mg-5CCA, and Mg-7.5CCA composites, respectively. The localized matrix deformation was constrained by the presence of hard and strong ball-milled CCA particles. The increasing trend of hardness can be due to the addition of an increasing amount of hard CCA reinforcement particles in the Mg matrix.

The compressive properties of pure Mg and Mg composites containing ball-milled CCA particles are listed in Table 2 and Figure 6. When compared to pure Mg, a significant improvement in compressive yield strength of 40%, 57%, and 63% in the Mg-2.5CCA, Mg-5CCA, and Mg-7.5CCA composites was achieved, respectively. In terms of ultimate compressive strength, 57%, 78%, and 79% increments over pure Mg were attained in the Mg-2.5CCA, Mg-5CCA, and Mg-7.5CCA composites, respectively. The results indicated an increasing trend of average compressive strengths with increased addition of ball-milled alloy particles from 2.5 to 7.5 wt % in the Mg matrix. Between the Mg-5CCA and Mg-7.5CCA composites, the strength increment was marginal and the strength level was the same if the standard deviation is taken into consideration. This can be explained based on the grain size measurement, which shows the occurrence of the same average grain size between the two composite compositions (Table 1). From the compressive failure strain results, the presence of CCA particles had no clear deteriorating effect on the ductility of pure Mg, which showed a similar failure strain to Mg and its composites (Table 2). However, the average failure strain in Mg-5CCA was found to be lower than that in Mg-2.5CCA and Mg-7.5CCA. From the grain size distribution result, a right-skewed distribution was observed with the influence of the small grain size in Mg-5CCA's composition. For materials with coarse grains, a higher compressive strain can be expected due to continuous twinning within the coarse grains. For materials with smaller grains, the compressive strain can be decreased due to less continuity of twinning within the fine grains. This phenomenon was reported in detail in a related paper based on magnesium-based composites [16]. In addition, the ductility increment in this composite can be attributed to the resultant microstructural homogeneity in terms of grain size and particle distribution. The resultant lower failure strain in the Mg-5CCA composite can be accounted for with the influence of small grains from the grain distribution measurement (Figure 4c) and microstructural observation (Figure 5c).

Table 2. Results of room temperature compressive properties.

Materials	0.2% Compressive Yield Strength (MPa)	Ultimate Compressive Strength (MPa)	Compressive Failure Strain (%)
Mg	91 ± 8	263 ± 16	12 ± 2
Mg-2.5 wt % CCA (1.6 vol %)	127 ± 5 (40%)	414 ± 6 (57%)	15 ± 1
Mg-5.0 wt % CCA (3.2 vol %)	143 ± 2 (57%)	469 ± 18 (78%)	10 ± 2
Mg-7.5 wt % CCA (4.9 vol %)	148 ± 4 (63%)	472 ± 19 (79%)	15 ± 2
Mg-6 vol % $Ni_{50}Ti_{50}$ [17]	89 ± 3	368 ± 8	15.1 ± 1.5
Mg-5 vol % $Ni_{60}Nb_{40}$ [18]	130 ± 11	320 ± 11	18.4 ± 1.3
AT81-5 vol % SiC [9]	127 ± 10	301 ± 20	11.4 ± 0.5
AZ91D-3 vol % TiC [10]	-	320 *	17 *

* Values approximated from the Compression graph.

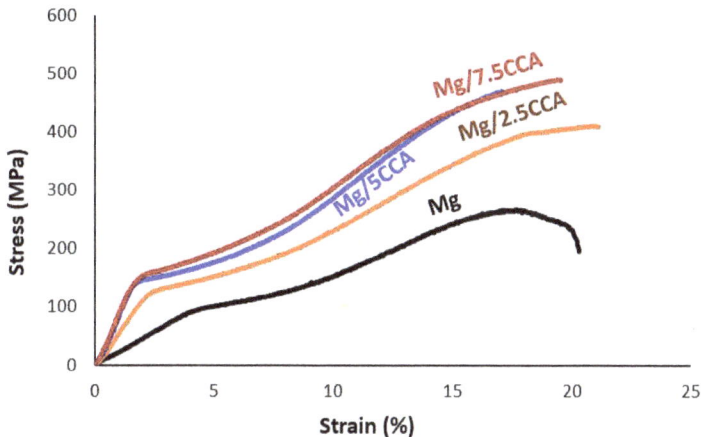

Figure 6. Compressive Stress-Strain curve of the Mg and Mg-CCA composites.

When compared to Mg composites containing ball-milled amorphous particles with a comparable amount of reinforcement addition, a significant improvement in compressive yield strength and ultimate compressive strength was observed in the Mg composite containing CCA particles while maintaining a similar compressive failure strain level (Table 2). In addition, significantly higher compressive strengths were achieved in the Mg-CCA composites when compared to the Mg-alloy-based composites containing micron size particles. This indicates the beneficial effect of CCA alloy particles for property enhancement in the composites under compressive loading. Furthermore, it shows the suitability and compatibility of this new type of CCA alloy particle as a reinforcement in an Mg matrix.

Fracture surface studies were done on Mg and its composites and the representative fractographs are shown in Figure 7. From the fractographs, the appearance of smooth fracture surfaces can be seen in the Mg, Mg-2.5CCA, and Mg-7.5CCA composites. In case of the Mg-5CCA composite, the appearance of ragged and rough fracture features was observed. The observed fracture features conform with the resultant compressive failure strain values presented in Table 2, indicating the reduced ductility attained in the Mg-5CCA composite's composition when compared to the Mg, Mg-2.5CCA, and Mg-7.5CCA composites (Figure 6).

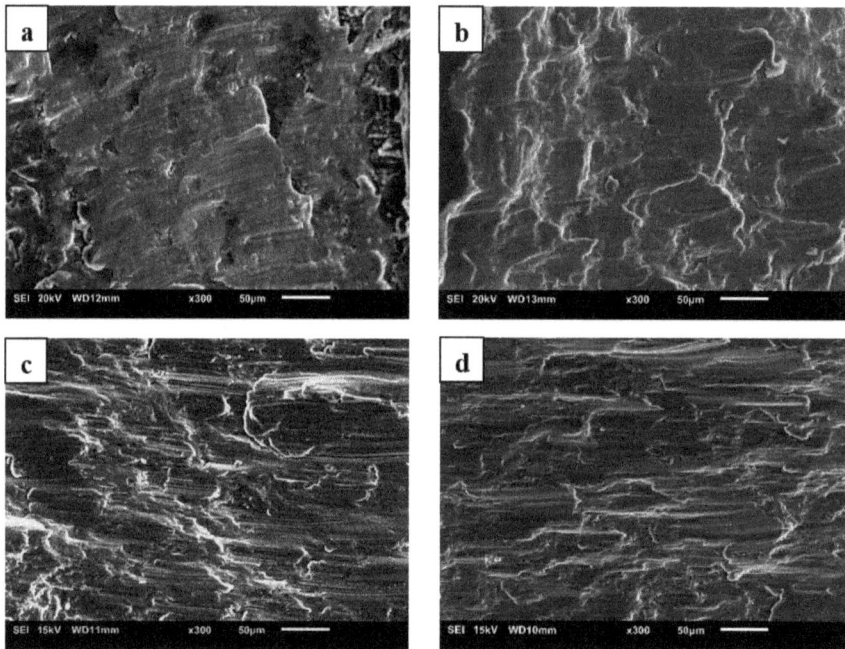

Figure 7. Representative fractographs of: (**a**) Mg; and the (**b**) Mg-2.5CCA; (**c**) Mg-5CCA; and (**d**) Mg-7.5CCA composites.

4. Conclusions

Based on the interrelation between the microstructural evolution and mechanical properties of the Mg-CCA composites developed in this work, conclusions are drawn as follows:

1. New Mg-CCA composites can be successfully developed using a powder metallurgy route incorporating microwave sintering and hot extrusion.
2. The addition of ball-milled CCA reinforcement particles assisted in a significant refinement of the matrix grain size. The measurement on the grain size distribution showed a normal distribution in the Mg, Mg-2.5CCA, and Mg-7.5CCA composite compositions while a right-skewed distribution was observed in the Mg-5CCA composite.

3. Hardness increased with an increasing amount of reinforcement addition in the Mg-CCA composites. The maximum microhardness of 80 HV was achieved in the Mg-7.5 wt % CCA composite.

4. The compressive yield strength and ultimate compressive strength were significantly enhanced in the Mg-CCA composites while maintaining the same ductility levels as unreinforced Mg. The newly developed Mg-CCA composites showed higher strength under compressive loading when compared to Mg composites containing ball-milled amorphous particles and Mg-alloy-based composites containing micron-size particle reinforcement. The achievement of enhanced mechanical properties in Mg-CCA composites highlighted the effectiveness of using ball-milled CCA particles as a reinforcement in Mg.

Acknowledgments: The authors would like to acknowledge the Ministry of Education Academic Research Funding (WBS# R-265-000-586-114) for the financial support in carrying out this research work.

Author Contributions: Manoj Gupta and Khin Sandar Tun proposed the original project and supervised the investigation. Yuming Zhang, Gururaj Parande, and Vyasaraj Manakari performed processing. Khin Sandar Tun, Yuming Zhang, Gururaj Parande, and Vyasaraj Manakari performed testing and characterization. Khin Sandar Tun performed data analysis and wrote the paper. Manoj Gupta contributed consultation, data analysis, and paper review.

Conflicts of Interest: The authors declare no conflict of interest.

References

1. Kulekci, M.K. Magnesium and its alloys applications in automotive industry. *Int. J. Adv. Manuf. Technol.* **2008**, *39*, 851–865. [CrossRef]
2. Caton, P.D. Magnesium—An old material with new applications. *Mater. Des.* **1991**, *12*, 309–316. [CrossRef]
3. International Magnesium Association. Available online: http://c.ymcdn.com/sites/intlmag.site-ym.com/resource/resmgr/docs/automotive/MgShowcase15_Feb2011.pdf (accessed on 28 January 2018).
4. Kainer, K.U.; Buch, F. The Current State of Technology and Potential for further Development of Magnesium Applications. In *Magnesium Alloys and Technology*; Wiley-VCH: Weinheim, Germany, 2003; pp. 1–22.
5. Housh, S.; Mikucki, B.; Stevenson, A. Selection and Application of Magnesium and Magnesium Alloys. In *ASM Handbook*, 10th ed.; ASM International: Materials Park, OH, USA, 1990; Volume 2, pp. 455–479.
6. Rohatgi, P.K. Metal Matrix Composites. *Def. Sci. J.* **1993**, *43*, 323–349. [CrossRef]
7. Lloyd, D.J. Particle reinforced aluminum and magnesium matrix composites. *Int. Mater. Rev.* **1994**, *39*, 1–23. [CrossRef]
8. Tjong, S.C. Novel nanoparticle reinforced metal matrix composites with enhanced mechanical properties. *Adv. Eng. Mater.* **2007**, *9*, 639–652. [CrossRef]
9. Luo, D.; Pei, C.-H.; Rong, J.; Wang, H.-Y.; Li, Q.; Jiang, Q.-C. Microstructure and mechanical properties of SiC particles reinforced Mg–8Al–1Sn magnesium matrix composites fabricated by powder metallurgy. *Powder Metall.* **2015**, *58*, 349–353. [CrossRef]
10. Cao, W.; Zhang, C.; Fan, T.; Zhang, D. In Situ Synthesis and Compressive Deformation Behaviors of TiC Reinforced Magnesium Matrix Composites. *Mater. Trans.* **2008**, *49*, 2686–2691. [CrossRef]
11. Gupta, M.; Nai, S.M.L. *Magnesium, Magnesium Alloys and Magnesium Composites*, 1st ed.; John Wiley & Sons: Hoboken, NJ, USA, 2011; pp. 113–205. ISBN 978-0-47-049417-2.
12. Perez, P.; Garces, G.; Adeva, P. Mechanical properties of a Mg–10 (vol %)Ti composite. *Comp. Sci. Technol.* **2004**, *64*, 145–151. [CrossRef]
13. Tun, K.S.; Gupta, M.; Srivatsan, T.S. Investigating influence of hybrid (yttria + copper) nanoparticulate reinforcements on microstructural development and tensile response of magnesium. *Mater. Sci. Technol.* **2010**, *26*, 87–94. [CrossRef]
14. Tun, K.S.; Gupta, M. Development of magnesium (yttria + nickel) hybrid nanocomposites using hybrid microwave sintering: Microstructure and tensile properties. *J. Alloy. Compd.* **2009**, *487*, 76–82. [CrossRef]
15. Tun, K.S.; Gupta, M. Role of microstructure and texture on compressive strength improvement of Mg/(Y_2O_3 + Cu) hybrid nanocomposites. *J. Comp. Mater.* **2010**, *44*, 3033–3050. [CrossRef]

16. Tun, K.S.; Gupta, M. Compressive deformation behavior of Mg and Mg/(Y_2O_3 + Ni) nanocomposite. *Mater. Sci. Eng. A* **2010**, *527*, 5550–5556. [CrossRef]
17. Sankaranarayanan, S.; Hemanth Shankar, V.; Jayalakshmi, S.; Nguyen, Q.B.; Gupta, M. Development of high performance magnesium composites using $Ni_{50}Ti_{50}$ metallic glass reinforcement and microwave sintering approach. *J. Alloy. Compd.* **2015**, *627*, 192–199. [CrossRef]
18. Jayalakshmi, S.; Sahu, S.; Sankaranarayanan, S.; Gupta, S.; Gupta, M. Development of novel Mg–$Ni_{60}Nb_{40}$ amorphous particle reinforced composites with enhanced hardness and compressive response. *Mater. Des.* **2014**, *53*, 849–855. [CrossRef]
19. Jayalakshmi, S.; Gupta, M. *Metallic Amorphous Alloy Reinforcements in Light Metal Matrices*, 1st ed.; Springer: Cham, Switzerland, 2015; pp. 85–105. ISBN 978-3-319-15015-4.
20. Tun, K.S.; Srivatsan, T.S.; Kumar, A.; Gupta, M. Synthesis of Light Weight High Entropy Alloys: Characterization of Microstructure and Mechanical Response. In Proceedings of the Twenty-Sixth International Conference on the Processing and Fabrication of the Advanced Materials (PFAM XXVI), Jeonju, Korea, 16–21 October 2017.
21. Inem, B. Dynamic recrystallization in a thermomechanically processed metal matrix composite. *Mater. Sci. Eng. A* **1995**, *197*, 91–95. [CrossRef]
22. Wang, X.J.; Wu, K.; Zhang, H.F.; Huang, W.H.; Chang, H.; Gan, W.M.; Zheng, M.Y.; Peng, D.L. Effect of hot extrusion on the microstructure of a particulate reinforced magnesium matrix composite. *Mater. Sci. Eng. A* **2007**, *465*, 78–84. [CrossRef]
23. Chan, H.M.; Humphreys, F.J. The recrystallisation of aluminium-silicon alloys containing a bimodal particle distribution. *Acta Metall.* **1984**, *32*, 235–243. [CrossRef]

![metals logo] *metals*

MDPI

Article

A Meso-Mechanical Constitutive Model of Particle-Reinforced Titanium Matrix Composites at High Temperatures

Weidong Song [1,*], Liansong Dai [1], Lijun Xiao [1], Cheng Wang [1], Xiaonan Mao [2] and Huiping Tang [2]

1   School of Mechatronical Engineering, Beijing Institute of Technology, Beijing 100081, China; dailiansong@yahoo.com (L.D.); xiaolijun20081016@gmail.com (L.X.); wangcheng@bit.edu.cn (C.W.)
2   Northwest Institute for Non-ferrous Metal Research, Xi'an 710016, China; maoxn2014@163.com (X.M.); hptang@c-nin.com (H.T.)
*   Correspondence: swdgh@bit.edu.cn; Tel.: +86-10-6891-4152

Academic Editor: Manoj Gupta
Received: 31 October 2016; Accepted: 26 December 2016; Published: 7 January 2017

Abstract: The elastoplastic properties of TiC particle-reinforced titanium matrix composites (TiC/TMCs) at high temperatures were examined by quasi-static tensile experiments. The specimens were stretched at 300 °C, 560 °C, and 650 °C, respectively at a strain rate of 0.001/s. scanning electron microscope (SEM) observation was carried out to reveal the microstructure of each specimen tested at different temperatures. The mechanical behavior of TiC/TMCs was analyzed by considering interfacial debonding afterwards. Based on Eshelby's equivalent inclusion theory and Mori-Tanaka's concept of average stress in the matrix, the stress or strain of the matrix, the particles, and the effective stiffness tensor of the composite were derived under prescribed traction boundary conditions at high temperatures. The plastic strains due to the thermal mismatch between the matrix and the reinforced particles were considered as eigenstrains. The interfacial debonding was calculated by the tensile strength of the particles and debonding probability was described by Weibull distribution. Finally, a meso-mechanical constitutive model was presented to explore the high-temperature elastoplastic properties of the spherical particle-reinforced titanium matrix composites by using a secant modulus method for the interfacial debonding.

Keywords: titanium matrix composite; constitutive model; interfacial debonding; high temperature; elastoplastic properties

1. Introduction

Titanium matrix composites (TMCs) become ideal materials for auto industry [1] and shipbuilding industry [2,3], with high specific strength, high specific modulus, and high temperature resistance. TMCs are mainly divided into two categories, continuously reinforced titanium matrix composites and particle-reinforced titanium matrix composites. Among them, particle-reinforced TMCs develop rapidly due to isotropic characteristics, high temperature properties, as well as low cost compared to the continuously-reinforced TMCs [4]. In order to achieve excellent properties, it is essential for reinforced particulates to have superior mechanical properties and also combine stably with the matrix materials [5]. Several ceramic particles were proposed as titanium reinforcements: SiC, B_4C, TiAl, TiB_2, TiN, TiC, and TiB [5–8]. Particularly, TiC was an excellent choice for its high modulus, strength, stiffness, hardness, and compatibility with titanium matrix [9,10]. According to the literature, TiC/Ti bulk nanocomposites have been significantly studied by Gu et al. [11,12], which systematically presented the influence of TiC on Ti matrix phase, densification, microstructure, and strengthening mechanisms.

A series of theory about particulate-reinforced composites taking account of particle size effects, damage evolution and debonding damage was carried out by K. Tohgo [13–16]. The coupled effects of the temperature and strain rate were studied by Song et al. [17,18], and a modified Johnson-Cook model was proposed to predict the dynamic behavior of TiCp/Ti. Recently, the combination of macroscopic and mesoscopic methods have been widely used to investigate dynamic mechanical behaviors and constitutive model. Meso-mechanical damage theory considers the variation of stiffness/compliance tensor as one measure of the damage, so how to determine the effective elastic modulus of the damaged materials becomes a key problem.

Main meso-mechanical theories are outlined as follows: Eshelby's equivalent inclusion theory [19,20], self-Consistent theory [21,22], Mori-Tanaka's theory [23], differential schemes [24], Hashin-Shtrikman Bounds [25,26], and so on. The elastoplastic behavior of the particle-reinforced composite with damage is widely explored by using the first-order stress moment, second-order stress moment, secant modulus method, and incremental method [27–29], the damage patterns include crack or hole in the matrix, interfacial debonding, particle fracture, and so on. In our previous work [30], a one-dimension dynamic constitutive model based on Eshelby's equivalent inclusion theory and Mori-Tanaka theory was established, by adding micro-crack nucleation and growth model. A three-dimensional interfacial debonding model to predict the stress-strain responses of weakly bonded composites was proposed by Lissenden [31], which was based on a modified Needleman type cohesive zone model. Considering progressively weakened interface, an elasto-plastic multi-level damage model was developed to predict the effective elasto-plastic behavior of particle-reinforced metal matrix composites in the work of Lee and Pyo [32]. According to Xia and Wang [33], a micromechanical model based on the analysis of localized deformation bands was provided to predict the toughening of dual-phase composites. However, there is little literature on the elastoplastic behavior of particle-reinforced composite at high temperatures. Whether the models widely used at room temperature still effective at high temperatures in new materials still have not identified by research.

In the current paper, the elastoplastic behavior of TiC particle-reinforced composite with interfacial debonding at high temperatures is discussed by means of Mori-Tanaka's mean field theory in conjunction with Eshelby's equivalent inclusion theory. A meso-mechanical constitutive model is proposed to predict the mechanical properties of the composite at high temperatures by considering the interfacial debonding.

2. Experimental Procedure

2.1. Materials

The material of titanium matrix composite reinforced with 3% TiCp was provided by Northwest Institute for Nonferrous Metal Research, which was manufactured by the pre-treatment melt process. The composition of the titanium matrix alloy was Ti-6Al-2.5Sn-4Zr-0.5Mo-1Nb-0.45Si, which could be used at high temperature ranging from 600 °C to 620 °C with excellent strength and oxidation resistance maintained above 600 °C. The reinforced particle dispersed homogeneously in the matrix which had an average diameter of about 5 μm [34] and no brittle phase existed. The interfacial reaction layers between the particle and the matrix were stable and the reaction zone width was below 3 μm, by which perfect ductility at room temperature and strength ratio above 650 °C were demonstrated.

2.2. Specimen Preparation

Specimens for quasi-static tensile tests were machined by linear cutting, which were in a shape of flat dumbbell with holes at both ends to be clamped. The thickness was 3 mm and the schematic of the specimens was presented in Figure 1.

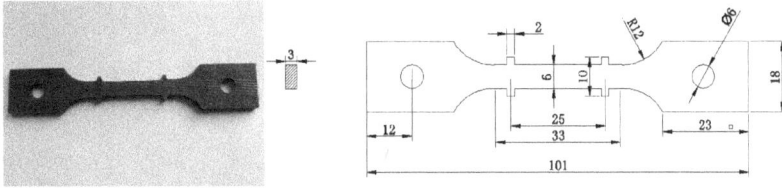

Figure 1. Specimen for quasi-static tensile test (unit: mm).

2.3. Quasi-Static Tensile Tests

The quasi-static tensile tests at elevated temperatures were carried on WDW-300 electronic universal testing machine (Jinan East Testing Machine Co., Ltd., Jinan, China). The temperature and measurement were controlled by GW-1200A controller and high-temperature furnace, respectively, during testing. The tests was conducted at deformation temperatures of 300 °C, 560 °C, and 650 °C with the same strain rate of 10^{-3} s^{-1}. The heat should be preserved for 5–10 min to ensure a uniform temperature in the test piece after the specimens were heated to the experimental temperature. Each experimental condition was repeated at least three times, and the average was taken from two valid experimental data of good reproducibility to be the final result. SEM tests were performed by a BCPCAS4800 scanning electron microscope (JEOL Co., Ltd., Tokyo, Japan) to observe the fracture morphology of each specimen stretched at different temperatures.

2.4. Experimental Results

2.4.1. Microstructure

The images in Figure 2 exhibit the fracture microstructure of TiC/TMCs composites samples tested at different temperatures. The results reveal that the failure of the composites is dominant by the interface debonding, particle cracking, and ductile fracture of the matrix. It can be seen that with the temperature rises up, the dimples of fracture surface tend to be more uniformly distributed. The sizes of dimples are getting to be larger and deeper when the experimental temperature increases from 300 °C to 650 °C. This demonstrates that the TiC/TMCs composites show better plasticity at elevated temperature.

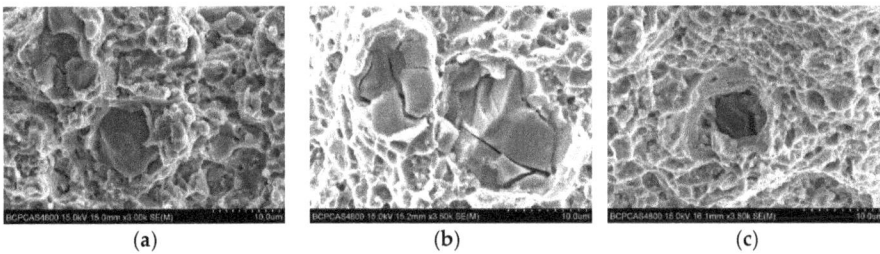

Figure 2. SEM images of TiC/TMCs composites: (**a**) 300 °C; (**b**) 560 °C; and (**c**) 650 °C.

2.4.2. Stress-Strain Relationship

Figure 3 shows the results of the quasi-static stress-strain curves for the titanium matrix composite at different temperatures. According to the tests, it is obvious that the TiC/TMCs composites demonstrate temperature sensitivity. The flow stress decreases with increasing experimental temperature at the same strain rate. From the data at 300 °C and 560 °C, the typical strain hardening curve can be obtained, but the flow stress of the latter rises more slowly than the former. For the stress-strain curve at 650 °C, the flow stress dropped with increasing strain which indicates that the temperature softening effect is greater than strain hardening. As for the ductile properties, the composite exhibits better ductility with the increasing temperature, which means elongation is positively correlated with temperature.

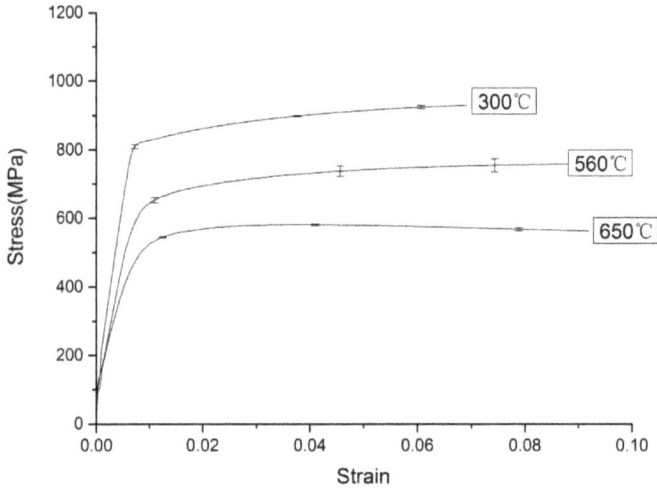

Figure 3. Stress-strain curves for the TiC/TMCs composites under quasi-static tensile at different temperatures.

3. Meso-Mechanical Theory for Multiphase Composite

3.1. Average Stress of the Reinforcement and the Matrix

Let us consider an n-phase composite, and we shall refer to the matrix as phase 0, the particle as phase 1, and damaged particle (void) as phase 2. Based on Mori-Tanaka theory, a uniform far-field stress σ is exerted on the composite and its boundary. Take the matrix as comparative material and then the average strain ε^0 of the comparative material will satisfy Equation (1) under the same external force:

$$\sigma = L_0 : \left(\varepsilon^0 - \varepsilon_0^p \right) \tag{1}$$

where L_0 and ε_0^p are the stiffness tensor the plastic strain of the matrix, respectively.

Due to the existence of the reinforcement, the average strain is actually different from ε^0 and perturbation stress $\tilde{\sigma}$ and perturbation strain $\tilde{\varepsilon}$ are generated by the interaction between the reinforcement, so the average stress of the matrix is:

$$\sigma^{(0)} = \sigma + \tilde{\sigma} = L_0 : \left(\varepsilon^0 + \tilde{\varepsilon} - \varepsilon_0^p \right) \tag{2}$$

Since the elastic property and the coefficient of thermal expansion of the reinforcement is different from that of the matrix, the average stress of the reinforcement is expressed as:

$$\begin{aligned} \sigma^{(1)} = \sigma + \tilde{\sigma} + \sigma_1^{pt} = & \; L_1 : \left(\varepsilon^0 + \tilde{\varepsilon} + \varepsilon_1^{pt} - \varepsilon_1^p - \alpha_1^* \right) \\ = & \; L_0 : \left(L_0^{-1} : \sigma + \tilde{\varepsilon} + \varepsilon_1^{pt} - \Delta\varepsilon_1^p - \alpha_1^* - \varepsilon_1^* \right) \end{aligned} \tag{3}$$

$$\sigma^{(2)} = \sigma + \tilde{\sigma} + \sigma_2^{pt} = L_2 : \left(\varepsilon^0 + \tilde{\varepsilon} + \varepsilon_2^{pt} - \varepsilon_2^p - \alpha_2^* \right) = 0$$

$$= L_0 : \left(L_0^{-1} : \sigma + \tilde{\varepsilon} + \varepsilon_2^{pt} - \Delta\varepsilon_2^p - \alpha_2^* - \varepsilon_2^* \right) \tag{4}$$

where, σ_r^{pt} and ε_r^{pt} indicate the average stress difference and the average strain difference between the rth phase and the matrix, ε_r^* is the eigenstrain of the rth phase, α_r^* is caused by the different coefficient of thermal expansion between the rth phase and the matrix:

$$\alpha_{ij}^{r*} = (\alpha^r - \alpha^0)\Delta T \delta_{ij} = \alpha^r \delta_{ij} \tag{5}$$

$$\varepsilon_r^{pt} = S : \left(\varepsilon_r^* + \Delta\varepsilon_r^p + \alpha_r^* \right) \qquad r = 1,2 \tag{6}$$

where α^r is the thermal expansion coefficient of the rth phase.

By using Equations (2), (3), (6), we have:

$$\sigma_r^{pt} = L_0 : (S - I) : \left(\varepsilon_r^* + \Delta\varepsilon_r^p + \alpha_r^* \right) \quad r = 1,2 \tag{7}$$

where S is the Eshelby tensor and I is the four order unit tensor.

The average stress of the composite meets the volume mixing ratio, namely:

$$\sigma = f_0 \sigma^{(0)} + f_p \sigma^{(1)} + f_v \sigma^{(2)} \tag{8}$$

where $f_0 + f_p + f_v = 1$.

Substitute Equations (2) and (3) into Equation (8), we have:

$$\tilde{\sigma} = -\left(f_p \sigma_1^{pt} + f_v \sigma_2^{pt} \right) \tag{9}$$

and:

$$\tilde{\varepsilon} = (I - S)[f_p(\varepsilon_1^* + \Delta\varepsilon_1^p + \alpha_1^*) + f_v(\varepsilon_2^* + \Delta\varepsilon_2^p + \alpha_2^*)] \tag{10}$$

Let $X = \varepsilon_1^* + \Delta\varepsilon_1^p + \alpha_1^*, Y = \varepsilon_2^* + \Delta\varepsilon_2^p + \alpha_2^*$ and substitute Equation (10) into Equation (4), we get:

$$Y = \frac{f_p X + (I - S)^{-1} L_0^{-1} \sigma}{1 - f_v} \tag{11}$$

Substituting Equation (10) into Equation (3), we have:

$$X = \left\{ L_0 + (L_1 - L_0) : \left[S - \frac{f_p}{1-f_v}(S - I) \right] \right\}^{-1} : \left[\frac{1}{1-f_v}(L_0 - L_1) : L_0^{-1} : \sigma + L_1 : \left(\Delta\varepsilon_1^p + \alpha_1^* \right) \right]$$

$$= \left\{ (1 - f_v - f_p)[(L_1 - L_0) : S + L_0] + f_p L_1 \right\}^{-1} : \left[(L_0 - L_1) : L_0^{-1} : \sigma + (1 - f_v)L_1 : \left(\Delta\varepsilon_1^p + \alpha_1^* \right) \right] \tag{12}$$

and:

$$\varepsilon_1^* = \left\{ (1 - f_v - f_p)[(L_1 - L_0) : S + L_0] + f_p L_1 \right\}^{-1} : \left[(L_0 - L_1) : L_0^{-1} : \sigma + (1 - f_v - f_p)(L_0 - L_1) : (S - I) : \left(\Delta\varepsilon_1^p + \alpha_1^* \right) \right] \tag{13}$$

$$Y = \left\{ (1 - f_v - f_p)[(L_1 - L_0) : S + L_0] + f_p L_1 \right\}^{-1} : \left\{ [(L_1 - L_0) : S + L_0] : (I - S)^{-1} : L_0^{-1} : \sigma + f_p L_1 : \left(\Delta\varepsilon_1^p + \alpha_1^* \right) \right\} \tag{14}$$

$$\varepsilon_2^* = \left\{ (1 - f_v - f_p)[(L_1 - L_0) : S + L_0] + f_p L_1 \right\}^{-1} : \left\{ [(L_1 - L_0) : S + L_0] : (I - S)^{-1} : L_0^{-1} : \sigma + f_p L_1 : \left(\Delta\varepsilon_1^p + \alpha_1^* \right) \right.$$

$$\left. - [(1 - f_v - f_p)[(L_1 - L_0) : S + L_0] + f_p L_1] : \left(\Delta\varepsilon_2^p + \alpha_2^* \right) \right\} \tag{15}$$

Substituting Equations (1), (10)–(12) into Equations (2) and (3), respectively, we get:

$$\sigma^{(0)} = \sigma + L_0 : (I - S) : (f_p X + f_v Y) = \frac{1}{1-f_v}\sigma + \frac{f_p}{1-f_v} L_0 : (I - S)$$

$$: \left\{ (1 - f_v - f_p)[(L_1 - L_0) : S + L_0] + f_p L_1 \right\}^{-1} : \left[(L_0 - L_1) : L_0^{-1} : \sigma + (1 - f_v)L_1 : \left(\Delta\varepsilon_1^p + \alpha_1^* \right) \right] \tag{16}$$

$$\sigma^{(1)} = \sigma + L_0 : (I - S) : (-(1 - f_p)X + f_v Y) = \frac{1}{1 - f_v}\sigma - \frac{1 - f_v - f_p}{1 - f_v}L_0 : (I - S)$$
$$: \left\{(1 - f_v - f_p)[(L_1 - L_0) : S + L_0] + f_p L_1\right\}^{-1} : \left[(L_0 - L_1) : L_0^{-1} : \sigma + (1 - f_v)L_1 : \left(\Delta \varepsilon_1^p + \alpha_1^*\right)\right] \tag{17}$$

3.2. Effective Stiffness Tensor of the Composite

The stress-strain relationship can be written as:

$$\varepsilon = L^{-1} : \sigma \tag{18}$$

The average strain of the composite is the sum of the strain:

$$\varepsilon = f_0 \varepsilon^{(0)} + f_p \varepsilon^{(1)} + f_v \varepsilon^{(2)}$$

Substitute $\varepsilon^{(0)} = \varepsilon^0 + \tilde{\varepsilon} - \varepsilon_0^p$, $\varepsilon^{(1)} = \varepsilon^0 + \tilde{\varepsilon} + \varepsilon_1^{pt} - \varepsilon_1^p - \alpha_1^*$ and $\varepsilon^{(2)} = \varepsilon^0 + \tilde{\varepsilon} + \varepsilon_2^{pt} - \varepsilon_2^p - \alpha_2^*$ into Equations (18), (6), (10), we have:

$$\varepsilon = L_0^{-1} : \sigma + f_p \varepsilon_1^* + f_v \varepsilon_2^* \tag{19}$$

Substituting Equations (13) and (15) into Equation (19), we get:

$$\varepsilon = L_0^{-1} : \sigma + f_p \varepsilon_1^* + f_v \varepsilon_2^* = L_0^{-1} : \sigma + f_p\{(1 - f_v - f_p)[(L_1 - L_0) : S + L_0]$$
$$+ f_p L_1\}^{-1} : [(L_0 - L_1) : L_0^{-1} : \sigma + (1 - f_v - f_p)(L_0 - L_1) : (S - I) :$$
$$\left(\Delta \varepsilon_1^p + \alpha_1^*\right)] + f_v\{(1 - f_v - f_p)[(L_1 - L_0) : S + L_0] + f_p L_1\}^{-1} :$$
$$\{[(L_1 - L_0) : S + L_0] : (I - S)^{-1} : L_0^{-1} : \sigma + f_p L_1 : \left(\Delta \varepsilon_1^p + \alpha_1^*\right)$$
$$-\{(1 - f_v - f_p)[(L_1 - L_0) : S + L_0] + f_p L_1\} : \left(\Delta \varepsilon_2^p + \alpha_2^*\right)\} \tag{20}$$

The stiffness tensor of the composite at high temperatures can be derived from Equation (20):

$$L^{-1} = L_0^{-1} + f_p\{(1 - f_v - f_p)[(L_1 - L_0) : S + L_0] + f_p L_1\}^{-1} : [(L_0 - L_1) : L_0^{-1}$$
$$+ (1 - f_v - f_p)(L_0 - L_1) : (S - I) : \alpha_1^* : \sigma^{-1}] + f_v\{(1 - f_v - f_p)[(L_1 - L_0) : S$$
$$+ L_0] + f_p L_1\}^{-1} : \{[(L_1 - L_0) : S + L_0] : (I - S)^{-1} : L_0^{-1} + f_p L_1 : \left(\Delta \varepsilon_1^p + \alpha_1^*\right) : \sigma^{-1}\}$$
$$= L_0^{-1} + \{(1 - f_v - f_p)[(L_1 - L_0) : S + L_0] + f_p L_1\}^{-1}\{f_p\left[(L_0 - L_1) : L_0^{-1}\right.$$
$$+ (1 - f_v - f_p)(L_0 - L_1) : (S - I) : \alpha_1^* : \sigma^{-1}] + f_v\{[(L_1 - L_0) : S + L_0] :$$
$$(I - S)^{-1} : L_0^{-1} + f_p L_1 : \alpha_1^* : \sigma^{-1}\}\} \tag{21}$$

4. Elastoplastic Analysis of the Composite

4.1. Constitutive Model of the Matrix

The elastoplastic relationship of the matrix can be described by modified Ludwik equation:

$$\sigma^{(0)} = \sigma_s^{(0)} + h\left(\varepsilon_0^p\right)^n \tag{22}$$

where, $\sigma_s^{(0)}$ is the yield stress of the matrix; h and n are material parameters determined by uniaxial tensile test.

Under monotonic loading, the secant modulus E_0^s of the matrix is expressed as:

$$E_0^s = \frac{\sigma^{(0)}}{\varepsilon_0^e + \varepsilon_0^p} = \frac{1}{\frac{\sigma_s^{(0)}}{E_0 \sigma^{(0)}} + \frac{\varepsilon_0^p}{\sigma_s^{(0)} + h\left(\varepsilon_0^p\right)^n}} \tag{23}$$

where, ε_0^e and ε_0^p are the elastic strain and the plastic strain of the matrix, respectively.

Under three-dimensional stress, by replacing $\sigma^{(0)}$ and ε_0^p with Mises effective stress $\sigma^{(0)*}$ and effective strain ε_0^{p*}, formula (22) can be rewritten as:

$$\sigma^{(0)*} = \sigma_s^{(0)} + h\left(\varepsilon_0^{p*}\right)^n \tag{24}$$

where:

$$\sigma^{(0)*} = \left(\frac{3}{2}\sigma_{ij}^{(0)'}\sigma_{ij}^{(0)'}\right)^{\frac{1}{2}}, \varepsilon_0^{p*} = \left(\frac{2}{3}\varepsilon_{ij}^{p(0)}\varepsilon_{ij}^{p(0)}\right)^{\frac{1}{2}} \tag{25}$$

$\sigma_{ij}^{(0)'}$ is the stress deviator of the matrix.

The secant bulk modulus and shear modulus of the matrix are expressed as:

$$k_0^s = \frac{E_0^s}{3\left(1-2v_0^s\right)}, \quad \mu_0^s = \frac{E_0^s}{2\left(1+v_0^s\right)} \tag{26}$$

where v_0^s indicates the secant Poisson's ratio.

Due to plasticity incompressibility, the secant bulk modulus k_0^s is equal to the elastic bulk modulus k_0, so, we have:

$$v_0^s = \frac{1}{2} - \frac{E_0^s}{E_0}\left(\frac{1}{2} - v_0\right) \tag{27}$$

In general, the elastoplastic behavior of the matrix under monotonic loading can be described by the secant Young's modulus E_0^s and two elastic constants E_0 and v_0.

4.2. Stress for the Reinforcement and the Matrix and the Secant Tensor of the Composite under Force Boundary Conditions

When the matrix is in the elastoplastic stage, the modulus changes with the deformation, so the modulus of the matrix takes the secant value indicating by superscript S. According to Equations (16) and (17), the stress for the matrix and the reinforcement can be written as:

$$\sigma^{(0)} = \frac{1}{1-f_v}\sigma + \frac{f_p}{1-f_v}L_0^s : (I-S) : \left\{(1-f_v-f_p)\right.$$
$$\left[(L_1-L_0^s):S+L_0^s]+f_pL_1\right\}^{-1} : \left[(L_0^s-L_1):L_0^{s-1}:\sigma+(1-f_v)L_1:\left(\Delta\varepsilon_1^p+\alpha_1^*\right)\right] \tag{28}$$

$$\sigma^{(1)} = \frac{1}{1-f_v}\sigma - \frac{1-f_v-f_p}{1-f_v}L_0^s : (I-S) : \left\{(1-f_v-f_p)\right.$$
$$\left[(L_1-L_0^s):S+L_0^s]+f_pL_1\right\}^{-1} : \left[(L_0^s-L_1):L_0^{s-1}:\sigma+(1-f_v)L_1:(\Delta\varepsilon_1^p+\alpha_1^*)\right] \tag{29}$$

According to Equation (21), the secant tensor of the composite is given:

$$(L^s)^{-1} = (L_0^s)^{-1} + \left\{(1-f_v-f_p)[(L_1-L_0^s):S+L_0^s]+f_pL_1\right\}^{-1}\{f_p\left[(L_0^s-L_1)\right.$$
$$: (L_0^s)^{-1} + (1-f_v-f_p)(L_0^s-L_1):(S-I):\alpha_1^*]+f_v\{[(L_1-L_0^s):S+L_0^s] \tag{30}$$
$$: (I-S)^{-1}:(L_0^s)^{-1}+f_pL_1:\alpha_1^*\}$$

where:

$$L_0^s = (2k_0^s, k_0^s-\mu_0^s, k_0^s-\mu_0^s, k_0^s+\mu_0^s, 2\mu_0^s, 2\mu_0^s) \tag{31}$$

$$L_1 = (2k_1, k_1-\mu_1, k_1-\mu_1, k_1+\mu_1, 2\mu_1, 2\mu_1) \tag{32}$$

$$L_2 = (0,0,0,0,0,0) \tag{33}$$

$$I = (1,0,0,1,1,1) \tag{34}$$

$$S = \left(\frac{2}{3}\alpha^s, \frac{\alpha^s}{3}-\frac{\beta^s}{2}, \frac{\alpha^s}{3}-\frac{\beta^s}{2}, \frac{\alpha^s}{3}+\frac{\beta^s}{2}, \beta^s, \beta^s\right) \tag{35}$$

$$\alpha^s = \frac{1 + v_0^s}{3(1 - v_0^s)}, \quad \beta^s = \frac{2(4 - 5v_0^s)}{15(1 - v_0^s)} \tag{36}$$

4.3. Interfacial Debonding Model

In order to descript the propagation of the interfaces, Weibull statistical distribution is introduced to discuss the cumulative probability of the interfacial debonding.

It is assumed that when the interfacial debonding happened between the matrix and the reinforcement (particle), the particle cannot bear any load and can be equivalent to a hole. By assuming that the tensile stress in the particle controls the debonding and the initial propagation strength along ii direction meets the Weibull statistical distribution P_{ii} [35], we have:

$$P_{ii}\left(\sigma_{11}^{(1)}\right) = 1 - \exp\left[-\left(\frac{\sigma_{ii}^{(1)}}{s}\right)^m\right], \quad i = 1, 2, 3 \tag{37}$$

where, $P_{ii}\left(\sigma_{11}^{(1)}\right)$ is the ratio of the damaged particles to all particles, i.e., the debonding probability. s and m are scale parameter and shape parameter of the Weibull function. Thus, the volume fraction of the damaged particle on the composite is:

$$f_1 P_{11}\left(\sigma_{11}^{(1)}\right) = f_1 \left\{ 1 - \exp\left[-\left(\frac{\sigma_{11}^{(1)}}{s}\right)^m\right]\right\} \tag{38}$$

The probability density of the damaged interface can be written as:

$$p_{11}\left(\sigma_{11}^{(1)}\right) = \frac{m}{s}\exp\left(\frac{\sigma_{11}^{(1)}}{s}\right)^{m-1}\left[-\left(\frac{\sigma_{11}^{(1)}}{s}\right)^m\right] \tag{39}$$

The relationship between the critical debonding strength of the interface σ_c and the two parameters s and m is given as [36]:

$$\sigma_c = \int_0^\infty \sigma_{11}^{(1)} p d\sigma_{11}^{(1)} = s \cdot \Gamma\left(1 + \frac{1}{m}\right) \tag{40}$$

When the critical debonding strength of the interface σ_c and the loading exerted on the particle are known, the volume fraction of the debonding particle can be obtained by Equation (37).

4.4. Elastoplastic Stress-Strain Relationship

The damage constitutive relation can be expressed by Equation (30). When the matrix is in the plastic stage, L_0^s is not a constant and changes with the deformation process. At the same time, volume fraction f_p and f_v are also changing. So, in order to obtain the stress-strain relationship, L_0^s and f_v of each stage should be calculated firstly. The numerical calculation is performed according to the following procedure: (1) calculate the effective elastic modulus L of the material and taking as the initial value; (2) for a given σ, determining $\sigma^{(0)}$ and $\sigma^{(1)}$ from Equations (28) and (29); (3) set σ_c and m, deriving f_v from Equation (37) ($f_p = f_1 - f_v$) and then obtaining the effective stress of the matrix from $\sigma^{(0)}$. If $\sigma^{(0)}$ is bigger than the elastic limit $\sigma_s^{(0)}$ of the matrix, L_s^0 should be calculated from Equations (23), (26), (27), (31); and (4) increasing σ and calculating L^s from the new L_s^0, then repeating the whole process.

It is assumed that the elastic modulus of the TiC particle does not change with the changes in temperature and the elastic modulus of the matrix decreases with the increase of temperature. In order to obtain the elastic of the matrix at different temperature, we assume that the elastic modulus of the matrix decreases linearly with increasing temperature. Based the known elastic modulus of the matrix, the elastic modulus at different temperatures can be obtained. The elastic modulus and yield stress of the matrix at room temperature was experimentally determined. By assuming that there are linear relationships between elastic modulus and temperature, and yield stress with temperature, respectively, the elastic modulus and yield stress of the matrix at different temperatures was calculated, which was shown in Table 1.

Table 1. The elastic modulus and yield stress of the matrix at different temperatures.

Temperature	R.T.	300 °C	600 °C	650 °C	700 °C
E_0/GPa	113	71	25	20	15
$\sigma_s^{(0)}$/MPa	1050	670	590	540	420

5. Comparison of Numerical Predictions with Experimental Results

When the uniaxial tensile stress σ_{11} is exerted on the particle-reinforced titanium matrix composite at high temperatures, the material parameters of the composite is listed as follows: $v_0 = 0.35$, $E_1 = 460$ Gpa, $v_1 = 0.188$, $f_1 = 0.03$, $h = 60$ Mpa, $n = 0.45$, $\sigma_c = 2.0\sigma_s^{(0)}$, $m = 5$. The stress-strain curves of the composite at three different temperatures ($T = 300$ °C, 560 °C, 650 °C) are shown in Figure 4.

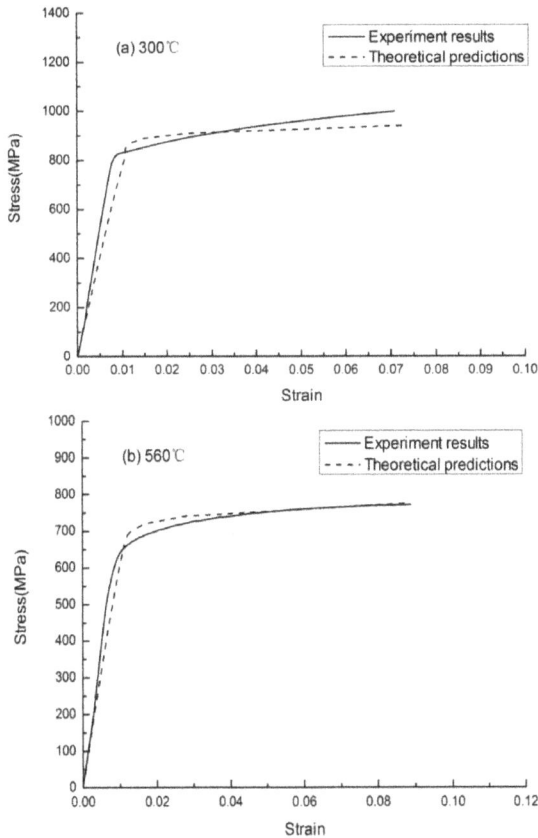

Figure 4. *Cont.*

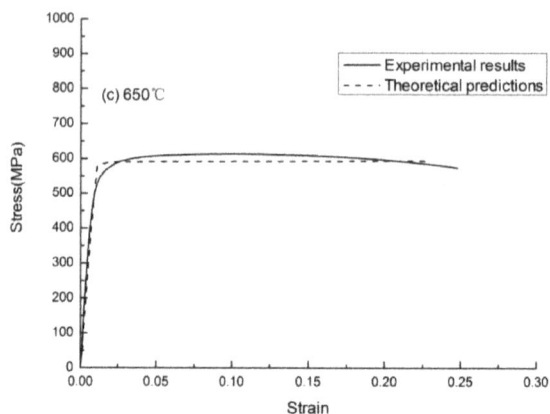

Figure 4. Comparisons of stress-strain curves of TiC/TMCs composites with the theoretical results and test results at different temperatures: (**a**) 300 °C; (**b**) 560 °C; and (**c**) 650 °C.

From Figure 4, it can be seen that the numerical predictions agree well with experimental results, which demonstrate the assumption that the elastic modulus decreases linearly with the increase of the temperature. The debonding model adopted in the current paper can be used to predict the elastoplastic behavior of the composite at elevated temperatures. There is little discrepancy of initial elastic modulus and yield stress, which may be caused by that the theoretical model did not consider the particle cracking and ductile fracture. Nevertheless, in the plastic section of the curve, the theoretical model prediction is not accurate enough, which needs further study.

6. Conclusions

Based on Eshelby's theory and Mori-Tanaka theory, the stress of the reinforcement and the matrix and the effective stiffness tensor of the composite under force boundary conditions are deduced. By using the assumption that the interfacial debonding is controlled by the tensile stress on the particle and the cumulative probability of the interfacial debonding is described by the Weibull function, a meso-mechanical constitutive model is proposed to investigate the elastoplastic properties of the particle-reinforced titanium matrix composite by using the secant modulus method. A good agreement between the numerical predictions and the experimental results is obtained, which demonstrate that the model and the method adopted in the current study is reliable and reasonable. When particle-reinforced titanium matrix composites were used at high temperatures, such as in aerospace and automobile industries, this model can be used to predict the mechanical properties so as to provide the theoretic basis for the design of structural parameters.

Acknowledgments: The author acknowledges the financial support of the National Nature Science Foundation of China (11672043, 11521062, 11325209), the Opening Project of State Key Laboratory for Strength and Vibration of Mechanical Structures (SV2015-KF-10), the Opening Project of State Key Laboratory of Traction Power (TPL1701) and the Project of State Key Laboratory of Explosion Science and Technology (YBKT16-19, KFJJ15-02M).

Author Contributions: Weidong Song conceived and designed the study; Liansong Dai and Lijun Xiao performed the experiments and analyzed the data; Weidong Song, Liansong Dai, Lijun Xiao and Cheng Wang made the theoretical analysis; Xiaonan Mao and Huiping Tang provided the specimens and essential parameters of the material; Weidong Song and Liansong Dai wrote the paper. All authors read and approved the manuscript.

Conflicts of Interest: The authors declare no conflict of interest.

References

1. Hunt, M. MMCs for exotic needs. *Compos. Mater. Sci.* **1992**, *104*, 53–62.

2. Gorynin, I.V. Titanium alloys for marine application. *Mater. Sci. Eng. A* **1999**, *263*, 112–116. [CrossRef]

3. Schutz, R.W.; Scaturro, M.R. An overview of current and candidate titanium alloy applications on U.S. navy surface ships. *Nav. Eng. J.* **1991**, 175–191. [CrossRef]

4. Luo, Y.Z. Recent development of titanium metallic matrix composites. *Rare Met. Mater. Eng.* **1997**, *26*, 1–7.

5. Vallauri, D.; Adrian, I.C.A.; Chrysanthou, A. TiC-TiB₂ composites: A review of phase relationships, processing and properties. *J. Eur. Ceram. Soc.* **2008**, *28*, 1697–1713. [CrossRef]

6. Leyens, C.; Peters, M. *Titanium and Titanium Alloys*; Wiley-VCH: Cologne, Germany, 2003.

7. Tjong, S.C.; Ma, Z.Y. Microstructural and mechanical characteristics of in-situ metal matrix composites. *Mater. Sci. Eng. R* **2000**, *29*, 49–113. [CrossRef]

8. Tjong, S.C.; Mai, Y.W. Processing-structure-property aspects of particulate- and whisker-reinforced titanium matrix composites. *Compos. Sci. Technol.* **2008**, *68*, 583–660. [CrossRef]

9. Wang, X.H.; Zou, Z.D.; Qu, S.Y.; Song, S.L. Microstructure and wear properties of Fe-based hardfacing coating reinforced by TiC particles. *J. Mater. Process. Technol.* **2005**, *168*, 89–94. [CrossRef]

10. Akhtar, F.; Guo, S.J. Microstructure, mechanical and fretting wear properties of TiC-stainless steel composites. *Mater. Charact.* **2008**, *59*, 84–90. [CrossRef]

11. Gu, D.D.; Hagedorn, Y.C.; Meiners, W.; Wissenbach, K.; Poprawe, R. Nanocrystalline TiC reinforced Ti matrix bulk-form nanocomposites by Selective Laser Melting (SLM): Densification, growth mechanism and wear behavior. *Compos. Sci. Technol.* **2011**, *71*, 1612–1620. [CrossRef]

12. Gu, D.D.; Meng, G.B.; Li, C.; Meiners, W.; Poprawe, R. Selective laser melting of TiC/Ti bulk nanocomposites: Influence of nanoscale reinforcement. *Scr. Mater.* **2012**, *67*, 185–188. [CrossRef]

13. Tohgo, K.; Weng, G.J. A Progressive Damage Mechanics in Particle-reinforced Metal-Matrix Composites under High Triaxial Tension, Trans. of ASME. *J. Eng. Mater. Technol.* **1994**, *116*, 414–420. [CrossRef]

14. Tohgo, K.; Chou, T.W. Incremental Theory of Particulate-Reinforced Composites Including Debonding Damage. *JSME Int. J. Ser. A* **1996**, *39*, 389–397. [CrossRef]

15. Tohgo, K.; Itoh, Y.; Shimamura, Y. A Constitutive Model of Particulate-Reinforced Composites Taking Account of Particle Size Effects and Damage Evolution. *Compos. A* **2010**, *41*, 313–321. [CrossRef]

16. Tohgo, K.; Fujii, T.; Shimamura, D.K.Y. Influence of Particle Size and Debonding Damage on an Elastic-Plastic Singular Field around a Crack-Tip in Particulate-Reinforced Composites. *Acta Mech.* **2014**, *225*, 1373–1389. [CrossRef]

17. Li, W.; Song, W.D.; Ning, J.G.; Mao, X.N. Effects of strain rate and temperature on mechanical properties of TP-650 composite. *Rare Mater. Eng.* **2010**, *39*, 1195–1199.

18. Song, W.D.; Ning, J.G.; Mao, X.N.; Tang, H.P. A modified Johnson-Cook model for titanium matrix composites reinforced with titanium carbide particles at elevated temperatures. *Mater. Sci. Eng. A* **2013**, *576*, 280–289. [CrossRef]

19. Eshelby, J.D. The determination of the elastic field of an ellipsoidal inclusion and related problems. *Proc. R. Soc. Lond.* **1957**, *241A*, 376–396. [CrossRef]

20. Eshelby, J.D. The elastic field outside an ellipsoidal inclusion. *Proc. R. Soc. Lond.* **1959**, *252A*, 561–569. [CrossRef]

21. Hill, R. A self-consistent mechanics of composite materials. *J. Mech. Phys. Solids* **1965**, *13*, 213–222. [CrossRef]

22. Budiansky, B. On the elastic model of some heterogeneous materials. *J. Mech. Phys. Solids* **1976**, *12*, 81–97.

23. Morit, T.; Tanaka, K. Average stress in matrix and average energy of materials with misfitting inclusions. *Acta Metall.* **1973**, *21*, 571–574. [CrossRef]

24. Roscoe, R.A. The Viscosity of suspensions of rigid spheres. *Br. J. Appl. Phys.* **1952**, *3*, 267–269. [CrossRef]

25. Hashin, Z.; Shtrikman, S. A variational approach to the theory of the elastic behaviour of multiphase materials. *J. Mech. Phys. Solids* **1963**, *11*, 127–140. [CrossRef]

26. Hashin, Z.; Shtrikman, S. A variational approach to the theory of the effective magnetic permeability of multi- phase materials. *J. Mech. Phys. Solids* **1962**, *10*, 335–342. [CrossRef]

27. Buryachenko, V. Elastic-plastic behavior of elastically homogeneous materials with a random field of inclusions. *Int. J. Plast.* **1999**, *15*, 687–720. [CrossRef]

28. Hu, G.K. Composite plasticity based on matrix average second order stress moment. *Int. J. Plast.* **1997**, *12*, 1007–1015. [CrossRef]

29. Tandon, P.; Weng, G.J. A theory of particle-reinforced plasticity. *ASME J. Appl. Mech.* **1991**, *55*, 126–135. [CrossRef]

30. Song, W.; Yang, Y.; Ning, J. A Meso-Mechanical Constitutive Model of Particle Reinforced Titanium Matrix Composites Subjected to Impact Loading. *J. Nano Res.* **2013**, *23*, 104–107. [CrossRef]
31. Lissenden, C.J. Fiber-matrix interfacial constitutive relations for metal matrix composites. *Compos. B* **1999**, *30*, 267–278. [CrossRef]
32. Lee, H.K.; Pyo, S.H. An elastoplastic multi-level damage model for ductile matrix composites considering evolutionary weakened interface. *Int. J. Solid Struct.* **2008**, *45*, 1614–1631. [CrossRef]
33. Xia, S.H.; Wang, J.T. A micromechanical model of toughening behavior in the dual-phase composite. *Int. J. Plast.* **2010**, *26*, 1442–1460. [CrossRef]
34. Song, W.D.; Ning, J.G.; Wang, J. Mechanical Behavior of Titanium Carbide Particle-Reinforced Titanium Matrix Composite under Quasi-Static and Dynamic Tension. *Adi. Sci. Lett.* **2011**, *4*, 635–640. [CrossRef]
35. Weibull, W. A statistical distribution function of side applicability. *J. Appl. Mech.* **1951**, *18*, 293–297.
36. Zhao, Y.H.; Weng, G.J. Plasticity of a two-phase composite with partially debonded inclusions. *Int. J. Plast.* **1996**, *12*, 781–804. [CrossRef]

![metals logo] *metals*

MDPI

Article

Microstructures and Tensile Properties of Al–Cu Matrix Composites Reinforced with Nano-Sized SiC$_p$ Fabricated by Semisolid Stirring Process

Feng Qiu [1,3], Xiang Gao [1], Jian Tang [1], Yu-Yang Gao [1], Shi-Li Shu [2], Xue Han [3], Qiang Li [1] and Qi-Chuan Jiang [1,*]

1   Key Laboratory of Automobile Materials, Ministry of Education and Department of Materials Science and Engineering, Jilin University, Renmin Street NO. 5988, Changchun 130025, China; qiufeng@jlu.edu.cn (F.Q.); xianggao15@mails.jlu.edu.cn (X.G.); xxe8198458@163.com (J.T.); gaoyy14@mails.jlu.edu.cn (Y.-Y.G.); liqiang15@mails.jlu.edu.cn (Q.L.)
2   State Key Laboratory of Luminescence and Applications, Changchun Institute of Optics, Fine Mechanics and Physics, Chinese Academy of Sciences, Changchun 130012, China; shushili@ciomp.ac.cn
3   Department of Mechanical Engineering, Oakland University, Rochester, MI 48309, USA; xhan@oakland.edu
*   Correspondence: jiangqc@jlu.edu.cn; Tel./Fax: +86-431-85094699

Academic Editor: Manoj Gupta
Received: 29 December 2016; Accepted: 3 February 2017; Published: 8 February 2017

Abstract: The nano-sized SiC$_p$/Al–Cu composites were successfully fabricated by combining semisolid stirring with ball milling technology. Microstructures were examined by an olympus optical microscope (OM), field emission scanning electron microscope (FESEM) and transmission electron microscope (TEM). Tensile properties were studied at room temperature. The results show that the α-Al dendrites of the composites were strongly refined, especially in the composite with 3 wt. % nano-sized SiC$_p$, of which the morphology of the α-Al changes from 200 μm dendritic crystal to 90 μm much finer equiaxial grain. The strength and ductility of the composites are improved synchronously with the addition of nano-sized SiC$_p$ particles. The as-cast 3 wt. % nano-sized SiC$_p$/Al–Cu composite displays the best tensile properties, i.e., the yield strength, ultimate tensile strength (UTS) and fracture strain increase from 175 MPa, 310 MPa and 4.1% of the as-cast Al–Cu alloy to 220 MPa, 410 MPa and 6.3%, respectively. The significant improvement in the tensile properties of the composites is mainly due to the refinement of the α-Al dendrites, nano-sized SiC$_p$ strengthening, and good interface combination between the SiC$_p$ and Al–Cu alloys.

Keywords: nano-sized SiC$_p$; aluminum matrix composites; mechanical properties; microstructures

1. Introduction

In the past decades, particulate reinforced aluminum matrix composites (AMCs) have attracted much attention in the field of structural and functional materials [1–4]. SiC$_p$ reinforced AMCs have been a hot research issue in recent years because of their excellent properties such as low density, high tensile strength, high elastic modulus and wear resistance, etc. [5–7]. For example, SiC$_p$ reinforced AMCs are used for engine piston, and heat sink [8,9]. Compared with traditional micron-sized SiC$_p$/Al composites, the higher tensile strength and good ductility of the nano-sized SiC$_p$/Al composites entitle them to have more competitive ability for advanced structural applications such as in automotive and aerospace industries and the military [10]. In the past decades, several processing techniques have been developed for fabricating Al matrix composites reinforced with nano-sized particles such as high-energy milling, powder metallurgy, and nano-sintering, and liquid-state solidification processing (e.g., stir casting) [11–19]. In these techniques, the semisolid stirring process has some important advantages such as low cost, capability of producing products with complex shapes, and processing simplicity [14].

Al–Si and Al–Mg alloys are the usually used matrix phase in nano-sized SiC_p/Al composites [10,16,20,21]. Xiong et al. [10] fabricated 14 vol. % nano-sized SiC_p/Al–Mg composites. They reported that the ultimate tensile strength increased from 223 MPa to 286 MPa, while the ductility decreased from 4.8% to 3.9%. Hamedan et al. [16] produced 1.0 wt. % nano-sized $SiC_p/Al356$ composite and reported that the ultimate tensile strength of the composite increased from 140 MPa to173 MPa, while the ductility decreased slightly from 6.1% to 5.38%. Compared to the Al–Si and Al–Mg alloys, the Al–Cu alloys can offer some good mechanical properties. For instance, a previous study by us has shown that the tensile strength and elongation increased by 26% and 50% respectively, for the modified Al–Cu alloys compared with unmodified alloy [22]. Moreover, we also found that the corrosion resistance of modified Al–Cu alloy had an improvement compared with the unmodified one [23]. However, to the best of our knowledge, so far, because there is no chemical affinity between the Cu element and SiC_p, and the Cu element can also not improve the wettability between the aluminum matrix and SiC_p [24], the Al–Cu alloys were rarely used as a matrix in the nano-sized SiC_p/Al composites in stir casting. It is believed that if the nano-sized SiC_p/Al–Cu composites with uniform distribution SiC_p and clean interface between the SiC_p and Al–Cu alloys could be successfully fabricated, the composites will exhibit excellent mechanical properties, which are very important for application in the automotive and aircraft industries.

In this paper, the nano-sized SiC_p/Al–Cu composites were fabricated by combining semisolid stirring with ball milling technology. Semisolid stirring can suppress the interfacial reaction due to the low stirring temperature [19]. The usage of precursor powders fabricated by the mix of the nano-sized SiC_p and alloy powders using mechanical ball milling is of benefit to the dispersion of the nano-sized SiC_p in the matrix due to the disruption of the agglomerate nano-sized SiC_p clusters in advance. The microstructures and tensile properties of the synthesized composites were investigated, and the strengthening mechanism was discussed. We expect that such knowledge would provide guidance for the fabrication and application of the nano-sized SiC_p/Al–Cu composites.

2. Experimental Procedure

The Al–Cu alloy with a composition of (wt. %): 5.0 Cu, 0.8 Mn, 0.7 Fe, 0.5 Mg, 0.5 Si, 0.25 Zn, 0.15 Ti, 0.1 Cr and Al (balance) was used as the matrix. The nano-sized SiC_p, with a purity of 99.9 wt. % and ~60 nm in diameter, were used as the reinforced particles. The morphology of the raw nano-sized SiC_p particles is shown in Figure 1a. If the agglomerate nano-sized SiC_p clusters are added into the melt directly, it is difficult for semisolid stirring to break the clustering and disperse the nano-sized particles uniformly. Figure 1b shows the Al–Cu alloy powders (99% pure) with average sizes of about 10 μm, their composition is the same as the Al–Cu alloy matrix. Figure 1c,d shows the precursor powders which are fabricated by the mix of the calculated nano-sized SiC_p and Al–Cu alloy powders using mechanical ball milling with ZrO_2 balls at the speed of 150 r/min for 50 h. Figure 1d is the a high magnification of the rectangular area in Figure 1c. It could be found that most of the nano-sized SiC_p display a relatively uniform distribution in each individual composite particle surface. The ball to powder weight ratio was 8:1. During melting, Al–Cu alloy was melted at 933 K in air using an electricity resistant furnace and then cooled to 873 K at which point the matrix alloy was in semi-solid condition. The temperature range for the Al–Cu alloy used in this study to be in the semi-solid condition is 813 K–903 K. Then, the precursor powder was added into the molten metal after stirring the molten metal with a graphite stirrer at the speed of 500 r/min. After that, the melt was poured into a preheated steel die. After the casting process, the Al–Cu alloy and the composites were homogenised for 10 h at 758 K in order to avoid segregation. The materials were extruded to the batten shaped samples with the help of a 200-ton hydraulic press at 773 K with the extrusion ratio of 16. Before the tensile test, all the extruded samples underwent the T6 heat treatment (solutionized at 773 K for 2 h and aged at 433 K for 18 h).

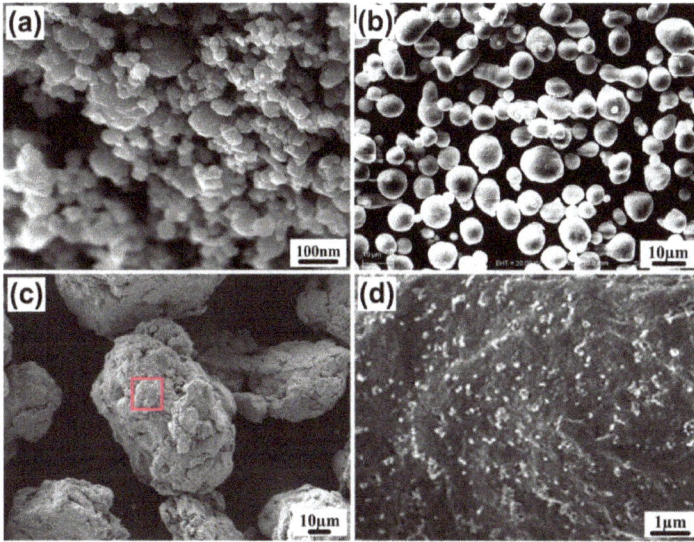

Figure 1. SEM images of morphologies of the raw (**a**) nano-sized SiC particles and (**b**) Al–Cu alloy powders; FESEM (field emission SEM) images of (**c**) nano-sized SiC$_p$/Al–Cu composite powders after ball milling; (**d**) high magnification of the area marked in (**c**).

Microstructures of the composites were examined by an Optical Microscope (Axio Imager A2m, Zeiss, Oberkochen, Germany) equipped with image analysis software and a camera; a computer was used for the OM observation and the quantitative measurements of microstructural features. The size of Al dendrites in every composite was measured from forty images taken at two magnifications, such as 50×, and 100×. Five samples of every composite were used to obtain the standard deviations (the error bars) plotted in Figure 2. Microstructures of the composites and morphologies of the raw nano-sized SiC particles and Al–Cu alloy powders were observed by field emission SEM (FESEM, JSM6700F, Tokyo, Japan) and SEM (Evo18, Carl Zeiss, Oberkochen, Germany).

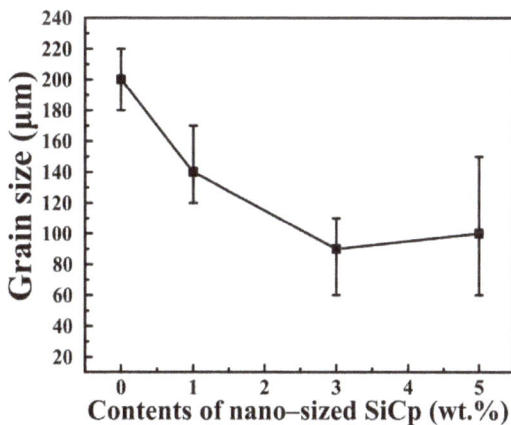

Figure 2. Grain sizes of α-Al in the cast Al–Cu alloy and nano-sized SiC$_p$/Al–Cu composites with different SiC$_p$ contents.

The extruded samples were machined into dog-bone shaped tensile samples with a gauge cross section of 5.0 mm × 2.5 mm and a gauge length of 30.0 mm. Tensile tests were conducted at room temperature by using a servo-hydraulic materials testing system (MTS, MTS 810, Minneapolis, MN, USA) at a constant strain rate of $3 \times 10^{-4} \text{ s}^{-1}$.

3. Results and Discussion

Figure 3 shows the as-cast microstructures of the Al–Cu alloy and nano-sized SiC_p/Al–Cu composites with the nominal content of 1 wt. %, 3 wt. %, 5 wt. % SiC_p. As shown in Figure 3a, the α-Al dendrites of the Al–Cu matrix alloy are coarse and their average size is about 200 μm. However, in the nano-sized SiC_p/Al–Cu composites, the α-Al dendrites are significantly refined by the addition of nano-sized SiC_p, as shown in Figure 3b–d. The refinement of the dendrite size is mainly due to some heterogeneous nucleation sties of the α-Al crystal provided by nano-sized particles during solidification, and the hindrance of the other added nano-sized SiC_p to the growth of α-Al dendrites during the solidification process. Figure 2 shows the size of dendrite in the nano-sized SiC_p/Al–Cu composites with different particle contents. In the 3 wt. % nano-sized SiC_p/Al–Cu composite, the morphology of α-Al changes from coarse dendritic grain to equiaxial grain with finer sizes of about 90 μm, which increases the boundary concentration in the Al matrix. The increase in the boundary concentration could be helpful to improve the tensile strength of metals or alloys due to the grain boundary playing a role as a barrier to the transmission of the dislocations. The as-cast microstructure of the composite with 5 wt. % SiC_p is similar to that with 3 wt. % SiC_p, although the sizes of the α-Al dendrites were uneven sizes of 60–150 μm (Figure 3d). In the composite with 5 wt. % SiC_p, the shape of Al dendrites became very non-uniform due to the agglomeration of nano-sized SiC_p particles. The hindrance effect of the nano-sized SiC_p on the α-Al dendrite growth is strong in the area of agglomeration of ceramic particles. On the contrary, the hindrance effect of the nano-sized SiC_p on the α-Al dendrite growth is weakened in the area of less ceramic particles. Thus, the difference in the size of the α-Al dendrites was probably due to the nonuniform dispersion of SiC_p when their contents reached 5 wt. %.

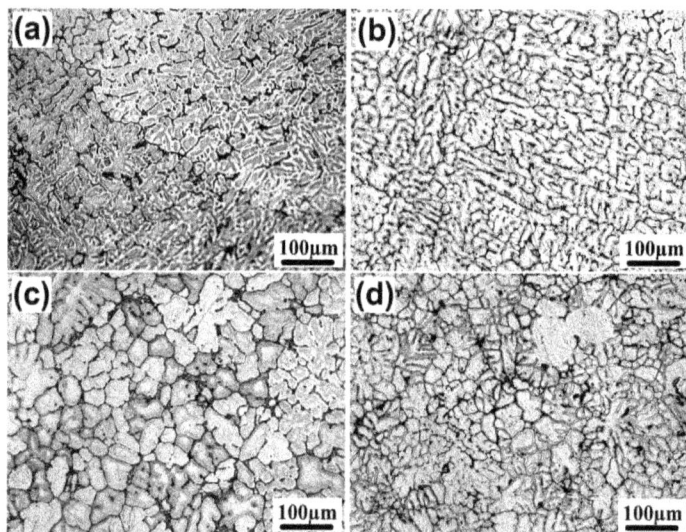

Figure 3. Cast microstructures of the nano-sized Al–Cu alloy and SiC_p/Al–Cu composites with different SiC_p contents; (**a**) Al–Cu alloy; (**b**) 1 wt. % SiC_p; (**c**) 3 wt. % SiC_p; (**d**) 5 wt. % SiC_p.

Figure 4 shows the engineering stress–strain curves of the cast Al–Cu alloy and nano-sized SiC$_p$/Al–Cu composites, and Table 1 lists the detailed data of the tensile properties. As indicated, the strength and ductility of the composites (1 wt. % and 3 wt. %) are improved synchronously which is quite rare for the composites reinforced with ceramic particles, because in most reported works [11,14,20,21], the composites had a higher strength and lower ductility than the matrix alloy. The yield strength ($\sigma_{0.2}$), ultimate tensile strength (UTS) and fracture strain (ε) of the nano-sized SiC$_p$/Al–Cu composites firstly increase and then decrease with the increase in the content of SiC$_p$. The 3 wt. % nano-sized SiC$_p$/Al–Cu composite possesses the best tensile properties. The yield strength, UTS and fracture strain of the 3 wt. % SiC$_p$/Al–Cu composite are 220 MPa, 410 MPa and 6.3%, which increase by 45 MPa (25.7%), 100 MPa (32.2%) and 2.2% (53.6%), respectively, compared to those of the as-cast Al–Cu alloy (175 MPa, 310 MPa and 4.1%).

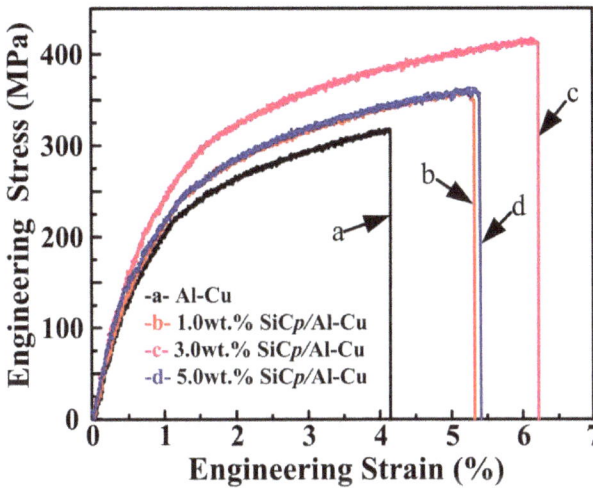

Figure 4. Tensile stress–strain curves of the cast Al–Cu alloy and nano-sized SiC$_p$/Al–Cu composites with different SiC$_p$ contents.

Table 1. Tensile properties of the as-cast Al–Cu alloy and nano-sized SiC$_p$/Al–Cu composites with different SiC$_p$ contents.

SiC$_p$ (wt. %)	$\sigma_{0.2}$ (MPa)	σ_b (MPa)	ε (%)
0	175^{+8}_{-6}	310^{+11}_{-10}	$4.1^{+1.2}_{-0.5}$
1	185^{+7}_{-8}	358^{+12}_{-11}	$5.3^{+0.8}_{-0.7}$
3	220^{+10}_{-6}	410^{+14}_{-8}	$6.3^{+0.7}_{-0.5}$
5	190^{+5}_{-8}	362^{+5}_{-13}	$5.4^{+1.3}_{-1.6}$

Figure 5a–d shows the FESEM images of the 3 wt. % and 5 wt. % nano–sized SiC$_p$/Al–Cu composite, and TEM micrographs of the 3 wt. % SiC$_p$/Al–Cu composite. As indicated in Figure 5a,b, more evenly distributed nano-sized SiC$_p$ particles in the 3 wt. % SiC$_p$/Al–Cu composite are observed compared with the 5 wt. % SiC$_p$/Al–Cu composite. However, in the 5 wt. % SiC$_p$/Al–Cu composite as shown in Figure 5b, although there are some uniform distribution zones of nano-sized SiC$_p$, the agglomeration of particles can still be easily found, as shown in Figure 5d. In other words, more and more particles aggregate to form the clusters with the increase in the content of SiC$_p$, resulting in the quite uneven α-Al dendrites sizes and higher concentration of defects. As shown in Figure 5c, it is clearly seen that the nano-sized SiC particles dispersed inside the α-Al dendrite and the interface

between the SiC particles and matrix are good and clean without any contaminations, indicating the superiority of this fabrication technology.

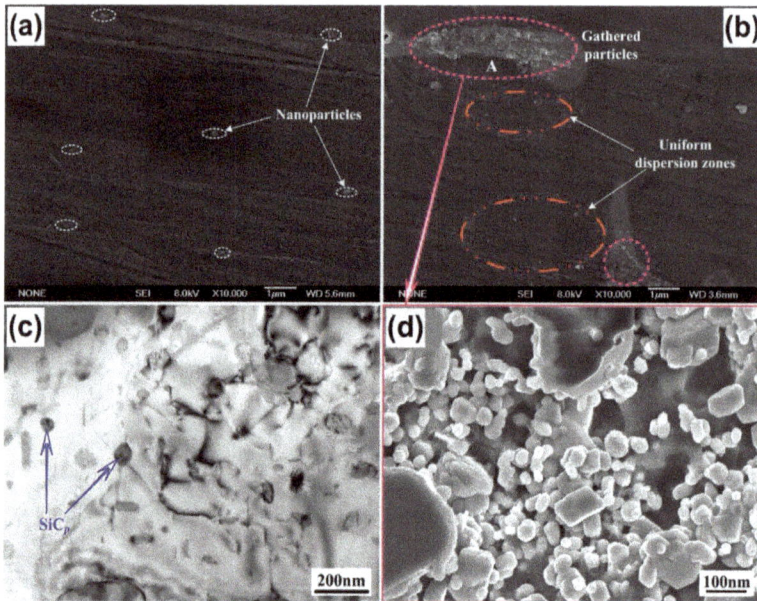

Figure 5. FESEM images of the (**a**) 3 wt. % and (**b**) 5 wt. % nano-sized SiC_p/Al–Cu composite; (**c**) TEM micrographs of the 3 wt. % SiC_p/Al–Cu composite; (**d**) the high magnification of area A in (**b**).

The significant improvement of the strength of the SiC_p/Al–Cu composites is mainly due to the refinement of the α-Al dendrites and the hindrance of the nano-sized SiC_p to the start and motion of dislocations in the matrix. Moreover, the significantly improved ductility of the nano-sized SiC_p/Al–Cu composites with a simultaneously increased tensile strength is derived from three factors:

(i) Nano-sized reinforcement. Compared with micron-sized ceramic particles, the nano-sized ceramic particles used as reinforcement can not only possess higher tensile strength but also maintain good ductility, especially in the low contents [21]. Large reinforcement particles could give rise to cleavage in the particle due to the fact that they are acting as concentrators of stress, and lead to the formation of pits or cavities due to the loss of interphase cohesion. However, the smallest reinforcement particles usually do not initiate pits or cavities at the particle and bond well to the metal matrix [11].

(ii) Dendrite refinement. The refinement of the α-Al dendrites will result in the increase in matrix dendrite boundaries. The finer the dendrite is, the more tortuous the grain boundaries are. Therefore, the crack propagation becomes more and more difficult and thus the composites can endure the larger plastic deformation before fracture.

(iii) Suppression of interfacial reaction. It is known that the reaction between molten Al and SiC_p takes place easily in the temperature range from 675 °C to 900 °C, producing Al_4C_3 which is a brittle and unstable phase [12]. The presence of Al_4C_3 degrades the mechanical properties through crack propagation. In the present work, low stirring temperature (600 °C) during the semisolid stirring process can suppress the interfacial reaction effectively, which will be helpful to restrict the formation of the Al_4C_3 phase. The improved strength and cracking resistance of the interface bonding make the occurrence of the crack source cracking become more difficult.

Nano-sized particle strengthening, microstructure refinement, and good interface between reinforcement and matrix with no brittle intermetallics can be responsible for the significant improvement in the mechanical properties of the nano-sized SiC$_p$/Al–Cu composites. Hence, the nano-sized SiC$_p$/Al–Cu composites showed high plasticity and strength. However, the properties of the composite with high content nano-sized SiC$_p$ particles could be weakened because this composite resulted in more agglomeration of the SiC$_p$ and higher concentration of defects. More severe agglomeration of SiC$_p$ could not lead to the matrix being completely wrapped up by the particles and thus result in the debonding of the interface. Moreover, micro-porosity and other defects around the SiC$_p$ clusters presented in the composites become the cracking source during the plastic deformation. The above analyses imply that the embrittlement of the composites resulting from micro-porosity and detects results in the decrease in strength and ductility of the 5 wt. % SiC$_p$/Al–Cu.

4. Conclusions

The nano-sized SiC$_p$/Al–Cu composites with contents of 1 wt. %, 3 wt. %, 5 wt. % SiC$_p$ were successfully fabricated by combining semisolid stirring with ball milling technology. The α-Al dendrites are significantly refined due to the addition of nano-sized SiC$_p$. The refinement of the dendrite size is mainly attributed to some nano-sized particles providing some heterogeneous nucleation sties of the α-Al crystal, and the hindrance of the other added nano-sized SiC$_p$ to the growth of α-Al dendrites during the solidification process. The strength and ductility of the composites are improved synchronously with the addition of nano-sized SiC$_p$ particles. The 3 wt. % nano-sized SiC$_p$/Al–Cu composite displays the best comprehensive tensile properties, i.e., the yield strength, UTS and fracture strain increase from 175 MPa, 310 MPa and 4.1% of the as-cast Al–Cu alloy to 220 MPa, 410 MPa and 6.3%, respectively. Nano-sized particle strengthening, microstructure refinement, and a good interface between reinforcement and the matrix with no brittle intermetallics can be responsible for the significant improvement in the mechanical properties of the nano-sized SiC$_p$/Al–Cu composites.

Acknowledgments: The National Natural Science Foundation of China (NNSFC, No. 51571101), the "Thirteenth Five-year Plan" Science & Technology Research Foundation of Education Bureau of Jilin Province, China (Grant No. 2015-479), NNSFC (No. 51501176) and the Project 985-High Properties Materials of Jilin University.

Author Contributions: Feng Qiu and Qi-Chuan Jiang conceived and designed the experiments; Feng Qiu, Xiang Gao, Jian Tang, Yu-Yang Gao and Qiang Li performed the experiments; Feng Qiu, Jian Tang, Shi-Li Shu, and Xue Han, analyzed the data; Feng Qiu wrote the paper.

Conflicts of Interest: The authors declare no conflict of interest.

References

1. Wang, L.; Qiu, F.; Ouyang, L.C.; Wang, H.Y.; Zha, M.; Shu, S.L.; Zhao, Q.L.; Jiang, Q.C. A novel approach of using ground CNTs as the carbon source to fabricate uniformly distributed nano-sized TiC$_x$/2009Al composites. *Materials* **2015**, *8*, 8839–8849. [CrossRef]
2. Zhao, Q.; Liang, Y.H.; Zhang, Z.H.; Li, X.J.; Ren, L.Q. Study on the impact resistance of bionic layered composite of TiC-TiB$_2$/Al from Al-Ti-B$_4$C System. *Materials* **2016**, *9*, 708. [CrossRef]
3. Shu, S.L.; Yang, H.Y.; Tong, C.C.; Qiu, F. Fabrication of TiCx-TiB$_2$/Al composites for application as a Heat Sink. *Materials* **2016**, *9*, 642. [CrossRef]
4. Zhao, Q.; Liang, Y.H.; Zhang, Z.H.; Li, X.J.; Ren, L.Q. Microstructure and dry-Sliding wear behavior of B$_4$C ceramic particulate reinforced Al 5083 matrix composite. *Metals* **2016**, *6*, 227. [CrossRef]
5. Sajjadi, S.A.; Ezatpoura, H.R.; Torabi Parizi, M. Comparison of microstructure and mechanical properties of A356 aluminum alloy/Al$_2$O$_3$ composites fabricated by stir and compo-casting processes. *Mater. Des.* **2012**, *34*, 106–111. [CrossRef]
6. Molina, J.M.; Prieto, R.; Narciso, J.; Louis, E. The effect of porosity on the thermal conductivity of Al–12 wt. % Si/SiC composites. *Scr. Mater.* **2009**, *60*, 582–585. [CrossRef]
7. Kerti, I.; Toptan, F. Microstructural variations in cast B$_4$C-reinforced aluminium matrix composites (AMCs). *Mater. Lett.* **2008**, *62*, 1215–1218. [CrossRef]

8. Du, X.F.; Gao, T.; Liu, G.L.; Liu, X.F. In situ synthesizing SiC particles and its strengthening effect on an Al-Si-Cu-Ni-Mg piston alloy. *J. Alloy. Compd.* **2017**, *695*, 1–8. [CrossRef]
9. Schöbel, M.; Altendorfer, W.; Degischer, H.P.; Vaucher, S.; Buslaps, T.; di Michiel, M.; Hofmann, M. Internal stresses and voids in SiC particle reinforced aluminum composites for heat sink applications. *Compos. Sci. Technol.* **2011**, *71*, 724–733. [CrossRef]
10. Xiong, B.; Xu, Z.; Yan, Q.; Cai, C.; Zheng, Y.; Lu, B. Fabrication of SiC nanoparticulates reinforced Al matrix composites by combining pressureless infiltration with ball-milling and cold-pressing technology. *J. Alloy. Compd.* **2010**, *497*, L1–L4. [CrossRef]
11. Kang, Y.C.; Chan, S.L.I. Tensile properties of nanometric Al_2O_3 particulate-reinforced aluminum matrix composites. *Mater. Chem. Phys.* **2004**, *85*, 438–443. [CrossRef]
12. Xiong, B.; Xu, Z.; Yan, Q.; Lu, B.; Cai, C. Effects of SiC volume fraction and aluminum particulate size on interfacial reactions in SiC nanoparticulate reinforced aluminum matrix composites. *J. Alloy. Compd.* **2011**, *509*, 1187–1191. [CrossRef]
13. Hsu, C.J.; Chang, C.Y.; Kao, P.W.; Ho, N.J.; Chang, C.P. Al–Al_3Ti nanocomposites produced in situ by friction stir processing. *Acta Mater.* **2006**, *54*, 5241–5249. [CrossRef]
14. Zhang, H.; Geng, L.; Guan, L.; Huang, L. Effects of SiC particle pretreatment and stirring parameters on the microstructure and mechanical properties of SiC_p/Al–6.8Mg composites fabricated by semi-solid stirring technique. *Mater. Sci. Eng. A* **2010**, *528*, 513–518. [CrossRef]
15. Mazahery, A.; Shabani, M.O. Characterization of cast A356 alloy reinforced with nano SiC composites. *Trans. Nonferr. Met. Soc. China* **2012**, *22*, 275–280. [CrossRef]
16. Dehghan Hamedan, A.; Shahmiri, M. Production of A356–1 wt % SiC nanocomposite by the modified stir casting method. *Mater. Sci. Eng. A* **2012**, *556*, 921–926. [CrossRef]
17. Nie, K.B.; Wang, X.J.; Wu, K.; Xu, L.; Zheng, M.Y.; Hu, X.S. Processing, microstructure and mechanical properties of magnesium matrix nanocomposites fabricated by semisolid stirring assisted ultrasonic vibration. *J. Alloy. Compd.* **2011**, *509*, 8664–8669. [CrossRef]
18. Mazahery, A.; Abdizadeh, H.; Baharvandi, H.R. Development of high-performance A356/nano-Al_2O_3 composites. *Mater. Sci. Eng. A* **2009**, *518*, 61–64. [CrossRef]
19. Tahamtan, S.; Halvaee, A.; Emamy, M.; Zabihi, M.S. Fabrication of Al/A206–Al_2O_3 nano/micro composite by combining ball milling and stir casting technology. *Mater. Des.* **2013**, *49*, 347–359. [CrossRef]
20. Amirkhanlou, S.; Niroumand, B. Effects of reinforcement distribution on low and high temperature tensile properties of Al356/SiC_p cast composites produced by a novel reinforcement dispersion technique. *Mater. Sci. Eng. A* **2011**, *528*, 7186–7195. [CrossRef]
21. Yang, Y.; Lan, J.; Li, X. Study on bulk aluminum matrix nano-composite fabricated by ultrasonic dispersion of nano-sized SiC particles in molten aluminum alloy. *Mater. Sci. Eng. A* **2004**, *380*, 378–383. [CrossRef]
22. Zhao, H.L.; Yao, D.M.; Qiu, F.; Xia, Y.M.; Jiang, Q.C. High strength and good ductility of casting Al-Cu alloy modified by Pr_xO_y and La_xO_y. *J. Alloy. Compd.* **2011**, *509*, L43–L46. [CrossRef]
23. Xia, Y.M.; Bai, Z.H.; Qiu, F.; Jin, S.B.; Jiang, Q.C. Effects of multi-modification of rare earth oxides Pr_xO_y and La_xO_y on microstructure and tensile properties of casting Al-Cu alloy. *Mater. Sci. Eng. A* **2012**, *558*, 602–606. [CrossRef]
24. Kobashi, M.; Choh, T. Effects of alloying elements on SiC dispersion in liquid aluminum. *Mater. Trans.* **1990**, *12*, 1101–1107. [CrossRef]

metals

MDPI

Article

Improved Compressive, Damping and Coefficient of Thermal Expansion Response of Mg–3Al–2.5La Alloy Using Y_2O_3 Nano Reinforcement

Amit Kumar [1], Khin Sandar Tun [1], Amit Devendra Kohadkar [2] and Manoj Gupta [1,*]

[1]   Department of Mechanical Engineering, National University of Singapore, 9 Engineering Drive 1,
    Singapore 117576, Singapore; amittonk@gmail.com (A.K.); mpekhst@nus.edu.sg (K.S.T.)
[2]   Department of Mechanical engineering, Visvesvaraya National Institute of Technology,
    South Ambazari Road, Nagpur 440010, India; amitkohadkar@gmail.com
*   Correspondence: mpegm@nus.edu.sg; Tel.: +65-6516-6358

Academic Editor: Daolun Chen
Received: 6 March 2017; Accepted: 14 March 2017; Published: 21 March 2017

Abstract: In the present study, the effects of the addition of Y_2O_3 nanoparticles on Mg–3Al–2.5La alloy were investigated. Materials were synthesized using a disintegrated melt deposition technique followed by hot extrusion. The samples were then characterized for microstructure, compression properties, damping properties, CTE (coefficient of thermal expansion) and fracture morphology. The grain size of Mg–3Al–2.5La was significantly reduced by the addition of the Y_2O_3 nano-sized reinforcement (~3.6 μm, 43% of Mg–3Al–2.5La grain size). SEM and X-ray studies revealed that the size of uniformly distributed intermetallic phases, $Al_{11}La_3$, Al_2La, and $Al_{2.12}La_{0.88}$ reduced by the addition of Y_2O_3 to Mg–3Al–2.5La alloy. The coefficient of thermal expansion (CTE) was slightly improved by the addition of nanoparticles. The results of the damping measurement revealed that the damping capacity of the Mg–3Al–2.5La alloy increased due to the presence of Y_2O_3. The compression results showed that the addition of Y_2O_3 to Mg–3Al–2.5La improved the compressive yield strength (from ~141 MPa to ~156 MPa) and the ultimate compressive strength (from ~456 MPa to ~520 MPa), which are superior than those of the Mg–3Al alloy (Compressive Yield Strength, CYS ~154 MPa and Ultimate Compressive Strength, UCS ~481 MPa). The results further revealed that there is no significant effect on the fracture strain value of Mg–3Al–2.5La due to the addition of Y_2O_3.

Keywords: Mg–Al–RE alloy; magnesium alloy; damping; $Al_{11}La_3$ phase; nanosize reinforcement; mechanical properties

1. Introduction

Mg–Al-based alloys are considered important lightweight alloys due to their low density, high strength, and stiffness with good casting and processing ability. Although Mg–Al alloys exhibit a superior combination of mechanical properties, they are not suitable for application in automobile engine components due to their poor creep resistance [1,2]. It is well reported that poor creep properties in Mg–Al alloys are due to the formation of the β-eutectic phase ($Mg_{17}Al_{12}$), which is unstable at high temperatures [3]. To improve the creep properties of Mg–Al, rare earth metals (RE) were used as alloying elements, as they can suppress the formation of the β-phase. In addition, RE also improved the grain refinement and strength while retaining the ductility, creep resistance, corrosion resistance and fatigue strength [4–7]. The addition of lanthanum (La) to Mg–4Al exhibited a good strengthening effect due to its precipitation hardening and grain refinement effects [3]. In our recent study on Mg–3Al–xLa (x = 1%, 2.5% and 4%), it was observed that the addition of La to Mg–3Al led to the consumption of most of the Al for the formation of $Al_{11}La_3$, Al_2La, and $Al_{2.12}La_{0.88}$ intermetallic phases and suppressed the formation of the $Mg_{17}Al_{12}$ phase [8]. Among all the compositions, the Mg–3Al–2.5La alloy exhibited the best tensile properties; Tensile Yield Strength, TYS ~160 MPa, Ultimate Tensile Strength, UTS ~249 MPa and fracture strain ~22%. However, the addition of La in Mg–3Al alloy caused a gradual decrease in the compressive strength and elongation [8].

On the other hand, nano-sized reinforcement (thermally stable ceramics such as Al_2O_3, ZrO_2, Y_2O_3) used in magnesium-based nanocomposites has already shown potential improvement in the mechanical properties and ductility without any significant increase in the density [9–13]. Many types of advanced metal matrix nanocomposites are now easily available, and they exhibit functional properties. Recently, a few particle-reinforced, self-lubricating and self-healing metal matrix nanocomposites were synthesized using solidification techniques [14–16]. Hassan et al. [17] showed that the addition of nano-sized yttrium oxide (Y_2O_3) particulates as a reinforcement in magnesium, synthesized by the disintegrated melt deposition (DMD) technique, enhanced the mechanical properties of the magnesium matrix. This work concluded that the addition of 1.9% Y_2O_3 by weight exhibits the best mechanical properties compared to 0.6% and 3.1% Y_2O_3 [17].

The present work addresses the further enhancement of the compression and damping response of Mg–3Al–2.5La alloy using Y_2O_3 nano particulates as a reinforcement. Mg–3Al–2.5La alloy, containing 1.9% Y_2O_3 by weight as reinforcement, is synthesized along with pure Mg, Mg–3Al and Mg–3Al–2.5La alloys, using the Disintegrated Melt Deposition (DMD) technique followed by hot extrusion. A detailed view of the effect of the Y_2O_3 addition on the microstructure, Coefficient of Thermal Expansion (CTE), compression and damping properties of Mg–3Al–2.5La is provided.

2. Materials and Characterizations

2.1. Materials

Magnesium turnings (99.9% purity) supplied by Acros Organics (Geel, Belgium) were used as the base material. Aluminium powder (99% purity) of size ~7–15 μm supplied by Alfa Aesar (Haverhill, MA, USA) and Mg–30%La master alloy supplied by Sunreiler Metal Co. Limited (Beijing, China) were used as alloying elements. Yttrium oxide (99.995% purity) of size 20–40 nm supplied by US Research Nanomaterials (Houston, TX, USA) was used as reinforcement in this study.

2.2. Processing

Four different compositions, pure Mg, Mg–3%Al, Mg–3%Al–2.5%La and Mg–3%Al–2.5%La–1.9%Y_2O_3 by weight were synthesized using disintegrated melt deposition technique [18]. Pure Mg turnings, Al powder, Y_2O_3 powder and Mg–30%La master alloy were placed in a multilayered sandwich fashion in a graphite crucible and superheated to 750 °C under an argon gas atmosphere using electrical resistance furnace (Dakin Engineering Pte Ltd., Singapore). For uniform distribution of reinforcement particulates within the alloy matrix, the superheated slurry was then stirred at 450 rpm for 5 min using a stainless steel impeller (Starlight Tool Precision Engineering, Singapore) with twin blade (pitch 45°). Stainless steel stirrer was used to avoid any iron contamination of the molten metal. After stirring, the molten melt was down poured through a nozzle of 10 mm diameter at the bottom of the crucible to the mould under the influence of gravity. Before entering the mold, the molten metal was disintegrated by two jets of argon gas, oriented normal to the melt stream. The flow of argon was maintained at 25 L/min [17]. An ingot of 40 mm diameter was then obtained. For synthesizing other compositions similar steps were followed. As cast ingot was later machined to 36 mm diameter and 45 mm length for the secondary processing.

Secondary processing involved the soaking of ingot at 400 °C for 1 h in a constant temperature furnace (Elite Thermal Systems Ltd., Market Harborough, Leicestershire, UK). Using a 150-ton hydraulic extrusion press, hot extrusion was carried out at 350 °C die temperature with an extrusion ratio of 20.25:1 to obtain rods of 8 mm diameter. Extruded rods were further used to prepare samples for different characterization studies.

2.3. Characterizations

2.3.1. Microstructural Characterization

The microstructure was characterized using an optical microscope (Olympus Corporation, Shinjuku, Tokyo, Japan) on polished and etched samples (etchant: 4.2 gm picric acid, 10 mL acetic acid, 70 mL ethanol and 10 mL distilled water). The grain size was measured on the longitudinal section of samples, with the help of Scion image analysis software (beta 4.0.2, Frederick, MD, USA, 2000). To observe intermetallic phase formation and distribution, scanning electron microscopes JEOL JSM-6010 (JEOL Ltd., Tokyo, Japan) and Hitachi FESEM-S4300 (Hitachi, Ltd., Tokyo, Japan) equipped with energy dispersive spectrometric analysis (EDS) were used. X-Ray diffraction analysis was conducted using an automated Shimadzu LAB-XRD-6000 (Shimadzu Corporation, Kyoto, Japan) (Cu Kα:λ = 1.54056 Å) spectrometer with a scan speed of 2°/min.

2.3.2. Physical Characterization

Density and Porosity: The density of extruded pure Mg, Mg–3%Al, Mg–3%Al–2.5%La and Mg–3%Al–2.5%La–1.9%Y$_2$O$_3$ was measured using a gas pycnometer (Micromeritics Instrument Corp., Norcross, GR, USA). Each sample was run for five cycles to measure the density more accurately. Pure helium gas was purged with a pressure of 19.5 Psig for all the five cycles with a cycle fill pressure of 19.5 Psig. The difference between theoretical density (calculated by the rule of mixture) and experimentally measured density was quantified as the porosity level in the material.

The Coefficient of thermal expansion: By using a thermo-mechanical analysis instrument LINSEIS TMA PT 1000LT (Linseis Thermal Analysis, Robbinsville, NJ, USA) the coefficient of thermal expansion (CTE) of pure Mg, Mg–3%Al, Mg–3%Al–2.5%La and Mg–3%Al–2.5%La–1.9%Y$_2$O$_3$ was determined. The heating rate of 5 °C/min was maintained with constant argon flow rate of 0.1 L per minute. The displacement of the test samples (each of 5 mm length and 8 mm diameter) was measured as a function of temperature (323 K to 673 K) using an alumina probe (Linseis Thermal Analysis, Robbinsville, NJ, USA).

Damping: The vibrational damping capacity of the materials was measured using the resonance frequency damping analyzer (RFDA), (IMEC, Genk, Belgium). The vibration signal of each material (8 mm diameter, 60 mm length) was measured as a function of amplitude vs. time.

2.3.3. Mechanical Characterization

Compression Properties: In accordance with ASTM E9-09, compressive properties of extruded pure Mg, Mg–3%Al, Mg–3%Al–2.5%La and Mg–3%Al–2.5%La–1.9%Y$_2$O$_3$ samples were determined at ambient temperature, using a fully automated servo-hydraulic mechanical testing machine, MTS-810 (MTS systems corporation, Eden Prairie, MN, USA). The compression properties were measured at a strain rate of 8.334 × 10^{-5} s^{-1}. The specimens of 8 mm diameter, with length to diameter ratio of one were used. At least five different samples of each composition were tested to ensure repeatability of results. Fractured surfaces of all samples were analyzed using Hitachi S-4300 FESEM (Hitachi, Ltd., Tokyo, Japan).

3. Results and Discussion

3.1. Microstructural Characterization

The microstructures of all the samples were initially characterized using SEM microscopy (JEOL Ltd., Tokyo, Japan) (Figure 1). Table 1 and Figure 2a show the grain size of different compositions after analysis. The results revealed that the addition of 2.5% La and 1.9% Y$_2$O$_3$ to Mg–3Al reduced the average grain size by ~50%. It was observed that the addition of Al in Mg (Figure 1a,b) significantly reduced the grain size from ~22.6 μm to ~7.74 μm. It is frequently reported that during the solidification of Mg–Al alloys, fine grains are nucleated as the primary-Mg solid solution, along with the $Mg_{17}Al_{12}$ eutectic mixture distributed along the grain boundaries [1,2,19,20]. Secondary processing or hot deformation during extrusion further breaks down the $Mg_{17}Al_{12}$ network structure into fine precipitates, which results in grain refinement as observed in Mg–3Al (Figure 1b). Hot extrusion was performed at 350 °C (which is >0.5 T_m of pure Mg), and therefore it resulted in recrystallization and the formation of nearly equiaxed grains.

As the solubility of La in Mg is very limited (~0.78 wt %) [21,22] and La is a grain refiner, therefore the addition of 2.5% La to Mg–3Al alloy further reduced the average grain size from ~7.74 μm to ~6.26 μm. These results obtained in this study are in good agreement with other available reports claiming La as an excellent grain refiner in Mg [23,24]. The reinforcement of nano-sized thermally stable 1.9% Y_2O_3 powder to Mg–3Al–2.5La further reduced the grain size as Y_2O_3 nanoparticles can act as the nucleation sites during solidification and recrystallization besides pinning the grain boundaries in the later stages. Figure 2a represents the change in the grain size of pure Mg with the addition of 3Al, 2.5La, and $1.9Y_2O_3$ subsequently.

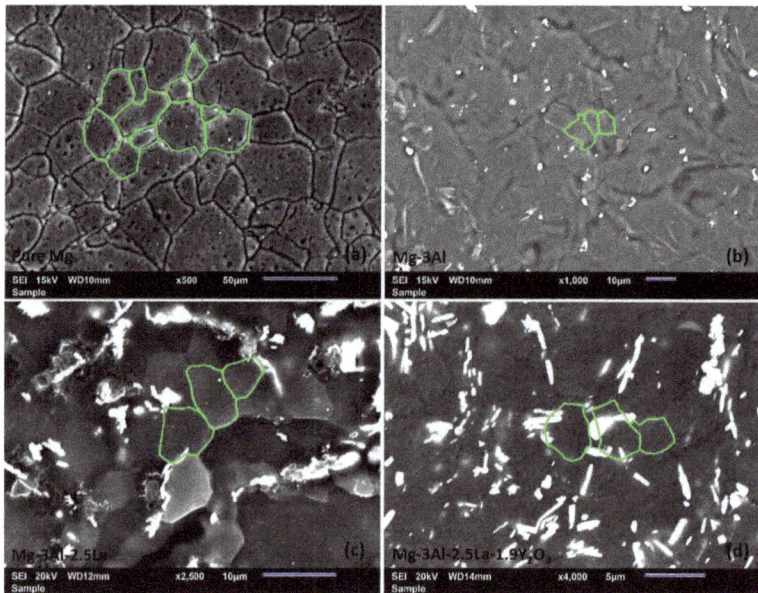

Figure 1. Scanning Electron Microscopic (SEM) micrographs of (**a**) pure Mg, (**b**) Mg–3Al, (**c**) Mg–3Al–2.5La, and (**d**) Mg–3Al–2.5La–1.9Y_2O_3 alloys, illustrating the grain structure.

Table 1. Results of average grain size, density, porosity and Coefficient of Thermal Expansion (CTE) measurements.

Material (wt %)	Average Grain Size (μm)	Density and Porosity Measurements			CTE ($\times 10^{-6}$/K)
		Theoretical Density (g/cc)	Experimental Density (g/cc)	Porosity (%)	
Pure Mg	22.6 ± 7.3	1.738	1.737	0.15	26.8 ± 3.9
Mg–3Al	7.74 ± 1.5	1.758	1.753	0.29	26.1 ± 2.6
Mg–3Al–2.5La	6.26 ± 1.1	1.791	1.788	0.17	25.3 ± 2.7
Mg–3Al–2.5La–1.9Y_2O_3	3.6 ± 0.5	1.818	1.813	0.16	25.0 ± 1.1

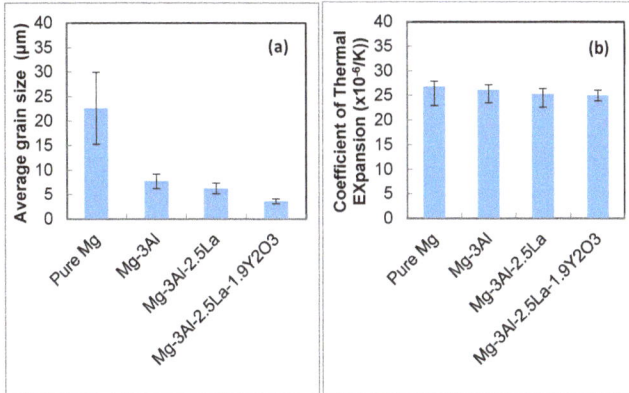

Figure 2. Change in (**a**) grain size and (**b**) coefficient of thermal expansion in pure Mg with the addition of 3Al, 2.5La and $1.9Y_2O_3$ subsequently.

Figure 3a–c are the SEM micrographs of Mg–3Al, Mg–3Al–2.5La, and Mg–3Al–2.5La–$1.9Y_2O_3$ alloys. Figure 3a–b show the SEM micrographs of extruded Mg–3Al and Mg–3Al–2.5La alloys. In Mg–3Al alloy, the dispersed $Mg_{17}Al_{12}$ phase is distributed inside the Mg matrix. A uniformly distributed bright white phase appeared in the Mg–3Al–2.5La alloy (Figure 3b) in rod-like ($Al_{11}La_3$) and polygon-type (Al_2La, $Al_{2.12}La_{0.88}$) shapes, which is consistent with earlier reports [3,21,25–29]. The brighter second phase in Figure 3c is broken into even finer shapes in the Mg–3Al–2.5La–$1.9Y_2O_3$ alloy, especially rod-like shapes, illustrating the ability of Y_2O_3 nanoparticles to refine the second phases. Similar findings were observed as a result of the addition of Al_2O_3 in the AZ31 alloy [30]. The uniform distribution of the second phase is due to the hot extrusion, which broke down these scattered rod-like and polygon shapes into small pieces throughout the microstructure.

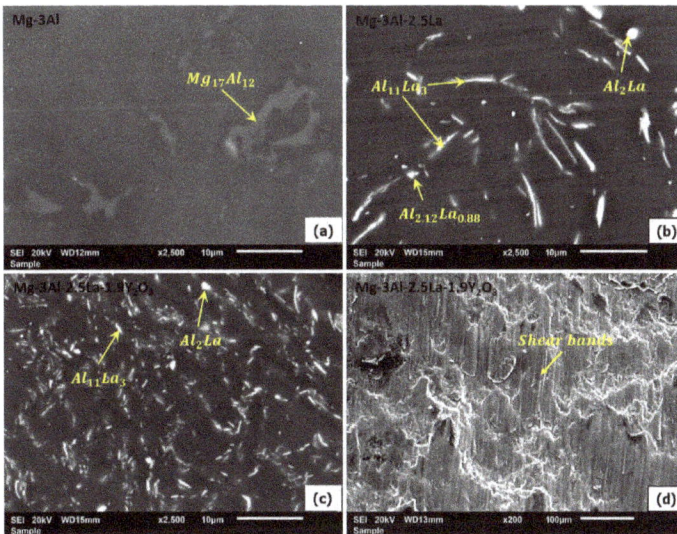

Figure 3. SEM micrographs of (**a**) Mg–3Al, (**b**) Mg–3Al–2.5La, and (**c**) Mg–3Al–2.5La–$1.9Y_2O_3$ alloys; (**d**) compressive fractograph of Mg–3Al–2.5La–$1.9Y_2O_3$ alloys.

X-ray diffraction (XRD) studies conducted in the longitudinal direction of the samples are shown in Figure 4. These diffractograms did not reveal the presence of any La phase with Mg, which is consistent with other available reports on Mg–Al–La alloys [8,21,22,25–29,31,32]. However, they revealed the strong presence of Mg peaks together with the phase comprised of $Al_{11}La_3$, Al_2La and $Al_{2.12}La_{0.88}$, which are also observed in the SEM micrographs. The formation of $Al_{11}La_3$, Al_2La and $Al_{2.12}La_{0.88}$ as intermetallic phases occurred due to the large difference in the electronegativity of Al and La when compared to Mg and Al [21,33,34]. It is well documented in the literature that dominating diffraction angles in extruded Mg rods corresponding to $2\theta = 32°$, $34°$ and $36°$, respectively, represent the prismatic $(1, 0, -1, 0)$ plane, the basal $(0, 0, 2, 0)$ plane and the pyramidal $(1, 0, -1, 0)$ plane of HCP Mg crystal [35]. From the intensity of these peaks at various diffraction angles, it is evident that the addition of La in Mg–3Al increased the I/I_{max} ratio for the basal plane but the pyramidal texture still dominated. The Mg–3Al–2.5La–1.9Y_2O_3 alloy showed that the peak corresponding to the basal plane becomes dominant. This indicates that the presence of Y_2O_3 clearly strengthens the basal texture in the Mg–3Al–2.5La alloy.

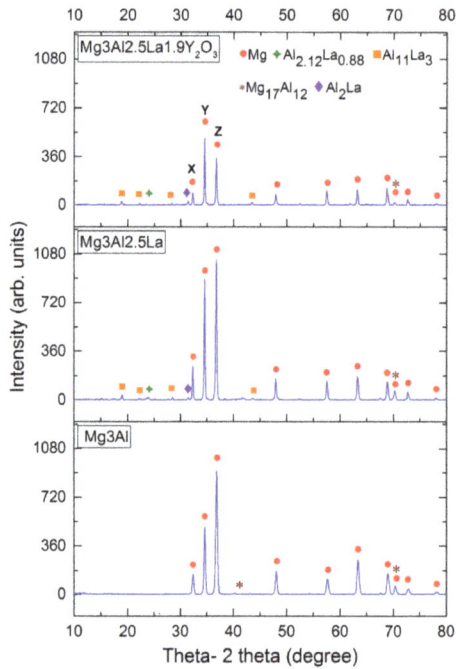

Figure 4. X-ray diffraction results of Mg–3Al, Mg–3Al–2.5La and Mg–3Al–2.5La–1.9Y_2O_3 alloys.

3.2. Physical Characterization

3.2.1. Density and Porosity

From Table 1, it is observed that near-dense Mg materials were synthesized utilizing the disintegrated melt deposition technique coupled with hot extrusion. The experimentally measured density values of the synthesized alloys and composite are closer to those of theoretically calculated density values. The increase in the density values of pure Mg was due to the addition of relatively high-density Al, La and Y_2O_3 elements when compared to pure magnesium.

The volumetric porosity results, which were calculated using theoretical and experimental density values, show that the addition of Y_2O_3 did not affect the porosity of the base Mg–3Al–2.5La alloy.

3.2.2. The Coefficient of Thermal Expansion

Table 1 and Figure 2b show the results of the coefficient of thermal expansion (CTE) measurements within the 25–400 °C temperature range. The results show that the CTE value of pure magnesium decreased with the alloying additions of Al, La and Y_2O_3. The gradual decrease in the CTE values of Mg, Mg–3Al, Mg–3Al–2.5La and Mg–3Al–2.5La–1.9Y_2O_3 alloy was due to the presence of the alloying addition of Al, La and Y_2O_3 which have lower CTE values (23.1 × 10^{-6}/K, 12.1 × 10^{-6}/K and 8.1 × 10^{-6}/K) as compared to pure Mg (26.8 × 10^{-6}/K) [36]. The results (see Figure 2b) suggest that the alloys and nanocomposites investigated in this study are more dimensionally stable with respect to temperature when compared to pure Mg.

3.2.3. Damping

The damping characteristics of extruded pure Mg, Mg–3Al, Mg–3Al–2.5La and Mg–3Al–2.5La–1.9Y_2O_3 alloys are presented in Table 2. The damping capacity of a material is defined as the ability to absorb vibration. The value of the damping capacity of a material depends on its properties such as density, microstructure, and elasticity [8,37]. The results show that the damping capacity of pure Mg decreased with the addition of Al in Mg–3Al, which was further enhanced by the addition of 2.5La in Mg–3Al–2.5La. The addition of Y_2O_3 further improved the damping capacity of the Mg–3Al–2.5La.

The damping loss rate represents how fast a material stops vibration. The results indicate that addition of 1.9% Y_2O_3 decreased the damping loss rate compared to the addition of 2.5% La. The significant change in the damping properties of alloys can be due to the damping mechanisms related to texture reorientation, thermal mismatch, defects, porosity, dislocation and grain boundary.

Table 2. Room-temperature compressive and damping properties of pure Mg, Mg–3Al, Mg–3Al–2.5La and Mg–3Al–2.5La–1.9Y_2O_3 alloys.

Material	0.2% CYS (MPa)	UCS (MPa)	Fracture Strain (%)	Damping Loss Rate	Damping Capacity
Pure Mg [8]	90 ± 6	333 ± 4	23 ± 0.74	8.00 ± 1.000	0.000456
Mg–3Al [8]	154 ± 2	481 ± 7	24 ± 0.5	6.16 ± 0.377	0.000204
Mg–3Al–2.5La [8]	141 ± 4	456 ± 3	18 ± 1	8.29 ± 0.827	0.000265
Mg–3Al–2.5La–1.9Y_2O_3	156 ± 5	520 ± 8	18 ± 0.70	7.60 ± 0.701	0.000272

3.3. Mechanical Characterization

Compression properties: Table 2 and Figure 5 show the room-temperature compression properties of extruded pure Mg, Mg–3Al, Mg–3Al–2.5La and Mg–3Al–2.5La–1.9Y_2O_3 samples under compression loading. As evident from the results, the addition of 3Al in pure Mg enhanced the compressive yield strength (CYS), the ultimate compressive strength (UCS) and the fracture strain (FS) from ~90 MPa, ~333 MPa and ~23% to a level of ~154 MPa, ~481 MPa and ~24%. This increase in compressive strength was due to the hall-patch effect as there was a tremendous (~73%, ~22.6 μm to ~7.76 μm) reduction in grain size of the pure Mg. Another possible reason is the presence of fine $Mg_{17}Al_{12}$ precipitates near the grain boundaries, which lead to precipitation hardening. The addition of 2.5La to Mg–3Al significantly reduced the CYS, UCS and failure strain values. In spite of the grain refinement (~7.74 μm to ~6.26 μm), the compression strength of Mg–3Al–2.5La decreased. This was due to the presence of intermetallic Al_2La and $Al_{2.12}La_{0.88}$ with fine $Al_{11}La_3$ phases, which are hard and exhibit sharp edges. Stress concentrates on these sharp edges and causes early crack initiation and subsequent crack propagation. The addition of nano-sized reinforcement particulates of Y_2O_3 further refined the grain size of Mg–3Al–2.5La (~6.26 μm to ~3.6 μm) and fragmented the second phases, resulting in the best improvement of the CYS and UCS.

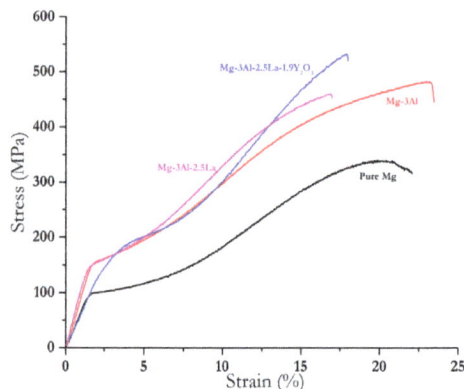

Figure 5. Stress-strain curves of extruded pure Mg, Mg–3Al, Mg–3Al–2.5La and Mg–3Al–2.5La–1.9 Y_2O_3 alloys under compression loading.

The presence of finer secondary phases assisted in restricting the motion of dislocations more effectively, leading to strength improvement in the case of the nanocomposite. Overall, the fracture strain remained unaffected (~18%) by the addition of the reinforcement when compared to the base alloy. Therefore, the addition of nano-sized Y_2O_3 compensated for the decrease in the compression strength of Mg–3Al due to the addition of 2.5La, while the presence of La suppressed the formation of the $Mg_{17}Al_{12}$ phase, which adversely affects the creep properties of Mg–3Al alloys.

The compression fracture morphology of the Mg–3Al–2.5La–1.9Y_2O_3 alloy is shown in Figure 3d. Compressive fractography studies (quasi-static) showed that the materials underwent the shear mode of deformation with the addition of the reinforcement. The ample split into two parts and the fracture surfaces of all samples were inclined at an angle of ~45°. The SEM fractograph of fractured surfaces revealed the presence of shear bands in the sample. Smooth fracture surfaces exhibited a ductile mode of fracture in the samples [38,39].

4. Conclusions

In this work, the effect of the addition of Y_2O_3 on the microstructural and mechanical properties of Mg–3Al–2.5La alloy was primarily investigated. The following conclusions can be drawn:

1. With the addition of the Y_2O_3 reinforcement, an even finer grain structure can be realized (~3.6 μm for Mg–3Al–2.5La–1.9 Y_2O_3 alloy, 43% less than that of Mg–3Al–2.5La at ~6.26 μm).
2. The microstructural characterization concluded that all intermetallic phases Al_2La and $Al_{2.12}La_{0.88}$ and $Al_{11}La_3$ were still present in dispersed form, but the sizes of these phases were refined by the addition of nanosize Y_2O_3 in the Mg–3Al–2.5La alloy.
3. The compressive results concluded that the addition of Y_2O_3 to Mg–3Al–2.5La significantly improved the compressive yield strength and the ultimate compressive strength (CYS from ~141 MPa to ~156 MPa and UCS from ~456 MPa to ~520 MPa), which are even better than those of the Mg–3Al alloy (CYS, ~154 MPa and UCS, ~481 MPa). There was no adverse effect on the fracture strain value recorded for Mg–3Al–2.5La with the addition of Y_2O_3.
4. The damping results concluded that the addition of nanosize Y_2O_3 to Mg–3Al–2.5La improved the damping capacity. The addition of the Y_2O_3 reinforcement also improved the CTE value of the Mg–3Al–2.5La alloy.

Acknowledgments: The authors would like to acknowledge the Ministry of Education Academic Research Funding (WBS# R-265-000-498-112) for the financial support in carrying out this research work.

Author Contributions: Amit Kumar and Amit Devendra Kohadkar performed processing; Amit Devendra Kohadkar performed mechanical testing; Amit Kumar and Khin Sandar Tun performed microstructure studies and data analysis; Amit Kumar wrote the paper; Manoj Gupta contributed consultation, data analysis and paper review.

Conflicts of Interest: The authors declare no conflict of interest.

References

1. Khomamizadeh, F.; Nami, B.; Khoshkhooei, S. Effect of rare-earth element additions on high-temperature mechanical properties of AZ91 magnesium alloy. *Metall. Mater. Trans. A* **2005**, *36*, 3489–3494. [CrossRef]
2. Braszczyńska-Malik, K.N. Precipitates of r–Mg17Al12 Phase in AZ91 Alloy. Available online: http://cdn.intechweb.org/pdfs/12741.pdf. (accessed on 14 March 2017).
3. Rokhlin, L.L. *Magnesium Alloys Containing Rare Earth Metals: Structure and Properties*; Crc Press: Florida, FL, USA, 2003.
4. Gupta, M.; Sharon, N.M.L. *Magnesium, Magnesium Alloys, and Magnesium Composites*; John Wiley & Sons: Hoboken, NJ, USA, 2011.
5. Mirzadeh, H. Constitutive analysis of Mg–Al–Zn magnesium alloys during hot deformation. *Mech. Mater.* **2014**, *77*, 80–85. [CrossRef]
6. Mirzadeh, H. Quantification of the strengthening effect of rare earth elements during hot deformation of Mg–Gd–Y–Zr magnesium alloy. *J. Mater. Res. Technol.* **2016**, *5*, 1–4. [CrossRef]
7. Hihara, L.H.; Adler, R.P.; Latanision, R.M. *Environmental Degradation of Advanced and Traditional Engineering Materials*; CRC Press: Florida, FL, USA, 2013.
8. Kumar, A.; Meenashisundaram, G.K.; Manakari, V.; Parande, G.; Gupta, M. Lanthanum effect on improving CTE, damping, hardness and tensile response of Mg–3Al alloy. *J. Alloy. Compd.* **2017**, *695*, 3612–3620. [CrossRef]
9. Hassan, S.; Gupta, M. Development of high performance magnesium nanocomposites using solidification processing route. *Mater. Sci. Technol.* **2004**, *20*, 1383–1388. [CrossRef]
10. Hassan, S.; Gupta, M. Effect of length scale of Al2O3 particulates on microstructural and tensile properties of elemental Mg. *Mater. Sci. Eng. A* **2006**, *425*, 22–27. [CrossRef]
11. Hassan, S.; Gupta, M. Effect of type of primary processing on the microstructure, CTE and mechanical properties of magnesium/alumina nanocomposites. *Compos. Struct.* **2006**, *72*, 19–26. [CrossRef]
12. Hassan, S.; Gupta, M. Development of high performance magnesium nano-composites using nano–Al2O3 as reinforcement. *Mater. Sci. Eng. A* **2005**, *392*, 163–168. [CrossRef]
13. Hassan, S.; Gupta, M. Effect of different types of nano-size oxide particulates on microstructural and mechanical properties of elemental Mg. *J. Mater. Sci.* **2006**, *41*, 2229–2236. [CrossRef]
14. Dorri Moghadam, A.; Ferguson, J.B.; Schultz, B.F.; Lopez, H.; Rohatgi, P.K. Direct Synthesis of Nanostructured in Situ Hybrid Aluminum Matrix Nanocomposite. *Ind. Eng. Chem. Res.* **2016**, *55*, 6345–6353. [CrossRef]
15. Moghadam, A.D.; Schultz, B.F.; Ferguson, J.B.; Omrani, E.; Rohatgi, P.K.; Gupta, N. Functional metal matrix composites: Self-lubricating, self-healing, and nanocomposites-an outlook. *JOM* **2014**, *66*, 872–881. [CrossRef]
16. Rohatgi, P.K.; Afsaneh, D.M.; Schultz, B.F.; Ferguson, J.B. Synthesis and Properties of Metal Matrix Nanocomposites (MMNCS), Syntactic Foams, Self Lubricating and Self-Healing Metals. Available online: https://www.researchgate.net/profile/Afsaneh_Dorri_Moghadam/publication/270510257_Synthesis_and_Properties_of_Metal_Matrix_Nanocomposites_MMNCS_Syntactic_Foams_Self_Lubricating_and_Self-Healing_Metals/links/54ac489a0cf2479c2ee7ac9a/Synthesis-and-Properties-of-Metal-Matrix-Nanocomposites-MMNCS-Syntactic-Foams-Self-Lubricating-and-Self-Healing-Metals.pdf (accessed on 18 March 2017).

17. Hassan, S.; Gupta, M. Development of nano–Y2O3 containing magnesium nanocomposites using solidification processing. *J. Alloy. Compd.* **2007**, *429*, 176–183. [CrossRef]
18. Sankaranarayanan, S.; Jayalakshmi, S.; Gupta, M. Effect of addition of mutually soluble and insoluble metallic elements on the microstructure, tensile and compressive properties of pure magnesium. *Mater. Sci. Eng. A* **2011**, *530*, 149–160. [CrossRef]
19. Braszczyńska-Malik, K. Some Mechanical Properties of Experimental Mg–Al–RE–Mn Magnesium Alloys. *Arch. Foundry Eng.* **2014**, *14*, 13–16. [CrossRef]
20. Greger, M.; Kocich, R.; Čížek, L.; Dobrzański, L.A.; Juřička, I. Possibilities of mechanical properties and microstructure improvement of magnesium alloys. *Arch. Mater. Sci. Eng.* **2007**, *28*, 83–90.
21. Moosa, A.A. Effect of Lanthanum Addition on the Microstructure of Mg–4Al Alloy. *Al-Khwarizmi Eng. J.* **2011**, *7*, 75–82.
22. Zhang, J.; Yu, P.; Liu, K.; Fang, D.; Meng, J. Effect of substituting cerium-rich mischmetal with lanthanum on microstructure and mechanical properties of die-cast Mg–Al–RE alloys. *Mater. Des.* **2009**, *30*, 2372–2378. [CrossRef]
23. Tekumalla, S.; Seetharaman, S.; Almajid, A.; Gupta, M. Mechanical properties of magnesium-rare earth alloy systems: A review. *Metals* **2014**, *5*, 1–39. [CrossRef]
24. Kainer, K.U.; Kaiser, F. *Magnesium Alloys and Technology*; John Wiley & Sons: Hoboken, NJ, USA, 2003.
25. Bai, J.; Sun, Y.; Xue, F. Microstructures and creep properties of Mg–4Al–(1–4) La alloys produced by different casting techniques. *Mater. Sci. Eng. A* **2012**, *552*, 472–480. [CrossRef]
26. Jain, C.C.; Koo, C.H. Creep behavior of extruded sheets of magnesium alloys containing La-rich Mischmetal. *Mater. Trans.* **2006**, *47*, 433–439. [CrossRef]
27. Kim, J.M.; Lee, S.J. Microstructure and Castability of Mg–Al–La Alloys for High Conductivity Applications. *Int. J. Met.* **2015**, *9*, 15–21. [CrossRef]
28. Zhang, J.H.; Zhang, M.L.; Meng, J.; Wu, R.Z.; Tang, D.X. Microstructures and mechanical properties of heat-resistant high-pressure die-cast Mg–4Al–xLa–0.3Mn (x = 1, 2, 4, 6) alloys. *Mater. Sci. Eng. A* **2010**, *527*, 2527–2537. [CrossRef]
29. Zhang, J.H.; Liu, S.J.; Zhe, L.; Zhang, M.L.; Meng, J.; Wu, R.Z. Structure stability and mechanical properties of high-pressure die-cast Mg–Al–La–Y–based alloy. *Mater. Sci. Eng. A* **2012**, *531*, 70–75. [CrossRef]
30. Nguyen, Q.; Gupta, M. Increasing significantly the failure strain and work of fracture of solidification processed AZ31B using nano–Al2O3 particulates. *J. Alloy. Compd.* **2008**, *459*, 244–250. [CrossRef]
31. Yang, Q.; Bu, F.; Zheng, F.; Liu, X.; Zhang, D.; Qiu, X.; Meng, J. Influence of trace Sr additions on the microstructures and the mechanical properties of Mg–Al–La–based alloy. *Mater. Sci. Eng. A* **2014**, *619*, 256–264. [CrossRef]
32. Zhang, J.; Zhang, D.; Tian, Z.; Wang, J.; Liu, K.; Lu, H.; Tang, D.X.; Meng, J. Microstructures, tensile properties and corrosion behavior of die-cast Mg–4Al–based alloys containing La and/or Ce. *Mater. Sci. Eng. A* **2008**, *489*, 113–119. [CrossRef]
33. Wei, L.; Dunlop, G. The solidification behaviour of Mg–Al–rare earth alloys. *J. Alloy. Compd.* **1996**, *232*, 264–268. [CrossRef]
34. Zou, H.; Zeng, X.; Zhai, C.; Ding, W. Effects of Nd on the microstructure of ZA52 alloy. *Mater. Sci. Eng. A* **2005**, *392*, 229–234. [CrossRef]
35. Sankaranarayanan, S.; Nayak, U.P.; Stbat, R.K.; Suwas, S.; Almajid, A.; Gupta, M. Nano–ZnO particle addition to monolithic magnesium for enhanced tensile and compressive response. *J. Alloy. Compd.* **2014**, *615*, 211–219. [CrossRef]
36. Goodfellowc. Metals and Materials for Research and Industry. Alloys, and Ceramics Compounds. Available online: http://www.goodfellow.com/E/M.html (accessed on 14 March 2017).

37. Nguyen, Q.B.; Nai, M.L.S.; Nguyen, A.S.; Seetharaman, S.; Jayalakshmi, S.; Leong, E.W.W.; Gupta, M. Microstructure and damping characteristics of Mg and its composites containing metastable Al85Ti15 particle. *J. Compos. Mater.* **2016**, *50*, 2565–2573. [CrossRef]

38. Batra, R.C.; Wei, Z.G. Instability strain and shear band spacing in simple tensile/compressive deformations of thermoviscoplastic materials. *Int. J. Impact. Eng.* **2007**, *34*, 448–463. [CrossRef]

39. Wang, T.S.; Hou, R.J.; Lv, B.; Zhang, M.; Zhang, F.C. Microstructure evolution and deformation mechanism change in 0.98C–8.3Mn–0.04N steel during compressive deformation. *Mater. Sci. Eng. A* **2007**, *465*, 68–71. [CrossRef]

metals

MDPI

Article

Characterization of In-Situ Cu–TiH2–C and Cu–Ti–C Nanocomposites Produced by Mechanical Milling and Spark Plasma Sintering

Nguyen Thi Hoang Oanh [1,*], Nguyen Hoang Viet [1], Ji-Soon Kim [2] and Alberto Moreira Jorge Junior [3,4,5,6,7]

[1] School of Materials Science and Engineering, Hanoi University of Science and Technology, No. 1 Dai Co Viet, Hanoi 100000, Vietnam; viet.nguyenhoang@hust.edu.vn
[2] School of Materials Science and Engineering, University of Ulsan, San-29, Mugeo-2 Dong, Nam-Gu, Ulsan 680-749, Korea; jskim@ulsan.ac.kr
[3] Department of Materials Science and Engineering, Federal University of São Carlos, Via Washington Luiz, km 235, São Carlos, SP 13565-905, Brazil; Jorge.Moreira@simap.grenoble-inp.fr
[4] University of Grenoble Alpes, Science et Ingénierie des Matériaux et Procédés (SIMAP), F-38000 Grenoble, France
[5] Centre National de la Recherche Scientifique (CNRS), Science et Ingénierie des Matériaux et Procédés (SIMAP), F-38000 Grenoble, France
[6] University of Grenoble Alpes, Laboratoire d'Electrochimie et de Physico-chimie des Matériaux et des Interfaces (LEPMI), F-38000 Grenoble, France
[7] Centre National de la Recherche Scientifique (CNRS), Laboratoire d'Electrochimie et de Physico-chimie des Matériaux et des Interfaces (LEPMI), F-38000 Grenoble, France
* Correspondence: oanh.nguyenthihoang@hust.edu.vn; Tel.: +84-4-3868-0409

Academic Editor: Manoj Gupta
Received: 5 February 2017; Accepted: 27 March 2017; Published: 29 March 2017

Abstract: This study focuses on the fabrication and microstructural investigation of Cu–TiH2–C and Cu–Ti–C nanocomposites with different volume fractions (10% and 20%) of TiC. Two mixtures of powders were ball milled for 10 h, consequently consolidated by spark plasma sintering (SPS) at 900 and 1000 °C producing bulk materials with relative densities of 95–97%. The evolution process of TiC formation during sintering process was studied by using X-ray diffraction (XRD), scanning electron microscopy (SEM), and high resolution transmission electron microscopy (HRTEM). XRD patterns of composites present only Cu and TiC phases, no residual Ti phase can be detected. TEM images of composites with (10 vol % TiC) sintered at 900 °C show TiC nanoparticles about 10–30 nm precipitated in copper matrix, most of Ti and C dissolved in the composite matrix. At the higher sintering temperature of 1000 °C, more TiC precipitates from Cu–TiH2–C than those of Cu–Ti–C composite, particle size ranges from 10 to 20 nm. The hardness of both nanocomposites also increased with increasing sintering temperature. The highest hardness values of Cu–TiH2–C and Cu–Ti–C nanocomposites sintered at 1000 °C are 314 and 306 HV, respectively.

Keywords: spark plasma sintering; Cu–TiC; in-situ composites; mechanical milling

---

1. Introduction

Metal matrix composites (MMCs) are advanced materials which combine ductility and toughness of metal and high strength and modulus of ceramic particles. The unique properties of MMCs are high specific strength, specific modulus, and good wear resistance compare to unreinforced metal [1]. In many type of MMCs, copper matrix composites (CMCs) have received a lot of interest because of super toughness and wear resistance which are used for structural application in wear industry [2].

Generally, there are two routes to produce particulate-reinforced CMCs, which are ex-situ and in-situ. In the ex-situ method, ceramic particles such as TiB2, TiC, and oxide are introduced into the metal matrix via powder metallurgy or conventional casting methods [3,4]. However, the CMCs fabricated by these methods revealed a drawback because of poor interfacial bonding between reinforcement particles and copper matrix [5]. In order to improve the wettability of the Cu matrix and reinforcement phase, nano-ceramic particles were used [6]. Nevertheless, ceramic nanoparticles have tendency to segregate into clusters in milling process leading decrease strength of composite. The distribution of reinforced particles are non-uniform in the copper matrix, the mechanical and electrical features of the composite will be affected negatively [7]. On the contrary, ceramic particles synthesized by the in-situ method were dispersed more homogeneously in the copper matrix. The interfaces between reinforcement particles and matrix are clean, and very fine reinforcement particles are formed. Among CMCs, Cu–TiC system is attracted more attention due to their potential applications as electrical sliding contacts, resistance welding electrodes [8]. In the in-situ method, TiC nanoparticles were produced by the reaction between Ti and C during sintering process. In order to prevent grain growth of reinforcement and copper particles occur at high sintering temperature a fast sintering process need to be carried out. Spark plasma sintering (SPS) has some advantages such as rapid sintering, uniform sintering, low running cost, easy operation proves a suitable sintering technique for consolidation nano-structure, nanocomposite, and amorphous materials. In SPS, very high temperature over melting temperature may be attained in the contact area of powder particles which enhances interparticle bonding without considerable grain growth occurring [9–13].

The replacement of Ti powder in Cu–Ti–C composite by another powder such as TiH2 is considerable because of high price of Ti powder. In addition to, dehydrogenation of TiH2 occurs during sintering process is always accompanied by formation of high concentration of lattice defects and the highly activated Ti atoms. Released hydrogen from TiH2 will react with oxygen on the surface of TiH2 powders in the form of H2O which affect positively on the electrical conductivity of the composite [14].

The objectives of the present work are to explore the possibility of synthesizing Cu–TiC in-situ composites made from Cu–TiH2–C and Cu–Ti–C powder mixtures by mechanical milling and SPS. The effect of reinforcement content and sintering temperature on microstructure and hardness properties of composites was investigated.

2. Experimental Procedure

The copper (with average particle size of 75 μm), titanium (average particle size of 45 μm), TiH_2 (average particle size of 40 μm) and graphite (average particle size of 5 μm) powder (≥99% purity, from HIGH PURITY CHEMICALS Co., Ltd., Chiyoda, Japan) were used as starting materials. The powder mixtures of two composites Cu–TiH2–C and Cu–Ti–C with mixing ratio of 10 and 20 vol % TiC were mechanically milled in a high-energy planetary ball mill (P100-Korea). Milling was operated for 10 h at the rotational speed of 500 rpm and 0.5 wt % stearic acid was used as the milling process control agent. Balls and vials are made of stainless steel, the diameter of the balls was 5 mm and the powder-to-ball ratio was 1:10. The vial was evacuated and subsequently filled with argon up to 0.3 MPa.

A 1.5 g amount of as-milled powder was loaded into a cylindrical graphite die with 10 mm-inner and was subjected to a pulsed current using a spark plasma sintering equipment, (SPS-515 apparatus Sumitomo Coal Mining, Tokyo, Japan). The chamber was pumped to low vacuum (<5 Pa). The composite powders were spark plasma sintered at 900 and 1000 °C under a pressure of 50 MPa for 5 min with a heating rate of 50 °C/min.

X-ray diffraction patterns of the composites were recorded by a SIEMENS D5000 diffractometer (Siemens Industry Inc., Karlsruhe, Germany) using Cu K_α radiation (λ = 1.5418 Å). Microstructural analysis of powders and composite samples was carried out by using Scanning Electron Microscopy (SEM/EDX-JEOL JSM-7600F, JEOL Ltd., Tokyo, Japan) and Transmission Electron Microscopy (TEM-JEOL JEM-2100, JEOL Ltd., Tokyo, Japan). Relative densities of bulk composites were determined by Archimedes method. The indexation of such selected area electron diffraction (SAED) patterns was performed using JEMS software [15].

Microhardness measurements were performed using a Vickers hardness instrument (Mitutoyo MVK-H1 Hardness Testing Machine, Mitutoyo, Japan) under a load of 100 g.

To analyze the surface of fracture, samples were simply fractured by gripping the halves of the composite with pliers and bending them apart.

3. Results and Discussion

3.1. Characterization of the Powders

Figure 1a,b shows SEM images of starting powders, respectively for TiH_2 and copper powders. TiH_2 particles have an irregular shape while Cu powder particles have a dendritic shape. Figure 1c–f show SEM images of composite powders formed after 10 h of milling of Cu–TiH2–C and Cu–Ti–C mixtures, with different amounts of reinforcement particle content. As one can observe, the increase of the reinforcement particle content the particle size of milled powders decreases for both composites. Cu–Ti–C composite presented finer particles than those for the Cu–TiH2–C composite, with the same reinforcement particle content. As can be seen from SEM images shown in Figure 1g,h. Some large particles were formed due to agglomeration of small particles reaching a size of 10–30 µm. EDS analyses (Figure 2) were performed on particles such as in Figure 1g,h which presents spectra relative to such analyses, of which it is worth noting that there was no contamination of Fe either from the milling tools (Table 1). However, such contamination was already observed even in the ex-situ method as reported in [16].

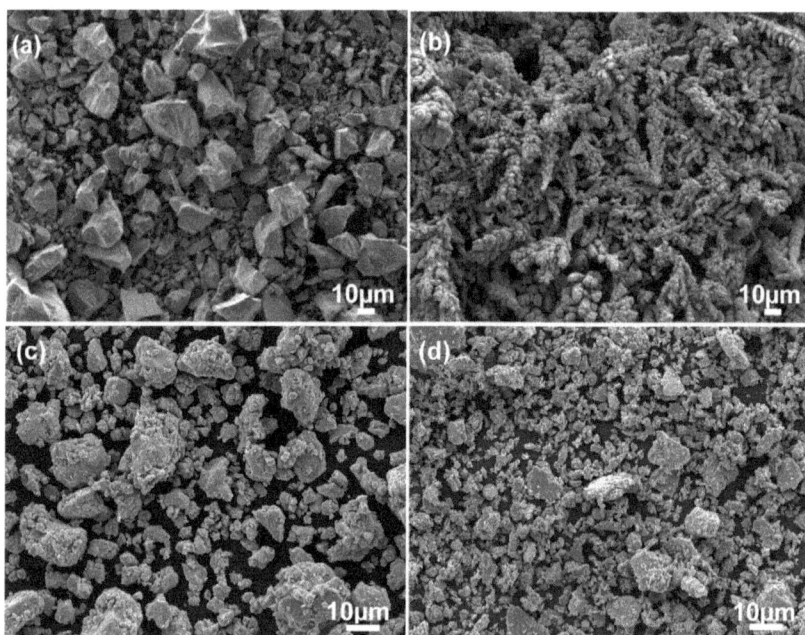

Figure 1. *Cont.*

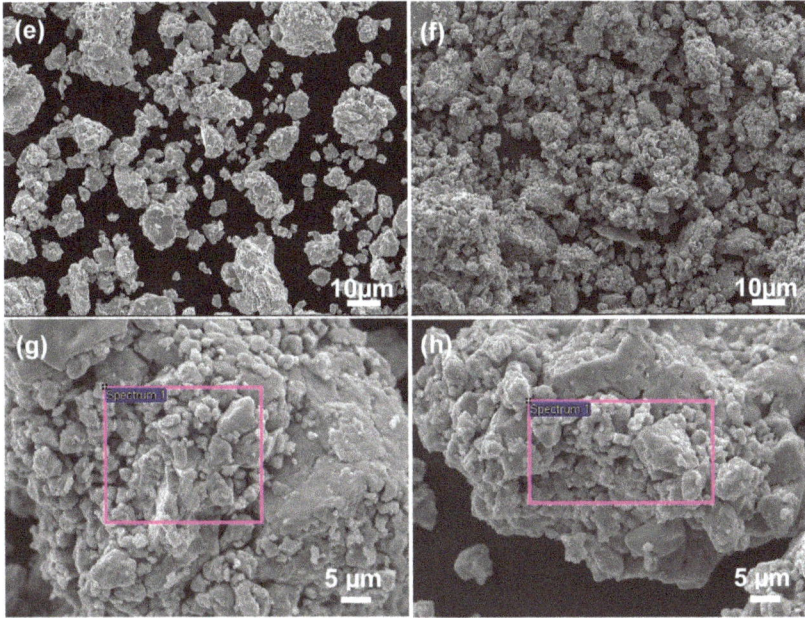

Figure 1. SEM of starting powders in (**a**) TiH$_2$ and (**b**) copper powders. SEM images of composite powders milled for 10 h, with 10 vol % TiC in (**c**) Cu–TiH2–C and (**d**) Cu–Ti–C; and with 20 vol % TiC in (**e**) Cu–TiH2–C and (**f**) Cu–Ti–C. Higher-magnification SEM images; the rectangles mark areas; from which EDS spectra were taken (**g**) Cu–TiH2–C and (**h**) Cu–Ti–C.

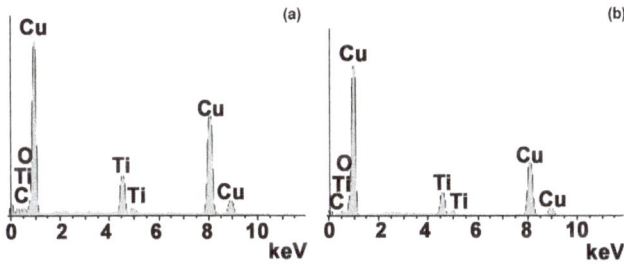

Figure 2. Typical EDS analyses acquired from particles presented in Figure 1e (Cu–Ti–C) in (**a**) and Figure 1f (Cu–TiH2–C) in (**b**), both with 20 vol % TiC.

Table 1. EDS analysis of the ball milled powders with 20 vol % of TiC reinforcement particles.

Composite	Concentration, wt %			
	Cu	Ti	C	O
Cu–TiH2–C	73.78	8.75	13.42	4.05
Cu–Ti–C	70.73	7.89	16.49	4.89

3.2. Characterization of Compacts after SPS

X-ray diffraction patterns of as-sintered nanocomposites are illustrated in Figure 3. After SPS at 900 °C (Figure 3a), XRD patterns presents only diffraction peaks related to pure copper. There is a shift peak of copper in sintered composite compare to starting Cu powder. This fact clearly evidences that most of the Ti and C have dissolved in the copper matrix and also that the sintering temperature was not high enough to precipitate the TiC phase. Conversely, by increasing the sintering temperature to 1000 °C, Ti reacts with C, and TiC precipitates, as it is clearly noticeable in the XRD patterns of Figure 3b for mixtures with 20 vol % TiC. However, there is reason to believe that the same has occurred for mixtures with 10 vol % of TiC because this is a thermodynamical condition. TiC weight percentages in bulk composites sintered at 1000 °C produced from starting powders was calculated by Rietveld refinement method as shown in Table 2. The weight percentages of in-situ TiC nanoparticles with 10 vol % reinforcement particles for Cu–TiH2–C and Cu-Ti–C composites are 1.43 and 2.86%, respectively. Furthermore, as it will be under mentioned, TEM analyses confirm the presence of TiC nanoparticles in mixtures with 10 vol % of TiC. At higher reinforcement particles of 20 vol %, the amount of TiC precipitated from Cu–TiH2–C and Cu-Ti–C composites also increases to 6.9 and 6.45 wt %, respectively. Additionally, it is important to observe that there was no precipitation of intermetallic phases during sintering process at any of the sintering temperatures. If one considers the Cu–Ti–C system, normally Cu, Ti, and C may interact to form several products through chemical reactions [17]. However, the Gibbs free energy of TiC formation at the temperature of 1273 K is about 84.4 kJ/mol, which is much lower than those to form other intermetallic phases of Ti and Cu, implying that the precipitation of TiC is thermodynamically preferred relative to other possible reactions.

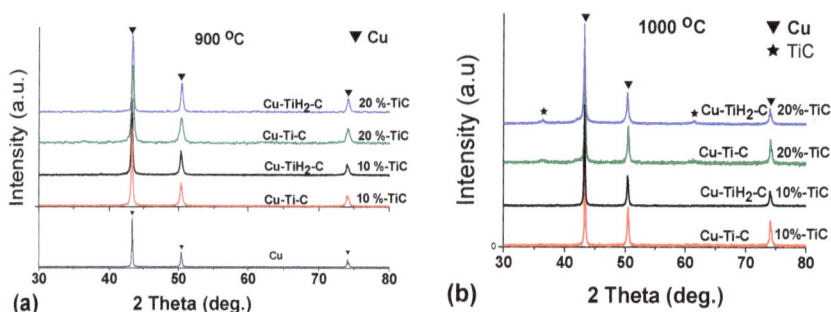

Figure 3. XRD patterns of nanocomposites (a) Spark Plasma Sintered at 900 °C; (b) Spark Plasma Sintered at 1000 °C.

Table 2. Fraction of phases of Cu–TiH2–C and Cu-Ti–C composites sintered at 1000 °C calculated by Rietveld refinement.

	Fraction of Phases (%)			
Phase	Cu–TiH2–C		Cu–Ti–C	
	10 vol % TiC	20 vol % TiC	10 vol % TiC	20 vol % TiC
TiC	2.86	6.9	1.43	6.45
Cu	97.14	97.14	98.57	93.55

Figure 4a–d present backscattered electron (BSE) SEM images of Cu–TiH2–C and Cu-Ti–C nanocomposites sintered at 1000 °C, showing details of the surface of samples. From these images, one can observe the presence of three gray tones, where black regions indicate the presence of some closed porosity on the surface, dark-gray regions correspond to the solid solution Cu(Ti, C), and, finally, white regions correspond to Cu-richer regions. It is interesting to note that the porosity is minimal for Cu–TiH2–C nanocomposites for any amount of TiC. The porosity increases for Cu-Ti–C nanocomposites but reduces by increasing the amount of TiC. However, apparently, Cu-richer regions are thinner and better distributed for Cu-Ti–C nanocomposites than for Cu–TiH2–C ones, diminishing

with the amount of TiC. Conversely, the thicker and worse distributed Cu-richer regions in Cu–TiH2–C nanocomposites augment with the amount of TiC.

Relative density measurements, presented in Table 3, confirm the above results regarding porosity. The relative density of the Cu–TiH2–C and Cu–Ti–C nanocomposites increases as the sintering temperature increases from 900 to 1000 °C. Therefore, one could infer that this fact may be due to the enhanced viscosity of Cu/Cu(Ti, C) matrix at the higher sintering temperature, which results in an efficient filling of pores due to high diffusion rates. In composites produced in-situ, the interfacial area is higher compared to ex-situ composites, detrimental effects of interfacial phenomena (decohesion, void formation) are more likely to prevail at high sintering temperatures compared to low sintering temperatures. Islak et al. [7] reported that in hot-pressed Cu–Ti–C nanocomposites, with 10 vol % TiC, the highest relative density obtained was about 86.4%, while in our samples, with the same TiC content, it is, on average, around 96.8 either for Cu–TiH2–C or Cu–Ti–C nanocomposites sintered at 1000 °C. The presence of porosity on the surface of nanocomposites is a natural result of the consolidation process. The relative density of Cu–TiH2–C has higher than that of Cu–Ti–C composite that why the amount of porosity can be seen in Figure 4b,d is more than in Figure 4a,c, respectively. The maximum relative density values were measured for Cu–TiH2–C and Cu–Ti–C nanocomposites at the sintering temperature of 1000 °C with 10 vol % TiC, which were 97% and 96.6%, respectively.

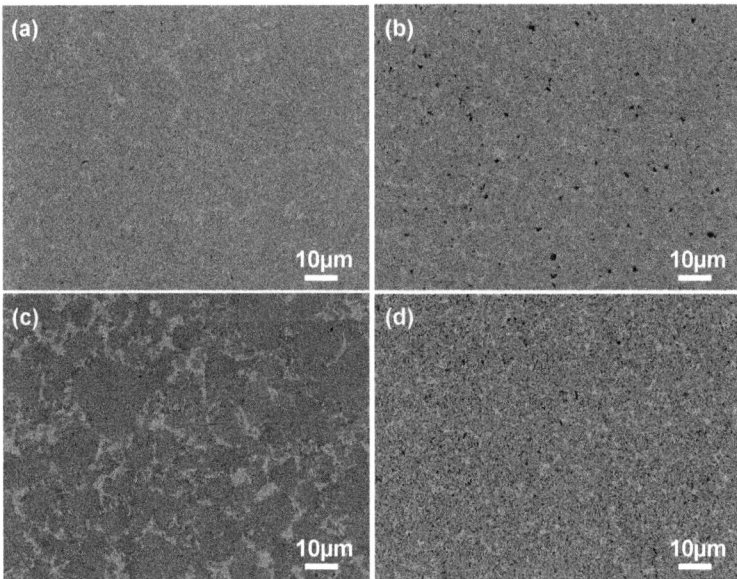

Figure 4. SEM images of Cu–TiH2–C and Cu–Ti–C nanocomposites Spark Plasma Sintered at 1000 °C: (**a**) 10 vol % TiC, Cu–TiH2–C nanocomposite; (**b**) 10 vol % TiC, Cu–Ti–C nanocomposite; (**c**) 20 vol % TiC, Cu–TiH2–C nanocomposite; (**d**) 20 vol % TiC, Cu–Ti–C nanocomposite.

Table 3. Relative density of Spark plasma sintered Cu–TiH2–C and Cu–Ti–C nanocomposites.

Sintering Temperature (°C)	Relative Density (%)			
	Cu–TiH2–C		Cu–Ti–C	
	10 vol % TiC	20 vol % TiC	10 vol % TiC	20 vol % TiC
900	95.8	95.3	95.4	95.2
1000	97.0	96.4	96.6	96.2

Figure 5 displays TEM images of Cu–TiH2–C nanocomposites with 10 vol % TiC and sintered at 900 °C. As one can observe, Ti still continues in the matrix, having a particle size of about 50 nm. Figure 5a shows some TiC nanograins of about 10–20 nm precipitated near Ti particles. As stated before, the reaction between Ti and C is not complete because the sintering temperature is not high enough to enhance the diffusion of C into Ti. Thus most of Ti and C remains in the composite matrix. Figure 5c–e presents examples of selected area electron diffraction (SAED) patterns of Cu–TiH2–C nanocomposite showing diffraction patterns for Cu, Ti, and TiC particles. There is a small misfit in the indexation of Cu due to the presence of Ti and C in solution into the copper. In addition, there are extra spots in SAEDs because of particle sizes. As the particles are smaller than the smallest aperture, the diffraction will be taken from the particle and regions around. At the sintering temperature of 1000 °C, TiC particles precipitate copiously on the nanocomposite matrix with particles size ranging between 10 and 30 nm as shown in Figure 6a,b. No other phase containing Ti could be detected under the TEM resolution.

Figure 6c,d present TEM images of Cu–Ti–C nanocomposites, with 10 vol % TiC, and sintered at 900 °C, showing few TiC nanoparticles with sizes ranging between 10 to 30 nm. As it occurred for Cu–TiH2–C nanocomposites, at the low sintering temperature of 900 °C, only a small fraction of Ti will react with C, and most of Ti and C continue dissolving in the composite matrix. At the sintering temperature of 1000 °C (Figure 6e,f), more TiC precipitates, whose particle size ranges from 10 to 20 nm. Nevertheless, Ti is still observed in the HRTEM image presented in Figure 6f.

The presence of TiC, even in mixtures with 10 vol % of TiC at high sintering temperature during XRD analyses (Figure 3), is confirmed by the observation that TiC precipitates in the TEM analysis.

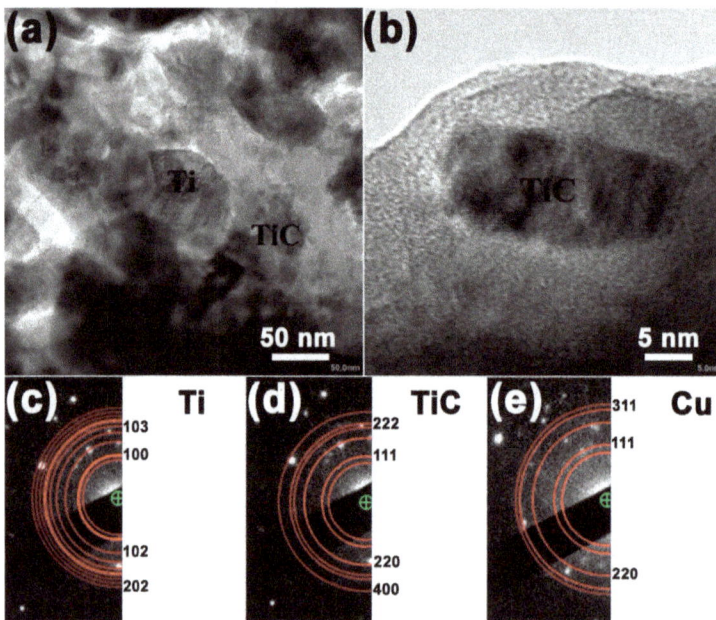

Figure 5. TEM images and SAED of Cu–TiH2–C nanocomposites of 10 vol % TiC sintered at 900 °C: (**a**) bright-field image; (**b**) HR-TEM image of TiC crystal on copper matrix; (**c**) SAED of Ti; (**d**) SAED of TiC; and (**e**) SAED of copper.

As it can be observed in Figures 5 and 6, the amount of TiC precipitated in the Cu–TiH2–C nanocomposite is higher than that for Cu–Ti–C after sintering at 1000 °C. Yang et al. [14] also prepared bulk TiC from TiH_2–C and Ti–C by self-propagation high-temperature synthesis method. By XRD and

DSC results, they found that for TiH_2–C, Ti and C reacted to produce TiC in the temperature range of 1130 to 1300 °C, very small amount of Ti remained after SHS process. For Ti–C, only an endothermic peak appears at 880 °C because of the α-Ti transformation to β-Ti and a lot of Ti and C remain. Little TiC is detected, but the amount is much smaller than TiH_2–C.

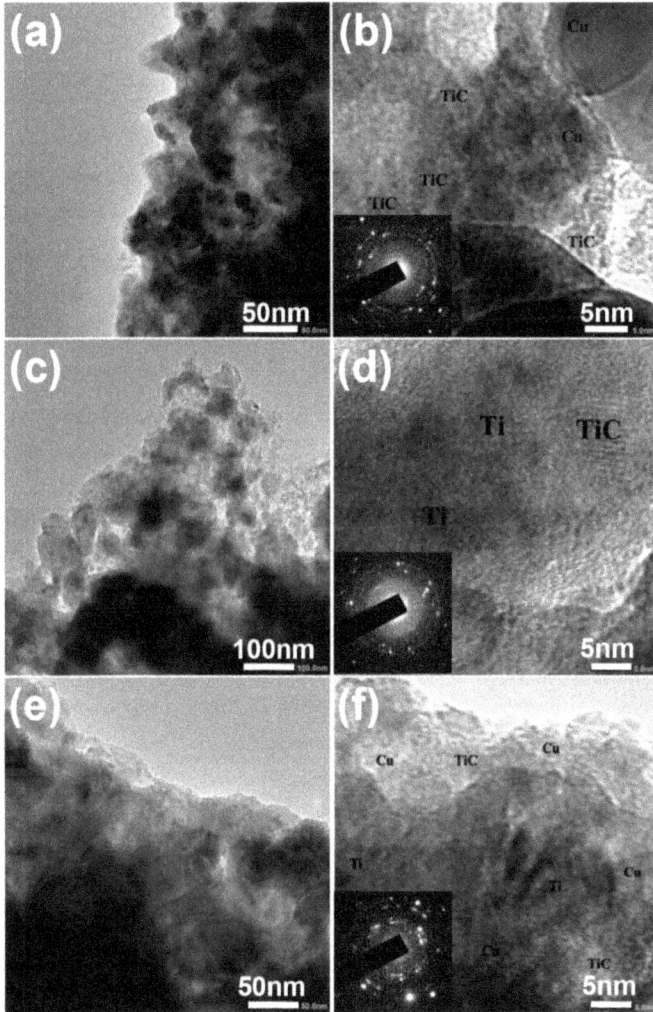

Figure 6. TEM images of: Cu–TiH2–C nanocomposite—10 vol %TiC and sintered at 1000 °C, bright-field image in (**a**) and HR-TEM image in (**b**); Cu–Ti–C nanocomposite—10 vol %TiC and sintered at 900 °C, bright-field image in (**c**) and HRTEM image in (**d**); Cu–Ti–C nanocomposite—10 vol %TiC and sintered at 1000 °C, bright-field image in (**e**) and HRTEM image in (**f**), SAEDs refer to TiC particles.

3.3. Hardness Results of Compacts

The hardness of a composite is an important parameter that determines their wear resistance for sliding contact applications. In this sense, hard TiC particles play an important role when reinforcing the soft Cu matrix. These particles lead to a higher hardness of the composite. Table 4 summarizes hardness values for all conditions studied in this work. From this table, it is possible to observe that, at the sintering temperature of 900 °C, hardness values of Cu–Ti–C nanocomposites are lower than those for Cu–TiH2–C nanocomposites, for any addition of TiC. The formation of Cu(Ti, C) solid solution enhances the hardness of the composites. As the sintering temperature increases, there is a significant increase in the hardness values because more TiC particles precipitate after the reaction between Ti and C. The highest hardness value of an ex-situ Cu–Ti–C composite with 10 vol % of TiC was found to be about 86.4 HV [7], much lower than those found for the in-situ produced nanocomposites of this work, which were 295 and 290 HV respectively for Cu–TiH2–C and Cu–Ti–C nanocomposites with the same TiC content and sintered at 900 °C. By increasing the volume of TiC, the hardness of nanocomposites also increases to 314 and 306 HV correspondingly for Cu–TiH2–C and Cu–Ti–C nanocomposites when sintered at 1000 °C.

Table 4. Hardness values of nanocomposites sintered at 900 and 1000 °C.

Sintering Temperature (°C)	Hardness Value, HV			
	Cu–TiH2–C		Cu–Ti–C	
	10 vol % TiC	20 vol % TiC	10 vol % TiC	20 vol % TiC
900	250	245	242	244
1000	295	314	290	306

3.4. Fracture Analysis

Figure 7 presents fracture morphologies of nanocomposites with 10 vol % TiC. The presence of dimples on the fracture surface suggests a ductile fracture mode in both nanocomposites Cu–TiH2–C and Cu–Ti–C. Fracture surface of the composites exhibits nano-void and micro-void coalescence in the matrix. Gray ultrafine particles are embedded in the copper matrix, which are fracture paths along grain boundaries of the matrix.

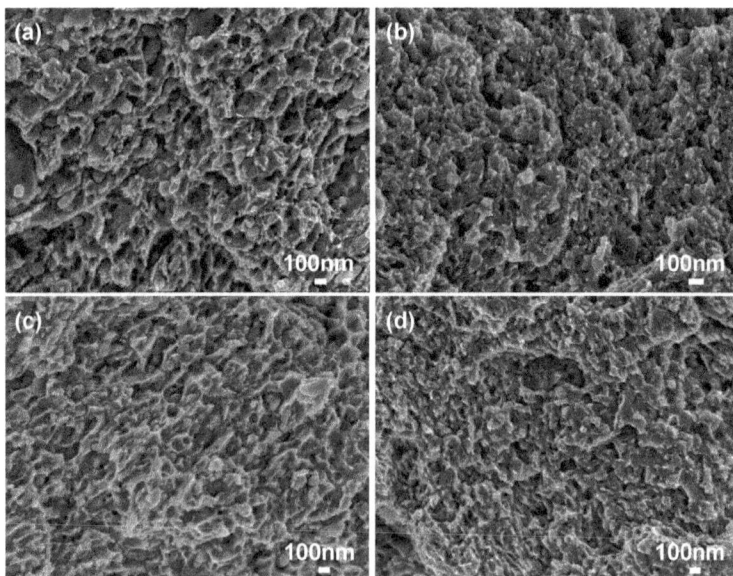

Figure 7. Fracture surfaces of 10 vol. %TiC (**a**) Cu–TiH2–C nanocomposite sintered at 900 °C; (**b**) Cu–Ti–C nanocomposite sintered at 900 °C; (**c**) Cu–TiH2–C nanocomposite sintered at 1000 °C; and (**d**) Cu–Ti–C nanocomposite sintered at 1000 °C.

At the highest sintering temperature of 1000 °C, one can observe a better contact between reinforcement particles and matrix, as presented in Figure 7c,d. The dimple size decreases with an increase in TiC content from 10 to 20 vol % as seen in Figure 8. The presence of more reinforcement particle content in the copper matrix prevents dimple growth. The fracture behavior of these nanocomposites also depicts a mixed mode, with the brittle mode of fracture being the predominant one. The fracture surface of both composites indicate nearly flat surface, minimal plastic deformation preceding the fracture occurs. The microscopic characteristics include faceted surface and transgranular fracture of Cu grains in the composites.

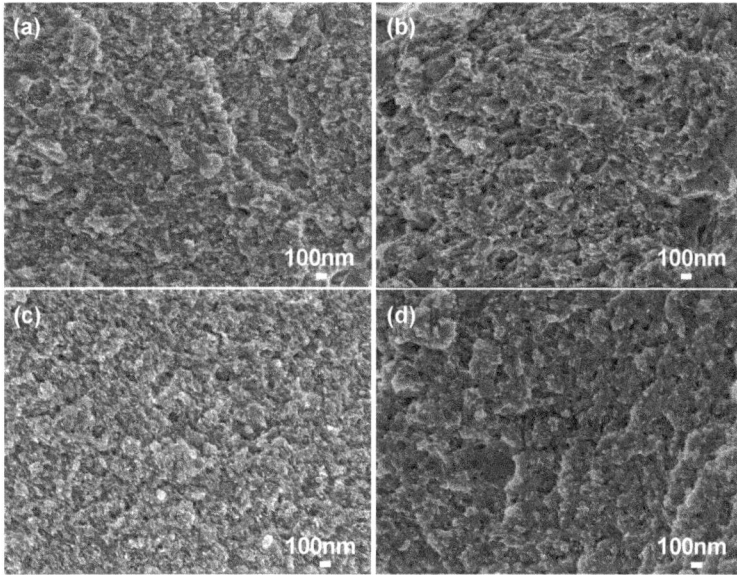

Figure 8. Fracture surfaces of 20 vol % TiC (**a**) Cu–TiH2–C nanocomposite sintered at 900 °C; (**b**) Cu–Ti–C nanocomposite sintered at 900 °C; (**c**) Cu–TiH2–C nanocomposite sintered at 1000 °C; and (**d**) Cu–Ti–C nanocomposite sintered at 1000 °C.

In order to investigate the distribution of reinforcing particulates in copper matrix, EDS analyses for the composite sintered at 1000 °C with 10 vol % TiC, is presented in Figure 9. The size of the "waves" on the Ti and C concentration profiles along the x-axis (distance) corresponds well to the size of the TiC nanoparticles introduced into the copper matrix through ball milling. The distribution of the Ti and C through copper matrix in Cu–TiH2–C is better than of Cu–Ti–C. The content of oxygen in Cu–TiH2–C is lower than Cu–Ti–C composite which means that the effect of hydrogen release from TiH$_2$ reacts with the oxygen lead to reduce the oxygen content in the composite after sintering. No contamination such as Fe from the balls and jar can be detected for both composites.

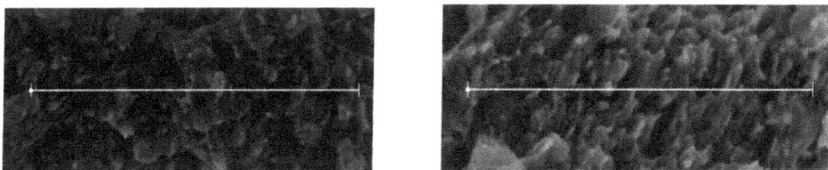

Figure 9. *Cont.*

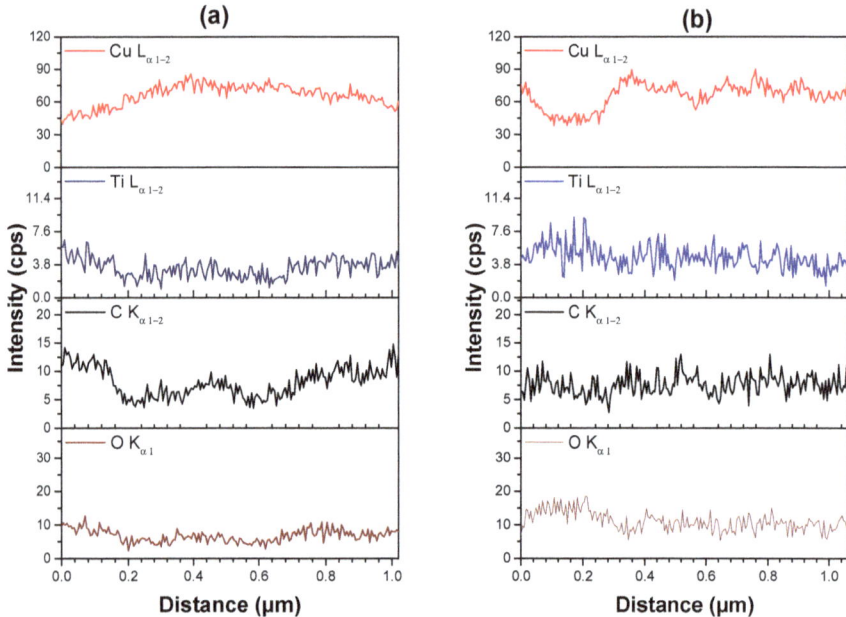

Figure 9. SEM images and EDS profiles for (**a**) Cu–TiH2–C and (**b**) Cu–Ti–C composites sintered at 1000 °C with 10 vol % reinforcement particles.

4. Conclusions

In-situ copper matrix composites strengthened with (10 and 20 vol %) TiC were prepared by SPS of ball milled mixtures of Cu–TiH2–C and Cu–Ti–C powders. The relative densities of composites sintered at 900 and 1000 °C are in the range of 95%–97%. The XRD of composites sintered at 900 °C showed only Cu(Ti, C) solid solution, no trace of TiC or oxide phases. Increasing sintering temperature to 1000 °C, diffraction peaks of TiC and pure Cu phases can be seen from XRD patterns for composites reinforced with 20 vol % TiC, while for composite with 10 vol % TiC reinforcement no diffraction peaks of TiC phase can be detected. The results of TEM analysis for these composites confirmed the presence of nanoparticle TiC about 10 to 20 nm disperse on Cu matrix near residual Ti particles whereas, under XRD resolution, it cannot detected. A larger amount of TiC nanoparticles precipitated in the Cu–TiH2–C composite is observed than for that of Cu–Ti–C from TEM images. These TiC nanoparticles enhance the hardness of composites prepared from the former. With the increasing the reinforcement particles content from 10 to 20 vol % the hardness values of composites increased while relative density slightly decreased. The hardness of composites also increased with increasing sintering temperature with any reinforcement particles addition. The highest hardness values of Cu–TiH2–C and Cu–Ti–C composites sintered at 1000 °C are 314 and 306 HV, respectively.

Acknowledgments: This research was funded by the Vietnam National Foundation for Science and Technology Development (NAFOSTED), grant number 103.02-2011.49.

Author Contributions: Nguyen Thi Hoang Oanh and Nguyen Hoang Viet designed the research, conducted the experiments, and wrote the paper. Ji-Soon Kim supervised the project and Alberto Moreira Jorge Junior revised the manuscript. All authors reviewed the manuscript.

Conflicts of Interest: The authors declare no conflict of interest.

References

1. Shirvanimoghaddam, K.; Hamim, S.U.; Karbalaei Akbari, M.; Fakhrhoseini, S.M.; Khayyam, H.; Paksersht, A.H.; Ghasali, E.; Zabet, M.; Munir, K.S.; Jia, S.; et al. Carbon fiber reinforced metal matrix composites: Fabrication processes and properties. *Compos. Part A Appl. Sci. Manuf.* **2017**, *92*, 70–96. [CrossRef]
2. Akhtar, F.; Askari, S.J.; Shah, K.A.; Du, X.; Guo, S. Microstructure, mechanical properties, electrical conductivity and wear behavior of high volume tic reinforced cu-matrix composites. *Mater. Charact.* **2009**, *60*, 327–336. [CrossRef]
3. Panda, S.; Dash, K.; Ray, B.C. Processing and properties of cu based micro- and nano-composites. *Bull. Mater. Sci.* **2014**, *37*, 227–238. [CrossRef]
4. Rathod, S.; Modi, O.P.; Prasad, B.K.; Chrysanthou, A.; Vallauri, D.; Deshmukh, V.P.; Shah, A.K. Cast in situ cu–tic composites: Synthesis by shs route and characterization. *Mater. Sci. Eng. A* **2009**, *502*, 91–98. [CrossRef]
5. Maity, P.C.; Panigrahi, S.K. Metal and intermetallic matrix in-situ particle composites. *Key Eng. Mater.* **1995**, *104–107*, 313–328. [CrossRef]
6. Casati, R.; Vedani, M. Metal matrix composites reinforced by nano-particles—A review. *Metals* **2014**, *4*, 65. [CrossRef]
7. Islak, S.; Kır, D.; Buytoz, S. Effect of sintering temperature on electrical and microstructure properties of hot pressed cu-tic composites. *Sci. Sinter.* **2014**, *46*, 15–21. [CrossRef]
8. Liang, Y.H.; Wang, H.Y.; Yang, Y.F.; Wang, Y.Y.; Jiang, Q.C. Evolution process of the synthesis of tic in the cu–ti–c system. *J. Alloys Compd.* **2008**, *452*, 298–303. [CrossRef]
9. Dudina, D.V.; Mukherjee, A.K. Reactive spark plasma sintering: Successes and challenges of nanomaterial synthesis. *J. Nanomater.* **2013**, *2013*, 12. [CrossRef]
10. Dudina, D.V.; Mali, V.I.; Anisimov, A.G.; Bulina, N.V.; Korchagin, M.A.; Lomovsky, O.I.; Bataev, I.A.; Bataev, V.A. Ti_3SiC_2-Cu composites by mechanical milling and spark plasma sintering: Possible microstructure formation scenarios. *Met. Mater. Int.* **2013**, *19*, 1235–1241. [CrossRef]
11. Ghasali, E.; Paksersht, A.H.; Alizadeh, M.; Shirvanimoghaddam, K.; Ebadzadeh, T. Vanadium carbide reinforced aluminum matrix composite prepared by conventional, microwave and spark plasma sintering. *J. Alloys Compd.* **2016**, *688 Pt A*, 527–533. [CrossRef]
12. Ghasali, E.; Shirvanimoghaddam, K.; Paksersht, A.H.; Alizadeh, M.; Ebadzadeh, T. Evaluation of microstructure and mechanical properties of al-tac composites prepared by spark plasma sintering process. *J. Alloys Compd.* **2017**, *705*, 283–289. [CrossRef]
13. Diouf, S.; Molinari, A. Densification mechanisms in spark plasma sintering: Effect of particle size and pressure. *Powder Technol.* **2012**, *221*, 220–227. [CrossRef]
14. Yang, Y.F.; Mu, D.K. Rapid dehydrogenation of tih2 and its effect on formation mechanism of tic during self-propagation high-temperature synthesis from tih2–c system. *Powder Technol.* **2013**, *249*, 208–211. [CrossRef]
15. Stadelmann, P. Java-Ems: Jems. C2004. Available online: Http://cimewww.Epfl.Ch/people/stadelmann/jemswebsite/jems.Html (accessed on 5 February 2017).
16. Zhuang, J.; Liu, Y.; Cao, Z.; Li, Y. The influence of technological process on dry sliding wear behaviour of titanium carbide reinforcement copper matrix composites. *Mater. Trans.* **2010**, *51*, 2311–2317. [CrossRef]
17. Wang, F.; Li, Y.; Wakoh, K.; Koizumi, Y.; Chiba, A. Cu–Ti–C alloy with high strength and high electrical conductivity prepared by two-step ball-milling processes. *Mater. Des.* **2014**, *61*, 70–74. [CrossRef]

Article

Characteristics of Cold and Hot Pressed Iron Aluminum Powder Metallurgical Alloys

Ahmed Nassef [1], Waleed H. El-Garaihy [2,3,*] and Medhat El-Hadek [1]

[1] Department of Production Engineering & Mechanical Design, Faculty of Engineering, Port-Said University, Port Fouad 42523, Egypt; nassef12@eng.psu.edu.eg (A.N.); melhadek@eng.psu.edu.eg (M.E.-H.)

[2] Mechanical Engineering Department, Unaizah College of Engineering, Qassim University, Unaizah 51911, Saudi Arabia

[3] Mechanical Engineering Department, Faculty of Engineering, Suez Canal University, Ismailia 41522, Egypt

* Correspondence: w.nasr@qu.edu.sa or wgaraihy@eng.suez.edu.eg; Tel.: +966-55-110-8490

Academic Editor: Manoj Gupta

Received: 4 March 2017; Accepted: 3 May 2017; Published: 12 May 2017

Abstract: Iron powders having average particle sizes of ~40 μm are mechanically mixed thoroughly with aluminum powders ranging from 1 to 10 in wt. %, with an average particle size of ~10 μm. Two different powder metallurgy (PM) techniques, cold and hot pressing, are used to study the effect of the additive element powder on the mechanical properties, wear properties, and the microstructure of the iron based alloys. The hot pressing technique was performed at a temperature reaching up to 500 °C at 445.6 MPa. The cold pressing technique was performed at 909 MPa at room temperature. By increasing the Al content to 10 wt. % in the base Fe-based matrix, the Brinell hardness number was decreased from 780 to 690 and the radial strength from 380 to 228 MPa with reductions of 11.5% and 40%, respectively. Improvement of the wear resistance with the increase addition of the Al powder to the Fe matrix up to five times was achieved, compared to the alloy without Al addition for different wear parameters: wear time and sliding speed.

Keywords: iron aluminum alloys; cold/hot PM; compressibility factor; wear resistance

---

1. Introduction

Some of the advantages of powder metallurgy (PM) alloys include minimum cost, high flexibility, ability to be shaped into complicated products, use metastable structures, and wide-ranging reinforcement levels [1,2]. Powder metallurgy technique can be used to reach homogeneity in the matrix distribution of the reinforcement without extreme reaction of matrix–reinforcement, which causes a problem in some techniques, such as stir casting or squeeze infiltration of the reinforcement in the molten matrix, under a dry and protective atmosphere. This is a result of the development of transient liquid phase throughout the sintering process. The liquid phase penetrates between the matrix grain boundaries, in the case of satisfactory wetting of the matrix by the melt.

Metal matrix composites (MMCs) containing nonmetallic particulates tend to ameliorate the wear and mechanical properties, through the creation of constraints to the distortion of the material during the mechanical working [3,4]. Perhaps the greatest advantage of PM techniques falls in the development of special material structures and the possibility of combining different components, widening the field of application of the PM materials [5,6]. Iron (Fe) is renowned for its strength and low price but it is very heavy in weight. To make use of it in scenarios that demand lightweight without resorting to buying expensive stronger materials such as titanium (Ti), it is often alloyed with aluminum (Al) which is light and cheap. The mixture of iron and aluminum usually includes a sprinkling of manganese to make it less brittle. Brittle intermetallic compounds can form poor ductility alloys at room temperature, which would limit their usage, as they are difficult to process into useful shapes, such as plates and tubes [7]. Using strengthening second phase control, Fe–Al alloys can be effectively hardened by controlling aluminum morphology and dispersion, in the iron based alloy [7]. Phase identification for Fe–Al alloys was initiated by the early effort of Koster and Tonn in 1933 [8] on the equilibrium phases. James [9] showed that subjected to the temperature and chemistry, there are three stated equilibrium phases for Fe–Al alloys: γ-austenite, α-ferrite, and β-Mn. The compound Fe–Al is present in a variety of compositions, mostly on the iron rich side of stoichiometry. It is characterized by its ordered body centered cubic (BCC) structure [10]. It has been an important subject of interest, commercially, due to its outstanding oxidation resistance, acceptable strength at high temperatures reaching $\approx$530 °C, and minimum density (5.76–6.32 g/cm^3 depending on Fe/Al ratio) when related to other iron based alloys utilized in market [11]. Iron aluminide intermetallic of Fe$_3$Al and FeAl possesses attractive properties for application, such as structural materials at high temperatures in aggressive environments [12,13]. Tanaka et al. [14] investigated Fe–Al superalloys with an 11.5% Al content; they have a relatively high solvus temperature with high hardness when aged at 600 °C. Although the agreement with the experimental results was not optimum, Carbon and Al contents had a positive influence on the constituent phase for the stability of the κ carbide phase; however, the experimental results were not in an optimum agreement with this finding [15].

Since the early 2000s, the interest in the Fe–Al alloys for automotive applications has been aggravated in Europe and Japan [16], as the demand arose in the light of producing durable automobiles. Due to the wide difference between the melting temperatures of iron (~1540 °C) and aluminum (~660 °C), the sintering of Fe–Al was, in fact, liquid phase sintering. As the sintering treatment is carried out above the melting point of the aluminum, the liquid phase may be stable or transient at the usual sintering temperatures, approximately 30 wt. % Al being soluble in α–Fe [16]. Although lightweight steel has an apparent shape of simplicity, it is very complex in its underlying metallurgical issues, which is associated to its potential to have structures, such as austenitic, ferritic, or even multiphase. TRIPLEX is a multiphase steel described in the literature by Frommeyer and Brux [17], which is a Fe–Al–Mn–C lightweight alloy characterized by having high ductility, strength, and a multiphase structure. The three major phases of TRIPLEX steel have a matrix phase composition of austenite with volume percentages of ferrite lying from 5 to 15; moreover, finely dispersed nano size κ-carbides all over the austenite with volume percentages less than 10 wt. %. Frommeyer and Brux [17] added Al up to 12 wt. % and reported a 1.5% linear relationship with the decrease in density per adding 1 wt. % of Al. Their distinctive of Fe–28Mn–12Al–1C alloy was tested mechanically in a uniaxial tensile test, which was accomplished at strain rate of 10^{-4} s^{-1} and room temperature. The results showed that ultimate tensile strength reached 1000 MPa, yield strength of 730 MPa, and total elongation of 55%. Sutou et al. [18] lately stated that by Al addition for more than 11 wt. %, the Fe–Al alloy cold workability has been depreciated to ~10%.

Material's wear resistance is vital for designers as it is linked to surfaces' interactions; particularly, mechanical action of the opposite surface leading to material's elimination and deformation. The wear analysis of Fe–Al alloys, with Fe being the base matrix, is deficient in the literature [19,20]. Perhaps one of the inimitable studies on the worn surfaces for Fe–Al alloys was reported by Xu et al. [21], where the Fe–Al alloys were found to exhibit high wear resistance. They found that with augmenting the load, a slight reduction in friction takes place, as a result of increased friction contact temperature and the bigger areas on the worn surface of the oxidized film, which act as a solid lubricant. At the deeper position below the surface, augmenting the load increases the shear stress; thus, aggravating the wear as aluminum affects the flexibility of the alloy [21]. Kim et al. [22] stated that the increase in wear rate of the iron-aluminides containing 25, 28 and 30 wt. % aluminum is associated with the increase of wear sliding speed and applied normal load; moreover, wear resistance of the aluminides decreased with augmenting the aluminum contents, where a ductile material's wear behavior linked to plastic deformation was discovered under SEM observations at worn surfaces of the iron-aluminides [22].

Mechanical alloying using two PM processing techniques, hot and cold pressed, were used for the production of strengthened Fe-based alloys, having additives of Al ranging within 1, 2, 3, 5 and

10 in wt. %. The procedure is based on a blending process in a hot and cold pressed attrition for elemental powders along with master alloy, which are ball milled in a protective and dry atmosphere. The present work aims to control and optimize the production of processing parameters, through suitable compaction die design to get dense PM parts, and study the influence of the production methods on the structure density, hardness and wear properties of the iron alloys containing aluminum as a dispersed.

2. Experimental Procedures

A commercial pure iron powder reduced by hydrogen, with 99.1% purity, average particle size of ~40 μm, molecular weight of 55.845 g/mol, and a density of 7.845 g/cm³, was mechanically mixed thoroughly with aluminum powders as reinforcement, with 99.9% purity, average particle size of ~10 μm, mesh No. 80, molecular weight of 26.98 g/mol, and a density of 2.9 g/cm³. The iron metallic powder was obtained from CNPC powder (Charlottetown, PE, Canada), whereas the aluminum metallic powder was obtained from ALDRICH (Darmstadt, Germany). The two metallic powders were mechanically mixed to study the effect of the aluminum addition on the mechanical and the wear properties of the iron based alloys. The weights of powders were calculated, to manufacture specimens of iron based alloys with Al additives ranging within 1, 2, 3, 5 and 10 in wt. %.

An arrangement of single acting piston cylinder, at room temperature, was adopted to perform the compaction process, as shown in Figure 1. The arrangement aimed obtaining dimensions of the green compact of 30 mm in diameter and 50 mm in height. The temperature was maintained at the desired level with a tolerance of ±5 °C. A pressing pressure of 909 MPa with cold pressing was calculated, with the assumption that the cross section area of the compact is equal to the cross section area of the die [8];

$$P = \frac{4\,F}{\pi D^2} \tag{1}$$

where P is the pressing pressure (MPa), F is the load (N) and D is the die diameter (mm). Various temperatures of molds were experienced with a maximum of 500 °C, while maintaining a pressure of 445.6 MPa with hot pressing and fixed cross head velocity of 2 mm/min. The setup was; consequently, heated then maintained for half an hour at the chosen temperature, for the sake of reaching temperature homogeneity all over the powder alloy. Afterwards, a reduction of the forming pressure was taken place for all tested hot components. Consequently after the compact operation, samples were enclosed using foils of aluminum surrounded with a graphite powder to shield its surface from the atmosphere during the sintering process, avoiding any potential reaction with oxygen and nitrogen. The cold and hot compacts were then sintered at 800 °C for one hour before being, finally, furnace cooled. The radial strength of the sintered pressed samples was determined using ASTM B939-15 standards [23,24], whereby annular compacts are crushed; the radial strength was calculated as:

$$RS = \frac{L_c(D - T)}{h \cdot T^2} \tag{2}$$

where RS is the radial strength (MPa), L_c is the crushing load at failure (N), T is the compact wall thickness (mm), D is the outer compact diameter (mm), and h (mm)is the height of the compact mass of specimen and its cross section were constant. The final compact height h_c was calculated as follow [25]:

$$h_c = \frac{h_0(RD)_i}{(RD)_C} \tag{3}$$

where h_0 is the initial compact height after 1 ton pressing, $(RD)_i$ is the initial relative density, the $(RD)_c$ is the compact relative density after sintering. The initial relative density can be calculated as follow:

$$(RD)_i = \frac{mass/\left(h_0 \cdot \frac{\pi}{4} \cdot D^2\right)}{\rho_c} \tag{4}$$

where ρ_c is the compact density (g/cm^3). The compact relative density after sintering can be calculated from Equation (5), and the compact relative density is calculated according to Equation (6);

$$(RD)_c = \frac{\rho_m}{\rho_c} \tag{5}$$

$$\rho_c = \frac{w_1\% + w_2\% + \ldots + w_n\%}{\sum_{i=1}^{n} (w\%/\rho)_i} \tag{6}$$

where ρ_m is the measured compact density, w_i % is the weight fraction of element i, ρ_i is the density of element i, and n is the element's number. The compressibility factor C_f of the metal powder was calculated using Equation (7) [26,27];

$$C_f = \frac{\rho_c - \rho_o}{P^{1/3}} \tag{7}$$

where ρ_o is the apparent density of the powder (g/cm^3), and P is the applied pressure. The porosity was calculated after the sintering process for each specimen according to $\theta = 1 - (RD)_c$. Metallographic specimens were then prepared by polishing and etching done mechanically in a solution of 20 mL HNO$_3$ + 10 mL H$_2$SO$_4$ + 20 mL H$_2$O then cleaned by a solution of 10 mL NaOH + 100 mL H$_2$O. Brinell hardness was measured using a 5 mm in diameter hardened ball, at 750 Kg load for 15 s.

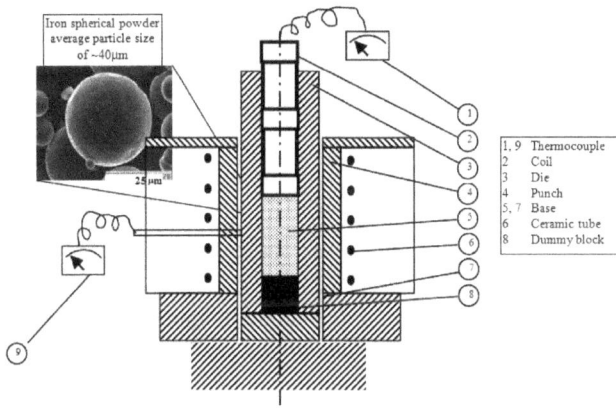

Figure 1. The setup of die of PM pressing technique.

3. Results and Discussion

3.1. Microstraultral Optical Investigations

The microstructure investigation on the Fe-based alloys was conducted using a Jeol 5400 scanning electron microscope (SEM) unit. It is connected to an EDS detector attachment used to detect various particle properties after the manufacture process. Particle size, morphology, shape, and agglomeration are obtained. The aluminum contents ranging within 1, 2, 3, 5 and 10 in wt. % were added to the iron powder and then produced using cold and hot pressing techniques. The measurements were carried out three times under the same conditions, to ensure repeatability. The achieved theoretical density after the sintering of the cold pressing samples was found to be 95% from the solid density.

3.1.1. Cold Pressed

Figure 2 shows the optical micrographs of the cold pressed (CP) Fe-based alloys with aluminum contents ranging within 1, 2, 3, 5 and 10 in wt. % at a pressing pressure of $P = 909$ MPa followed by a sintering process at 800 °C for one hour followed by furnace cooling process. The pure iron specimens were sintered at higher temperature up to 900 °C for one hour followed by furnace cooling process. The sintering process resulted in reducing the specimen's porosity due to coalescence processes between powders and eliminating the specimen's pores. It creates new distinctive solid–solid interface, and reduces the free energy occurring during sintering on the 1 μm particles to a 1 cal/g decrease. On a microscopic scale, varying pressure and free energy across the curved surface affect obviously the material transfer. In other words, the increase of small particle size with high curvature results in augmenting the differences in free energy across the curved surface [28–32].

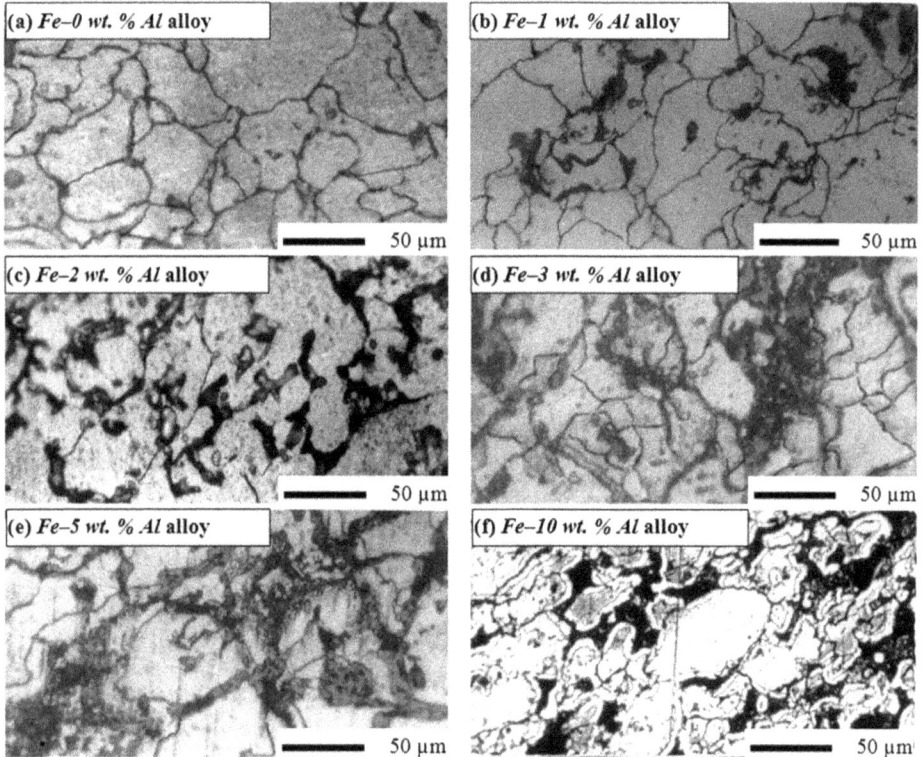

Figure 2. Optical micrographs of (**a**) Fe-0 wt. % Al, (**b**) Fe-1 wt. % Al, (**c**) Fe-2 wt. % Al, (**d**) Fe-3 wt. % Al, (**e**) Fe-5 wt. % Al, (**f**) Fe-10 wt. % Al alloys cold pressed at pressure of 909 MPa followed by sintering process of at 800 °C per one hour at constant furnace cooling conditions.

It is noticed in Figure 2c,d that a fine dispersion of precipitates occurs at the grain surface. The uniform distribution of the Fe and Al in the homogenous structures was noticed in alloys with Al additions of 2 or 3 wt. % compared to the rest of the Fe-based matrix alloys. Fast growth of the iron grains during sintering was observed as one of the limitations of the cold pressed PM method. In addition, it should be reported that surface cracks in the Fe–10 wt. % Al alloy were encountered.

3.1.2. Hot Pressed

Hot pressing (HP) technique followed by heat treatment was performed to avoid the limitations observed under the cold compaction technique. The Fe–5 wt. % Al and Fe–10 wt. % Al alloys were selected to conduct the hot pressing process. The specimens were pressed at 445.6 MPa, under a pressing temperature ranging from 200 to 500 °C, followed by a heat treatment for one hour duration, at a temperature of 800 °C, going to furnace cooling conditions of 1 °C/min [6]. The temperature was then held at the same level with a tolerance range of ±5 °C. Figure 3 shows the microstructure of the hot pressed Fe–5 wt. % Al alloys samples with various pressing temperatures (TP).

The hot pressed samples followed by heat treatment showed very fine and uniformly dispersed Fe–5 wt. % Al alloys, precipitated around the grain boundaries, with an average particle size of ~5 μm in the Fe-based matrix. Figure 4 shows a homogenous structure of Fe–10 wt. % Al alloy, which was obtained with heat treatment at 800 °C for one hour and furnace cooling as the cracks after the sintering process had disappeared.

Figure 3. Microstructure of Fe–5 wt. % Al alloys hot pressed at (**a**) 200 °C, (**b**) 300 °C, (**c**) 400 °C, (**d**) 500 °C.

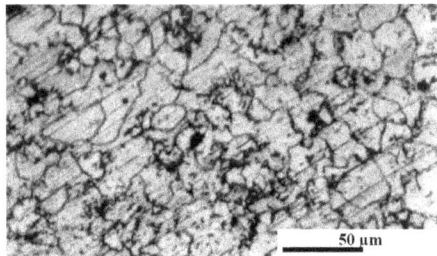

Figure 4. Microstructure of hot pressed Fe–10 wt. % Al alloy with heat treatment at 800 °C for one hour and furnace cooling conditions.

3.2. Hardness and Radial Crushing Strength

A minimum of ten readings were taken for the different cases to guarantee consistency of the Brinell hardness values crosswise the surface of material; the average of the readings is presented in Table 1. Radial strength tests were performed using an Instron 8562 universal mechanical tester under quasi static loading and a strain rate of $8 \times 10^{-5} \pm 5\%$ s^{-1}, at laboratory temperature. Specific dimensions of cylindrical specimens were set at a 30 mm diameter and a height of 50 mm. The samples went through deformation until crashed. Three similar samples were arranged for every test situation and subjected to the same loading conditions, to guarantee homogeneity and consistency. The average test value of all the three samples of the radial crushing strength was reported in Table 1. By increasing the Al content to 10 wt. % in the base Fe-based matrix, the Brinell hardness number was reduced from 780 to 690 and the radial strength from 380 to 228 MPa, as shown in Table 1. The hardness and radial strength of Fe–10 wt. % Al alloy was considerably lower than that of the Fe–0 wt. % Al alloy with a reduction of 11.5% and 40%, respectively. This reduction was due to a combined effect of the Fe-base matrix with the Al powder, which causes strengthening of the alloy matrix.

Table 1. Brinell hardness number and the radial strength for Fe-based alloys with various Al additions.

Radial Strength MPa	Brinell Hardness Number	Al Content in Fe-Based Alloys wt. %
380.44	780	0
369.78	751	1
344.67	735	1.5
310.11	722	2
270.23	714	2.5
245.11	705	3
235.14	698	5
230.21	691	7.5
228.43	690	10

The Al phase was dispersed in many pools or lakes, present in the cylindrical crashed Fe-base alloy samples. The cold pressed Fe–based alloys with the addition of Al were detected to go through fine crystallization, as presented in the SEM in Figure 5a for Fe–5 wt. % Al and Figure 5b for Fe–10 wt. % Al alloys. Fine crystal precipitation in the Fe–Al alloys was detected within vein protrusions, on the surface of compression fracture, and all along the paths of the rough crack propagation; additionally, within shear bands resulting from bending [33].

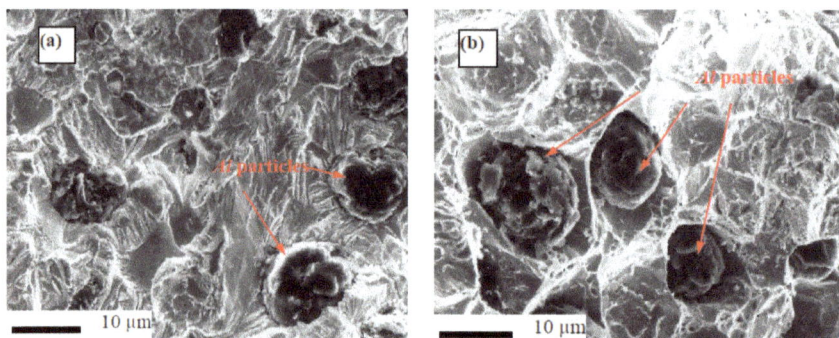

Figure 5. SEM micrographs of the rough fracture surface morphology of the CP Fe-based alloys: (a) Fe–5wt. % Al; and (b) Fe–10 wt. % Al.

The effect of the pressing temperature on the density, relative density, and compressibility factor of the Fe–5 wt. % Al alloy is shown in Figure 6a. The compact density and porosity are affected by the pressing temperature, as shown in Figure 6b; in other words, the density increases by increasing the temperature, which leads to porosity minimization. Higher relative density (RD) was reached for hot pressing Fe–5 wt. % Al alloy samples to ~95% from the solid density. Compressibility was used to indicate that the density of the powder had increased by a given pressure. By increasing the pressing temperature, the compressibility factor of the Fe–5 wt. % Al alloy was increased, as shown in Figure 6b, where augmented relative density and porosity reduction of the compact were detected. The porosity has already been presented in the green compacts. The pores formed during the sintering process were causing a different type of porosity, due to the formation of transient liquid phase during sintering, where the liquid phase penetrates into the matrix grain boundaries. Due to satisfactory wetting of the matrix by the melt, the alloying element particles replaced the pores and the homogenization leads to an overall expansion of the material during sintering.

The hot pressed compact quality was identified by the correlation with the three process parameters: time, pressure, and temperature. These variables had a major effect on the microstructure, physical properties, dimensional accuracy, and surface condition of the product. For the powders heated to low temperatures, pressure had the same effect as in cold pressing. Particles were brought closer together and were rotated and deformed, sheared, or fractured. At elevated temperatures, plastic deformation becomes the dominant mechanism. The liquid phase was formed at the hot pressing temperature and consolidation was further enhanced, by the isostatic action of the compressive stresses on the compact inside the dies [26,34,35]. Additionally, diffusion rates were increased through enhancing the liquid phase and densification, by good wetting between liquid and solid components of the alloy system.

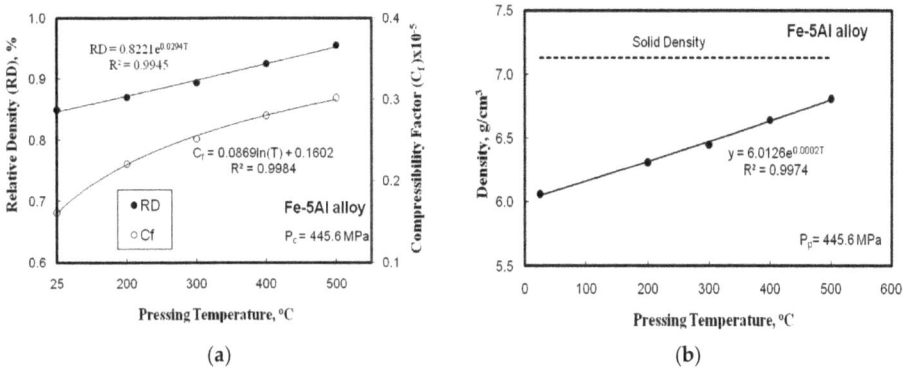

Figure 6. (a) The effect of pressing temperature on the compact relative density *(RD)* and compressibility factor *(C$_f$)*; and, (b) the effect of the pressing temperature on the compact density and porosity, for Fe–5 wt. % Al alloys.

3.3. Wear Resitance

Dry sliding wear tests were done using pin-on-disc apparatus for the Fe-based alloys with aluminum contents of 1, 2, 3, 5 and 10 in wt. % compared to stainless steel. Cylindrical specimens with 30 mm in diameter and height of 50 mm were slid against a rotating steel disc at various sliding speeds as follow: 0.33, 0.51, and 0.82 m/s. The severities of the harder material surface of the stainless steel using a pin-on-disc device was a turn over act on the surface of the Fe-based alloys. The applied wear pressure ranging from 2.2 to 3.56 MPa and duration tests from 15 to 60 min were used. The wear rate was calculated using mass losses divided by the product of the density and the sliding distance. Mass loss was measured using a precise digital balance of 170 g capacity with an accuracy of ±0.05 g. Figure 7 shows the effect of wear parameters on the wear rate *(WR)* for cold and hot pressing Fe-Al alloys using abrasive wear tester [21,22,36–38], the wear rate *(WR)* was calculated using the following equation:

$$WR = \frac{Wl}{\rho \cdot S_s \cdot t} \tag{8}$$

where *Wl* is the weight of the different alloy, ρ is the production density of compact alloy, S_S is the sliding speed, and *t* is the wear time.

The wear rate of Fe-base alloys with Al powder addition was much lower than the one without addition of Al element to the iron base matrix, at various wear parameters. The wear rate at higher speed was much lower than the one at lower sliding speed as shown in Figure 7b. Comparison of the wear behavior for the alloys produced, using different powder metallurgy techniques, is shown in Figure 7a,b. A significant reduction in wear rate was obtained using Fe-Al alloys in comparison with the Fe–0 wt. % Al alloy. While the addition of Al element powder to the Fe-base matrix causes a reduction in the wear rate, which increased the wear resistance up to five times compared with the alloy without Al addition, as shown in Figure 7a,b.

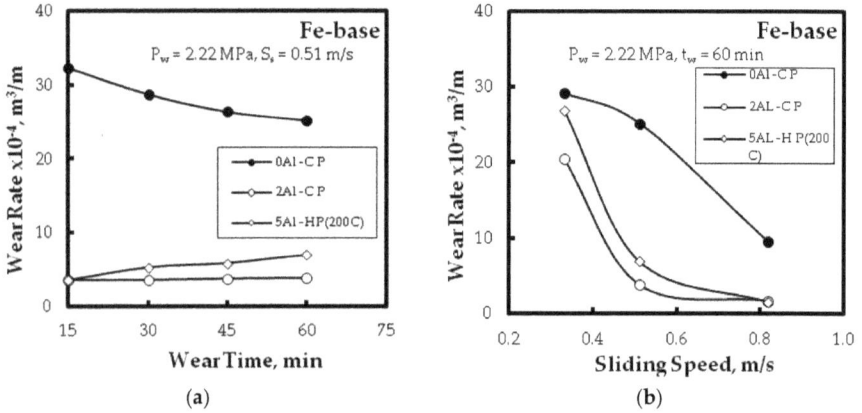

Figure 7. The effect of wear parameter for Fe-base alloys different PM technique on: (**a**) wear time; and (**b**) sliding speed for Fe–0 wt. % Al (CP), Fe–2 wt. % Al (CP), and Fe–5 wt. % Al (HP at 200 °C) alloys, respectively.

Figure 8 shows the wear behavior of Fe–2 wt. % Al alloy (cold pressing) under different wear pressures. It is clear that increasing the wear pressure resulted in increasing the materials loss, which is a regular known behavior for many metals and all alloys, regarding wear reactions. Hardness plays the predominant role at light wear process, while plasticity has big role at heavy operating conditions [39]. Heat generated due to the friction at high speed affects the microstructure, reduces hardness, and causes thermal stresses in the alloy matrix. As a result, a change in the wear mechanism takes place from abrasive to adhesive wear. Therefore, the worn surface layer is removed and the friction surface is protected from more wear and motion is facilitated, due to the relatively low value of shear strength [40,41].

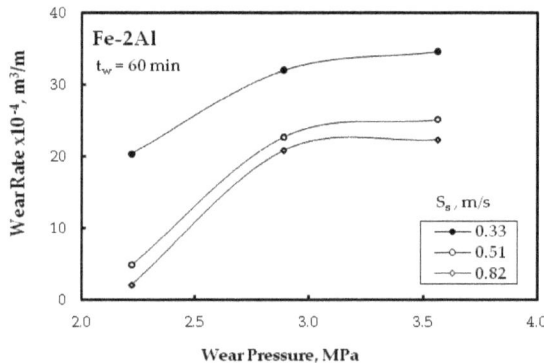

Figure 8. The effect of wear pressure for Fe–2 wt. % Al alloy (CP) on the wear rate.

The relation between the wear rate and wear time of the Fe–5 wt. % Al alloy pressed at different pressing temperatures is shown in Figure 9a, while Figure 9b shows the relation between the wear rate and the sliding speed of the Fe–5 wt. % Al alloy pressed at the same pressing temperatures. The highest wear rate of hot pressed specimen (300 °C) was due to its coarsening structure and the pressing temperature, which was not enough to precipitate the Fe-Al alloys hard phase around the grain boundaries. The lower wear rate of hot pressed specimens at 400 or 500 °C could be attributed to its finer grain structure and spheroidization of the pores after the sintering process. In general, there was a decrease in the wear rate for the hot pressed Fe-Al alloy sampled at 200 °C and an increase in the wear rate with increasing the hot pressing temperature, under the same wear parameters, as presented in Figure 9a,b. This finding could be attributed to the decrease of the residual porosity by increasing the pressing temperature.

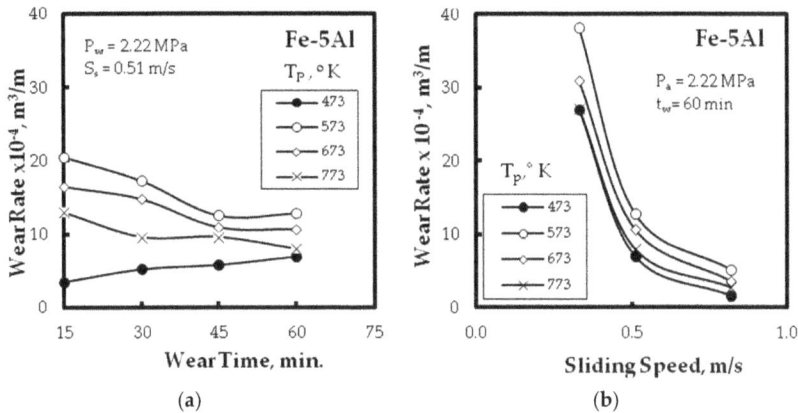

Figure 9. The effect of wear parameters for Fe–5 wt. % Al alloy at different pressing temperature on: (**a**) wear time; and (**b**) sliding speed.

4. Conclusions

Based on the findings of this study, the following conclusions could be summarized:

1. Under laboratory conditions and without the use of lubricant or binder, it was possible to produce Fe-Al with Al additives ranging within 1, 2, 3, 5 and 10 in wt. % alloys, using two powder metallurgy techniques, namely, cold and hot pressed, having comparable theoretical density and properties of the solid metals. Uniform distribution of dispersed phase inside the alloy structure was obtained.
2. By augmenting the Al content to 10 wt. % in the base Fe-based matrix, the Brinell hardness number was reduced from 780 to 690 MPa and the radial strength from 380 to 228 MPa, with a reduction of 11.5%, and 40%, respectively. The reduction of the compressibility factor of hot pressed powder alloys was detected with increasing the hot pressing temperature up to 500 °C.
3. Improvement of the wear resistance was observed, with augmenting the Al powder to the Fe matrix up to five times compared with the alloy without Al additions for different wear parameters: wear time, and sliding speed. This was also detected for the different PM technique.
4. The hot compact of Fe-Al alloys of about 95% theoretical density can be obtained from the metal powders, by employing a pressure of about 445.6 MPa and temperature of 500 °C. These alloys had higher density, better wear resistance, and homogenous structure than the parts produced by separate compaction and sintering obtained after elemental powders were added.

Author Contributions: All authors contributed equally in designing and manufacturing of the studied alloys, conceiving, designing and performing the experiments, collecting and analyzing the data, and finally discussion association of the results. I would like to also indicate that the corresponding author was responsible for addressing all reviewers' remarks and journal correspondence.

Conflicts of Interest: The authors also declare no conflict of interest.

References

1. Caballero, E.S.; Cintas, J.; Cuevas, F.G.; Montes, J.M.; Ternero, F. Influence of Milling Atmosphere on the Controlled Formation of Ultrafine Dispersoids in Al-Based MMCs. *Metals* **2016**, *6*, 224. [CrossRef]
2. Luka, F.; Vilemova, M.; Nevrla, B.; Klecka, J.; Chraska, T. Properties of Mechanically Alloyed W-Ti Materials with Dual Phase Particle Dispersion. *Metals* **2017**, *7*, 3. [CrossRef]
3. Chen, C.L.; Lin, C.H. A Study on the Aging Behavior of Al6061 Composites Reinforced with Y_2O_3 and TiC. *Metals* **2017**, *7*, 11. [CrossRef]
4. El-Hadek, M.A. Numerical simulation of the inertia friction welding process of dissimilar materials. *Metall. Trans. B* **2014**, *45*, 2346–2356. [CrossRef]
5. Nassef, A.; El-Hadek, M. Mechanics of hot pressed aluminum composites. *Int. J. Adv. Manuf. Technol.* **2015**, *76*, 1905–1912. [CrossRef]
6. Nassef, A.; El-Hadek, M. Microstructure and Mechanical Behavior of Hot Pressed Cu-Sn Powder Alloys. *Adv. Mater. Sci. Eng.* **2016**, *2016*, 9796169. [CrossRef]
7. Kim, S.H.; Kim, H.; Kim, N.J. Brittle intermetallic compound makes ultrastrong low-density steel with large ductility. *Nature* **2015**, *518*, 77–79. [CrossRef] [PubMed]
8. Köster, W.; Tonn, W. The Iron Corner of the Iron-Manganese-Aluminium System. *Arch. Eisenhuettenwes* **1933**, *7*, 365–366.
9. James, P.J. Precipitation of the Carbide-FEMN-3 ALC IN AN Iron-Aluminium Alloy. *J. Iron Steel Inst.* **1969**, *207*, 54–57.
10. Furushima, R.; Katou, K.; Shimojima, K.; Hosokawa, H.; Mikami, M.; Matsumoto, A. Effect of η-phase and FeAl composition on the mechanical properties of WC–FeAl composites. *Intermetallics* **2015**, *66*, 120–126. [CrossRef]
11. Amaya, M.; Romero, J.M.; Martinez, L.; Pérez, R. Mechanical Properties of Spray-Atomized FeAl40 at. % Al Alloys. In *Materials Characterization*; Springer International Publishing: Cham, Switzerland, 2015; Volume 5, pp. 199–207.
12. Trotter, G.; Baker, I. The effect of aging on the microstructure and mechanical behavior of the alumina-forming austenitic stainless steel Fe–20Cr–30Ni–2Nb–5Al. *Mater. Sci. Eng. A* **2015**, *627*, 270–276. [CrossRef]
13. Zamanzade, M.; Barnoush, A.; Motz, C. A Review on the Properties of Iron Aluminide Intermetallics. *Crystals* **2016**, *6*, 10. [CrossRef]
14. Tanaka, Y.; Kainuma, R.; Omori, T.; Ishida, K. Alloy Design for Fe-Ni-Co-Al-based Superelastic Alloys. *Mater. Today Proc.* **2015**, *2*, S485–S492. [CrossRef]
15. Ikeda, O.; Ohnuma, I.; Kainuma, R.; Ishida, K. Phase equilibria and stability of ordered BCC phases in the Fe-rich portion of the Fe–Al system. *Intermetallics* **2001**, *9*, 755–761. [CrossRef]
16. Kim, H.; Suh, D.W.; Kim, N.J. Fe–Al–Mn–C lightweight structural alloys: A review on the microstructures and mechanical properties. *Sci. Technol. Adv. Mater.* **2013**, *14*, 014205. [CrossRef] [PubMed]
17. Frommeyer, G.; Bruex, U. Microstructures and mechanical properties of high-strength Fe-Mn-Al-C light-weight TRIPLEX steels. *Steel Res. Int.* **2006**, *77*, 627–633. [CrossRef]
18. Sutou, Y.; Kamiya, N.; Umino, R.; Ohnuma, I.; Ishida, K. High-strength Fe-20Mn-Al-C-based alloys with low density. *ISIJ Int.* **2010**, *50*, 893–899. [CrossRef]
19. Greer, A.L.; Rutherford, K.L.; Hutchings, I.M. Wear resistance of amorphous alloys and related materials. *Int. Mater. Rev.* **2002**, *47*, 87–112. [CrossRef]
20. Maupin, H.E.; Wilson, R.D.; Hawk, J.A. Wear deformation of ordered Fe-Al intermetallic alloys. *Wear* **1993**, *162*, 432–440. [CrossRef]
21. Xu, B.; Zhu, Z.; Ma, S.; Zhang, W.; Liu, W. Sliding wear behavior of Fe–Al and Fe–Al/WC coatings prepared by high velocity arc spraying. *Wear* **2004**, *257*, 1089–1095. [CrossRef]

22. Kim, Y.S.; Kim, Y.H. Sliding wear behavior of Fe 3 Al-based alloys. *Mater. Sci. Eng. A* **1998**, *258*, 319–324. [CrossRef]
23. Chonglin, W. Discussion on radial crushing strength testing. *Powder Metall. Technol.* **1996**, *8*, 206–211.
24. Candela, N.; Plaza, R.; Rosso, M.; Velasco, F.; Torralba, J.M. Radial crushing strength and microstructure of molybdenum alloyed sintered steels. *J. Mater. Process. Technol.* **2001**, *119*, 7–13. [CrossRef]
25. Larker, H.T.; Larker, R. Hot isostatic pressing. *Mater. Sci. Technol.* **1991**. [CrossRef]
26. Leuenberger, H.; Rohera, B.D. Fundamentals of powder compression. I. The compactibility and compressibility of pharmaceutical powders. *Pharm. Res.* **1986**, *3*, 12–22. [CrossRef] [PubMed]
27. Hryha, E.; Dudrova, E.; Bengtsson, S. Influence of powder properties on compressibility of prealloyed atomised powders. *Powder Metall.* **2008**, *51*, 340–342. [CrossRef]
28. El-Hadek, M.A.; Kaytbay, S.H. Fracture properties of SPS tungsten copper powder composites. *Metall. Trans. A* **2013**, *44*, 544–551. [CrossRef]
29. Kaytbay, S.; El-Hadek, M. Wear resistance and fracture mechanics of WC–Co composites. *Int. J. Mater. Res.* **2014**, *105*, 557–565. [CrossRef]
30. El-Hadek, M.A.; Kassem, M. Failure behavior of Cu–Ti–Zr-based bulk metallic glass alloys. *J. Mater. Sci.* **2009**, *44*, 1127–1136. [CrossRef]
31. El-Hadek, M.; Kaytbay, S. Characterization of copper carbon composites manufactured using the electroless precipitation process. *Mater. Manuf. Process.* **2013**, *28*, 1003–1008.
32. El-Katatny, S.M.; Nassef, A.E.; El-Domiaty, A.; El-Garaihy, W.H. Fundamental Analysis of Cold Die Compaction of Reinforced Aluminum Powder. *Int. J. Eng. Tech. Res.* **2015**, *3*, 180–184.
33. Ahari, F. Flexible High Radial Strength Stent. U.S. Patent 6,264,685, 24 July 2001.
34. Chtourou, H.; Guillot, M.; Gakwaya, A. Modeling of the metal powder compaction process using the cap model. Part I. Experimental material characterization and validation. *Int. J. Solids Struct.* **2002**, *39*, 1059–1075. [CrossRef]
35. Bocchini, G.F. Warm compaction of metal powders: Why it works, why it requires a sophisticated engineering approach. *Powder Metall.* **2013**, *42*, 171–180. [CrossRef]
36. Wang, J.; Xing, J.; Cao, L.; Su, W.; Gao, Y. Dry sliding wear behavior of Fe$_3$Al alloys prepared by mechanical alloying and plasma activated sintering. *Wear* **2010**, *268*, 473–480. [CrossRef]
37. Sharma, G.; Limaye, P.K.; Ramanujan, R.V.; Sundararaman, M.; Prabhu, N. Dry-sliding wear studies of Fe$_3$Al-ordered intermetallic alloy. *Mater. Sci. Eng. A* **2004**, *386*, 408–414. [CrossRef]
38. Dhokey, N.B.; Rane, K.K. Wear behavior and its correlation with mechanical properties of TiB$_2$ reinforced aluminium-based composites. *Adv. Tribol.* **2011**, *2011*, 837469. [CrossRef]
39. Tong, C.J.; Chen, M.R.; Yeh, J.W.; Lin, S.J.; Chen, S.K.; Shun, T.T.; Chang, S.Y. Mechanical performance of the Al x CoCrCuFeNi high-entropy alloy system with multiprincipal elements. *Metall. Trans. A* **2005**, *36*, 1263–1271. [CrossRef]
40. Hsu, C.Y.; Yeh, J.W.; Chen, S.K.; Shun, T.T. Wear resistance and high-temperature compression strength of Fcc CuCoNiCrAl0. 5Fe alloy with boron addition. *Metall. Trans. A* **2004**, *35*, 1465–1469. [CrossRef]
41. Chen, M.R.; Lin, S.J.; Yeh, J.W.; Chuang, M.H.; Chen, S.K.; Huang, Y.S. Effect of vanadium addition on the microstructure, hardness, and wear resistance of Al$_{0.5}$CoCrCuFeNi high-entropy alloy. *Metall. Trans. A* **2006**, *37*, 1363–1369. [CrossRef]

![metals logo] *metals*

MDPI

Article

Mechanical and Corrosion Behavior of Al-Zn-Cr Family Alloys

Ahmed Nassef [1], Waleed H. El-Garaihy [2,3,]* and Medhat El-Hadek [1]

1   Department of Production Engineering & Mechanical Design, Faculty of Engineering, Port-Said University, Port Fouad 42523, Egypt; nassef12@eng.psu.edu.eg (A.N.); melhadek@eng.psu.edu.eg (M.E.-H.)
2   Mechanical Engineering Department, Unaizah College of Engineering, Qassim University, Unaizah 51911, Saudi Arabia
3   Mechanical Engineering Department, Faculty of Engineering, Suez Canal University, Ismailia 41522, Egypt
*   Correspondence: W.Nasr@qu.edu.sa; Tel.: +966-55-110-8490

Academic Editor: Manoj Gupta
Received: 9 February 2017; Accepted: 3 May 2017; Published: 12 May 2017

Abstract: Aluminum base alloys containing chromium (Cr) and zinc (Zn) were produced using extrusion and powder metallurgy techniques. Cr additions ranged between 5 to 10 wt. %, while Zn was added in an amount between 0 and 20 wt. %. Heat treatment processes were performed during powder metallurgy process, at different temperatures, followed by water quenching. Similar alloys were extruded with an extrusion ratio of 4.6 to get proper densification. Optical microscopy was used for microstructure investigations of the alloys investigated. The element distribution microstructure study was carried out using the Energy Dispersive X-ray analysis method. Hardness and tensile properties of the investigated alloys have been examined. Wear resistance tests were carried out and the results were compared with these of the Al-based bulk alloys. Results showed that the aluminum base alloys, containing 10 wt. % Cr and heat treated at 500 °C for one hour followed by water quenching, exhibited the highest wear resistance and better mechanical properties.

Keywords: Al-Zn-Cr alloys; powder metallurgy; strengthening; extrusion; dry sliding wear

1. Introduction

Aluminum chrome matrix composites (ACMC) are used in a wide range of industries including transportation, electronics, leisure, and others [1–4]. ACMC are usually produced by mechanical alloying of different kinds of metal powders using compression, at a temperature near room temperature, which results in nanocrystalline and supersaturated alloy powders at low temperatures [3]. The large mechanical energies introduced into the powder particles are nanocrystalline interfaces, amorphous state, and promoted reactions between the metals in solid state [5–7]. Studies on the trinity Al-Zn-Cr alloys are very limited throughout the literature. Recently, Kurtuldu et al. have investigated the effect of trace addition on the aging behavior of aluminum alloys containing Cr; moreover, the effect of Cr addition on Zn diffusions in molten of Al-Zn-Cr alloys [8,9]. It has been a subject of interest that minor and trace additives have a noticeable effect on these alloys, as far as structure and tensile properties are concerned [10]. In view of the outstanding corrosion and oxidation resistance, the strengthening and toughening of these alloys have been of recent concern [7–11]. Running experiments on the Al-Cr-Zn system at 600 °C, eleven three-phase regions had been identified [12]. Li et al. studied the effect of grain structure on quench sensitivity of an Al-Zn-Mg-Cu-Cr alloy. The results showed that the decrease of the quenching rate from 960 °C/s to 2 °C/s revealed a decrease in the hardness after aging, for the homogenized and solution heat treated alloy with large equiaxed grains. For the extruded and solution heat treated alloy, elongated grains and subgrains had dominated [13]. The Al-based alloys, with additives of magnesium (Mg) and silicon

(Si), belong to wrought aluminum alloys which are heat treated. It is worth mentioning that there is an increase in the mechanical strength of the alloys with the increase of Mg and Si contents to the solubility limits; on the other hand, the ductility is reduced. Small amounts of Mg, Cr, Zn, Zr or Ti were added to modify the microstructure to improve the mechanical properties, formability, and corrosion resistance of the alloy [14–17]. Enhanced wear and corrosion resistance is considered to be one of the most important attributes of AMC that contain ceramic particles for engineering applications [18]. A progressive degradation of material occurs when two sliding surfaces come in contact; this process is known as wear. Such interaction between surfaces gives rise to friction. Wang et al. [19] studied the effect of additives, such as copper (Cu) and silicon carbide (SiC), on Al-based alloys. Aluminum alloy matrices, with particulates of Al_2O_3 or SiC reinforcements, possess higher strength and stiffness as well as, greater wear resistance and improved high properties [20,21]. On the other hand, the effects of other reinforcement particles such as Cr, Fe, and so on have been studied.

In the present work, the control and optimization of the production processing parameters were implemented through suitable die design compaction with dense powder metallurgy (PM) parts. Using a cold pressed powder metallurgical process, the alloys were manufactured with controlled alloy constitutions using two techniques. The first is based on adopting cold compaction followed by heat treatment at 500 °C. In the second technique, and in order to reduce the grain coarsening resulting from the heat treatment process, similar cold compacted and heat treated alloys were subjected to an extrusion process followed by solution heat treatment and quenching in room temperature water. The influence of adding Zn and various percentages of Cr to aluminum alloys on the density, tensile behavior, hardness, corrosion, and wear resistance of the Al-based alloys was examined; in addition, the effect of production methods on those properties was also inspected.

2. Materials and Methods

Materials with desirable characteristics were attained due to the strong ionic interatomic bonding of Cr and Zn. Powders with purity greater than 99%, with an average particle size less than 50 μm in diameter and manufactured by CNPC powder (Charlottetown, PE, Canada), were used as the starting source materials. The aluminum metallic powder was obtained from ALDRICH (Darmstadt, Germany). Al-based alloys with different compositions were prepared using a mechanical mixer to achieve the needed composition. Six different compositions were prepared namely, Al-5Cr, Al-7.5Cr, Al-10Cr, Al-20Zn, Al-20Zn-5Cr, and Al-20Zn-10Cr. Two techniques were implemented to perform this work. The first uses cold compaction then heat treatment, and the second adopt extrusion followed by solution heat treatment to reduce the grain coarsening resulted from heat treatment process. In the first technique, each powder mixture was cold pressed using a compaction pressure of 425 MPa on the 30 mm diameter billets. The material properties are expected to be diverse since the melting temperatures of Al is around 660 °C, Cr 1900 °C and Zn 420 °C. The powder density of Al is 7190 kg/m^3, Cr is 2700 kg/m^3, and Zn is 7140 kg/m^3. A hydraulic testing machine of 40 Tons capacity was used to perform the compaction of the alloy powder, with constant cross head velocity of 0.002 m/min. The height of green compact was directly measured before and after die ejection. The final height is calculated from the load-displacement curve. After unloading, the elastic recovery of the compacts is neglected [4]. A compacting pressure ranging from 227 to 909 MPa was calculated, by assuming that the cross-section of the compact is equal to that of the die. The temperature of the die was measured using NiCr-Ni thermocouple, which is inserted into the die and maintained at the die cavity. After cold pressing, all of the MMCs alloys were heat–treated at about 500 °C to allow the atoms to diffuse randomly into a uniform solid solution, as liquid phase sintering [4,7]. The tin melts to form a thin film surrounding the copper particles enhancing the alloying element bonding [7]. All alloys were then heat treated to 500 °C for duration of 60 min.

On the other hand, the second technique uses the same alloy powders with different manufacturing steps. It starts with an extrusion process for alloy powders, with an extrusion ratio of 4.6 to get proper densification. The behavior of the extruded alloy was consequently studied.

The extruded bars were subjected to solution treatment temperature of 500 °C for 60 min and then quenched in room temperature water. As the size of the small particle with high curvature increases, the differences in free energy across the curved surface increases, as discussed in [22,23]. The setup was heated to the predefined temperature level and maintained constant for 30 min, to guarantee homogenous distribution within the powder alloy. The forming pressure was then lowered for all tested hot components. After the compact operation, the samples were covered with aluminum foil and embedded in a graphite powder to protect their surface from oxygen and nitrogen from atmosphere, during the sintering process. The specimens were sintered at a steady heating rate of 20 °C/min up to 500 °C for one hour. The temperature was maintained at that level with a tolerance of ±5 °C.

3. Results and Discussion

3.1. Densification Behavior

The relative density densification behaviors of the investigated Al-based alloys as affected by sintering temperature for different percentage of Cr and Zn are shown in Figure 1. From Figure 1 it is clear that the relative density increases by increasing the sintering temperature for the aluminum alloy reinforced with different percentages of Cr and Zn. Three classes of mechanisms contribute to densification: plastic yielding, low power creep, and various sorts of diffusion. In the absence of an external pressure, only the diffusion mechanisms exist as sintering is applied. Pore dragging and pore separation from the pores tend to inhabit the diffusion contribution. In order to take the initial pressed density into consideration, the compact sinterability was computed in terms of densification parameter (ΔD) as a proportional function of (sintered − green)/(theoretical − green) densities. The theoretical density was estimated using theory of mixtures [4,7]. Theoretical solid densities, normalized by after extrusion density of all produced parts, are shown in Figure 2. Higher relative densities (RD) were achieved for hot extruding samples to around 99% from the solid density values. Zn particles, having a distinctive lubricating effect, will enhance to a great extent the packing loss of the powders and retard the sealing for the individual pores. Consequently, density will decrease with Zn addition. This may be liquid Zn diffused inside the aluminum matrix and leaves pore inside the matrix. For the Al-based alloys, Cr particles were observed in the microstructure after heat treatment at 500 °C for one hour followed by water quenching. No difference in the microstructures was observed for these alloys that are heated treated at 500 °C for one hour followed by water quenching, as shown in Figure 3. From the microstructures of the heat treated samples, Al-5Cr homogeneous structure was observed at a temperature of 500 °C followed by water quenching. It should be noted that, with the decrease of the aluminum content in the Al-based alloys and the increase of the Cr and Zn content, the density increases accordingly.

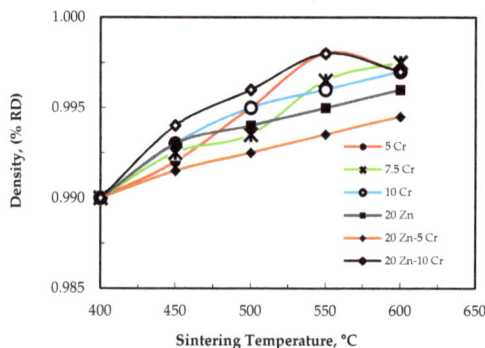

Figure 1. The effect of the sintering temperature on the relative density for different Al-base alloys.

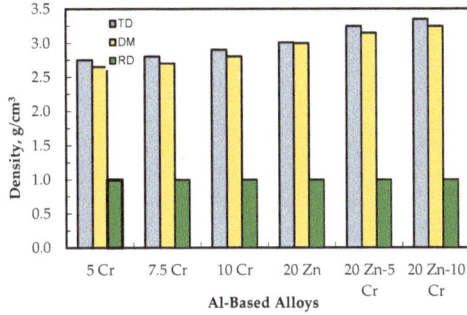

Figure 2. The density of as extruded Al-base alloys (TD theoretical density, DM density measured, and RD relative density).

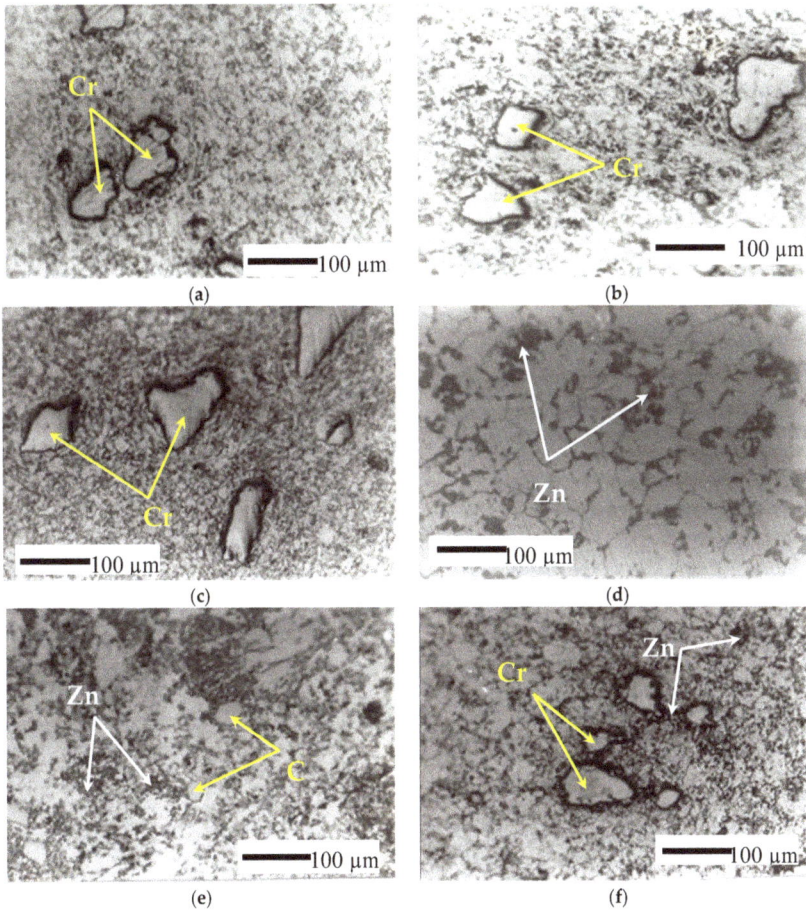

Figure 3. The microstructure of the different heat treated (**a**) Al-5Cr (**b**)Al-7.5Cr (**c**) Al-10Cr (**d**) Al-20Zn (**e**) Al-20Zn-5Cr and (**f**) Al-20Zn-10Cr alloys.

3.2. Microstructural Optical Investigations

There is a difficulty in revealing the grain boundaries in aluminum alloys, especially under heat treatment conditions; therefore, anodizing is required for the lower alloy content grades. Anodizing is an electrolytic etching procedure, which deposits a film on the specimen surface, revealing the grain structure when viewed with crossed polarized light. Using either the Keller's or the Grafet-Sargent reagent for the more highly alloyed grades would lead to a successful etching.

The microstructure of Al-based alloys, which is subjected to heat treatment temperature of 500 °C for 60 min followed by quenching in water at room temperature, is shown in Figure 3. On the other hand, the microstructure of Al-based alloys subjected to extrusion, with an extrusion ratio of 4.6 and solution heat treatment temperature of 500 °C for 60 min followed by quenching in water at room temperature, is shown in Figure 4. The uniform distributions of the Cr and Zn in the homogenous structures were noticed in alloys with Al-based alloys. In addition, it should be reported that surface cracks in the Al-based alloys were very limited. From Figure 3 it is clear that the cold compaction followed by heat treatment technique yielded homogenous microstructure with few pores. In the other hand, the Cr contents in the Al-Zn-Cr ternary system were investigated in Al-20Zn-5Cr and Al-20Zn-10Cr extruded alloys. It is noticed in the micrographs of the extruded billet with heat treatment shown in Figure 4a–f that a fine dispersion of precipitates occurs on the grain surface. In addition, multi-phases were observed in the microstructure in the as-extruded sample, as shown in Figure 4e,f.

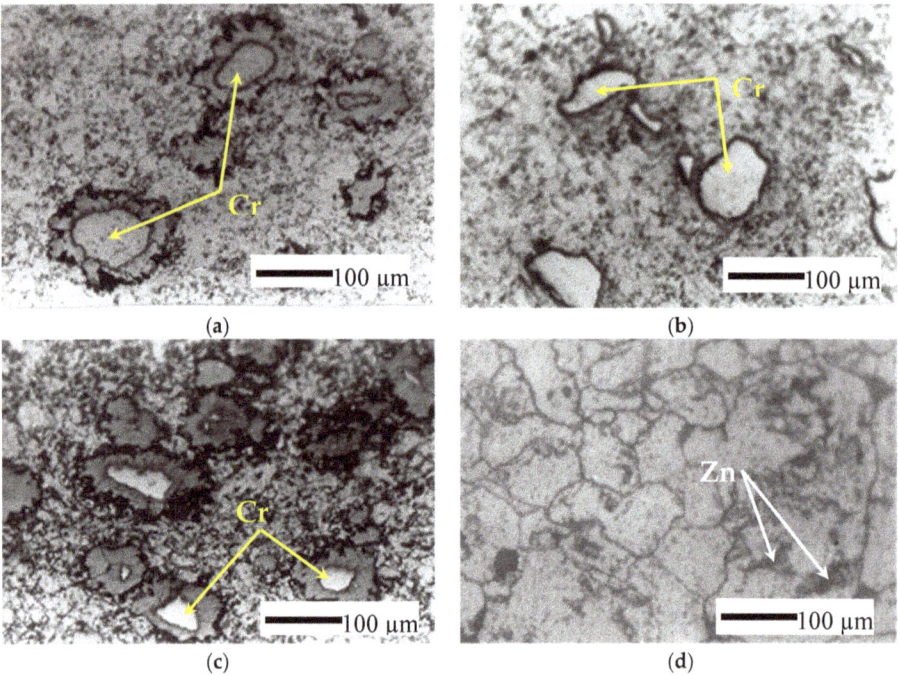

Figure 4. *Cont.*

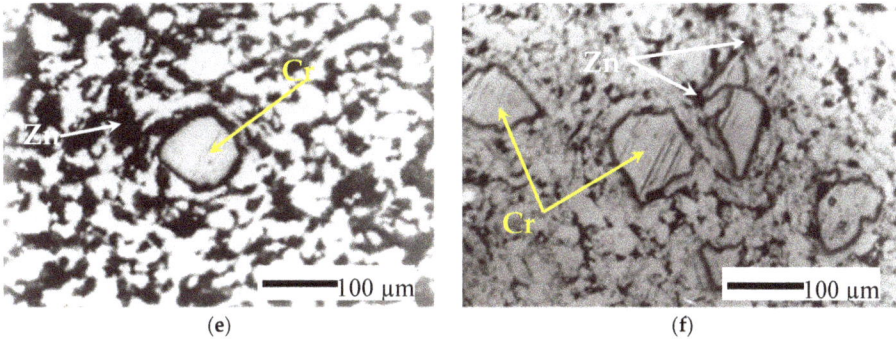

Figure 4. The microstructure of the different as-extruded (**a**) Al-5Cr (**b**)Al-7.5Cr (**c**) Al-10Cr (**d**) Al-20Zn (**e**) Al-20Zn-5Cr and (**f**) Al-20Zn-10Cr alloys.

The formation of lots of finely distributed polygonal type of the Cr and Zn particles, in the interface of liquid solid earlier during solidification, may lead to refinement. The process reduces the porosity by the formation of necks between powders, to achieve final elimination of the small pores at the end of the process. It forms new but lower energy solid-solid interfaces with a decrease in the free energy occurring on sintering, where 1 μm particles decreases with 1 cal/g [4]. On a microscopic scale, material transfer was affected by the change in the pressure and the differences in free energy across the curved surface.

It is worth mentioning that the liquid phase was formed at the hot pressing temperature and consolidation was further enhanced, by the isostatic action of the compressive stresses on the compact inside the dies [4,7]. Additionally, diffusion rates were increased by the liquid phase; densification was enhanced by good wetting between liquid and solid components of the alloy system.

3.3. Energy-Dispersive X-ray (EDX) Analysis

The Al-based alloy samples for microscopic examination were prepared using the standard metallographic procedures and were examined at a magnification of 100× by scanning electron microscopy (SEM, Jeol, Peabody, MA, USA). Quantitative energy dispersed X-ray (EDX, Jeol, Peabody, MA, USA) analysis was also performed to analyze the element composition for the different heat treated samples.

After the fabrication process, a Jeol 5400 SEM unit with a link energy dispersive X-ray spectroscopy (EDS) detector (Jeol, Peobody, MA, USA) is attached to observe the particle morphology, size, shape, and agglomeration. The EDX analysis emphasizes this behavior, as shown in Figure 5. The magnified image attached to Figure 5e,f shows that a white large phase was being embedded in a matrix of Al-Zn, as seen from EDX analysis region. The small white phases were mainly Zn particles. After the heat treatment process mentioned above, more homogeneous structure was obtained. The magnified image attached to Figure 5f shows white and dark areas, and the analysis of these areas was given as in EDX analysis region. It can be observed that the white area was Cr rich while the dark was mainly Zn. Figure 3e,f shows micrographs of Al-20Zn-5Cr and Al-20Zn-10Cr alloys, respectively, with heat treatment for one hour followed by water quenching. The liquid Zn during heating indicates homogenous structure, as the Cr particles become smaller than before heating. It shows that the nucleus was enriched with the addition of Cr. The data from EDS spectrum shows that the nature of nucleus was facilitated, as the foreign particle displays a small lattice mismatch in the al-based alloys. The results of calculation for some possible crystallographic orientations on the Cr particles shows that the planar disregistry was low; therefore, Cr can act as the heterogeneous nucleation for Al-based alloys.

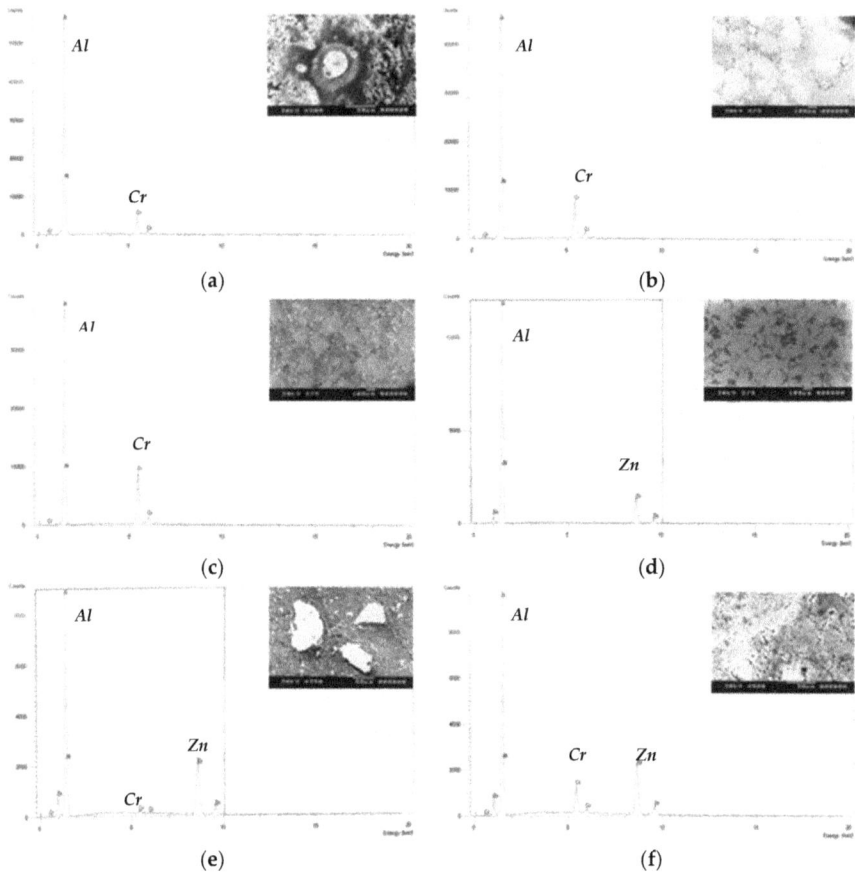

Figure 5. The EDX analysis at the indicated regions with higher magnification for of the different heat treated (**a**) Al-5Cr (**b**) Al-7.5Cr (**c**) Al-10Cr (**d**) Al-20Zn (**e**) Al-20Zn-5Cr and (**f**) Al-20Zn-10Cr alloys.

3.4. Mechanical Properties

Tensile specimens were prepared from the heat treated materials and as extruded states. The gauge length and diameter of the specimens were 30 mm and 6 mm; respectively. Tensile tests were performed at a cross head speed of 2.0 mm/min, which corresponded to an initial strain rate of $1.1 \times 10^{-3}\,\mathrm{s}^{-1}$. Tensile tests were conducted at room temperature using materials testing system (MTS) testing machine (model 610), fitted with a 160 kN load cell, operating in the displacement control mode under quasi static loading. The samples were deformed until crashed. For consistency and homogeneity, three identical samples were prepared for each test case and exposed to the same loading conditions. The average test value of all the three samples of the radial crushing strength was reported in Table 1.

Table 1. Average tensile data of Al-base alloys as extrusion and heat treatment with standard deviation SD.

Alloy Composition	As Extruded State				Heat Treated State			
	UTS (σ_u) MPa	SD	Fracture Strain (ε_f) %	SD	UTS (σ_u) MPa	SD	Fracture strain (ε_f) %	SD
Al-5Cr	75	0.1	12	1.2	105	1.4	10	1
Al-7.5Cr	70	0.3	10	1	101	1.4	8	0.78
Al-10Cr	65	0.3	8	0.78	96	0.7	6	0.7
Al-20Zn	188	5.8	18	1.7	185	5.7	16	1.5
Al-20Zn-5Cr	95	0.7	14	1.3	160	4.8	12	1.2
Al-20Zn-10Cr	80	1.4	10	1	143	3.4	8	0.78

Table 1 presents the tensile properties of the materials in both cases of the as-extruded and the heat treated state. For the alloys of Al-Cr without Zn, the ultimate tensile strength and tensile ductility decreases with increasing Cr amount. Comparing the tensile properties of the materials containing Zn with the Zn free counterpart, it is noticed that the addition of Zn resulted in improving both the tensile strength and ductility of Al-20Zn-5Cr and Al-20Zn-10Cr alloys, in the as-extruded state. Thus, the Zn addition appears to be quite favorable, as far as the mechanical properties are concerned. The heat treatment applied to the Zn-containing alloys does not improve their tensile ductility. This was probably related to an improved bonding between the matrix and the Cr as a result of the released local stresses around the Cr particles induced during extrusion.

Rockwell hardness measurements are performed for different produced materials using digital Rockwell hardness tester (SUN-TEC, Novi, MI, USA) at 60 kg load (HRA-60). At least five readings were taken for each case; the averages were recorded to insure consistency and homogeneity throughout the material surface. It was found for all alloys that the the bright phase, which corresponds to δ–phase, had Vickers hardness measurements about two times greater than the corresponding matrix, as presented in Table 2.

Table 2. Rockwell average hardness values Kg/mm^2 for the different Al-base alloys after extrusion and heat treatment with standard deviation SD.

Chemical Composition	Rockwell Hardness As Extruded	SD	Rockwell Hardness Heat Treated at 500 °C	SD
Al-5Cr	8.9	0.9	9.0	0.9
Al-7.5Cr	9.5	1	12.5	1.4
Al-10Cr	9.9	1	14.6	1.4
Al-20Zn	10.4	1	27.6	2.3
Al-20Zn-5Cr	17.4	1.6	30.6	2.5
Al-20Zn-10Cr	16.9	1.5	32.1	2.7

The effect of heating temperatures on hardness for both of the as-extruded and heat treated Al-base alloys were listed in Table 2. The heat treated Al-Zn-Cr alloys demonstrated improvement in hardness values compared with specimen without heat treatment. The heat treatment of Al-20Zn-10Cr alloys at 500 °C had the highest hardness value of 32.1, which represents an increase in hardness for more than 90% compared to that of the specimen without heat treatment. On the other hand, the increase of Cr particles in the Al-base matrix results in an increase of hardness values, as shown in Table 2.

3.5. Corrosion Rate

Corrosion rate were measured for the Al-based alloys using electrochemical polarization. The tests were performed in a corrosion cell, which contained 250 mL of 3.5 wt. % NaCl solutions, at room temperature and scan rate of 0.5 mV/s saturated air. All electrochemical measurements were conducted using a potentiostat AUTOLAB PGSTATE 30 (Artisan, Champaign, IL, USA) and analyzed using Galvanostat M352 software, at room temperature. Platinum gauze was used, as a counter electrode, and silver/silver chloride was the referenced. The exposed area was 1 cm^2 of all the heat treated and as-extruded Al-based alloys.

An activation-controlled cathodic process occurred in the cathodic branch; the main reaction was hydrogen evolution during the measurements. As the applied potential increased, an activation controlled anodic process was observed. The electrochemical parameters, such as polarization resistance (R_p), were measured. Corrosion current density (I_{corr}) was measured using the linear polarization resistance technique and obtained as a function of R_p, with β_c as the cathodic and β_a anodic Tafel slopes, as $I_{corr} = \beta/R_p$, where β is a constant value and can be calculated by following equation:

$$\beta = \beta_c \times \beta_a/2.3 \times (\beta_a + \beta_c) \tag{1}$$

The corrosion rate (C_R) expressed in mm per year was obtained from I_{corr} in air saturated sodium chloride solution, according to the Equation (2):

$$C_R = 0.13 \times I_{corr} \times (eq_{wt})/\rho \tag{2}$$

where eq_{wt} is the equivalent weight, ρ is the density in g/cm^3, and I_{corr} is the corrosion current density determined by the linear polarization method, using the Stern-Geary equation [24]. The corrosion parameters of Al-based alloys are presented in Table 3 for both of the as-extruded and heat treated Al-based alloys.

Table 3. The electrochemical corrosion data for both as extruded and heat treated Al-based alloys.

Chemical Composition	As Extruded		Heat Treated at 500°C	
	C_R mm/year	SD	C_R mm/year	SD
Al pure	0.031	1	–	–
Al-5Cr	0.030	0.9	0.028	0.9
Al-7.5Cr	0.029	0.9	0.027	0.8
Al-10Cr	0.029	0.9	0.026	0.7
Al-20Zn	0.028	0.8	0.024	0.7
Al-20Zn-5Cr	0.026	0.7	0.022	0.6
Al-20Zn-10Cr	0.022	0.6	0.020	0.5

It is noticed that with increase of the Cr and Zn additives in the Al-based alloys, the corrosion rate decreases. As Al-20Zn-10Cr alloys have corrosion rate of 30% less than pure aluminum, the heat treated Al-based alloys had an approximate corrosion rate of 7% less than as-extruded Al-based alloys.

3.6. Wear Resistance

The materials were also subjected to pin-on-disk wear tests under an unlubricated condition, in air and at room temperature. The counter face (the disk) was made of steel with a hardness of RC 32. Before each test, the disk surface was polished by standard metallographic procedure [4,7,25]. The tests were made for different lengths of running times up to 20 min under a constant pressure applied with about 0.127 MPa. Weight loss was calculated using a precise digital analog weight balance (accuracy of ±0.005 g). Results of the wear resistance for the different Al-based alloys compositions and heat treatment conditions are plotted in Figures 6–8. Figure 6. It shows that increasing Cr content in the binary Al-Cr alloys, for the extruded conditions, improved the wear resistance. Heat treatment of these alloys at 500 °C for one hour decreased the weight loss by about 1.5 times.

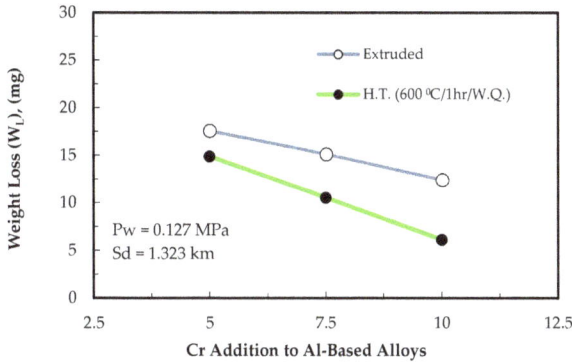

Figure 6. The effect of heat treatment on the weight loss for Al-Cr alloys.

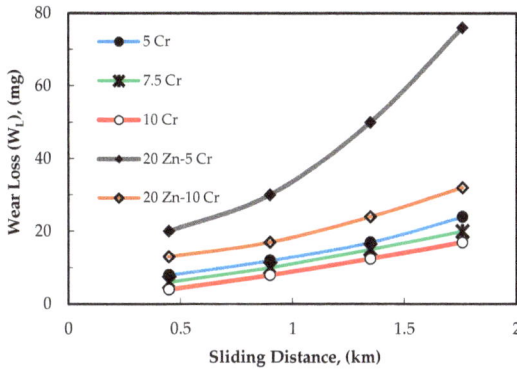

Figure 7. The weight loss versus sliding distance of Al-base alloys in the as-extruded condition.

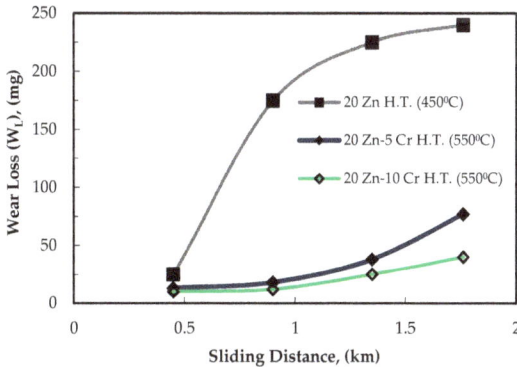

Figure 8. The effect of Cr addition on the weight loss of heat treated Al-20Zn alloys.

The microstructure shown in Figure 3a indicates that heat treatment of the binary Al-Cr alloys led to a diffusion of aluminum in the Cr particles, resulting in gradual alloying of the Cr particle. It is possible that some Cr particles were dissolved in Al-based alloys, which resulted in alloy hardening and; consequently, an increase in the wear resistance. In the ternary alloys of Al-Zn-Cr, the addition of Zn increases the weight loss for the alloys containing 5 or 10 wt. % Cr, as shown in Figure 7. This effect was clearer for the alloys containing 5 wt. % Cr. The weight loss of these alloys was about 2 times higher compared to the 10 wt. % Cr ternary alloys. In the as-extruded condition, the addition of Zn to the Al-Cr alloys resulted in a higher degree of homogeneity in the microstructure, as shown in Figure 3e,f; however, Cr is a hard metal and is expected to improve wear resistance.

The wear resistance results for the heat treated Al-20Zn alloys and Al-20Zn alloys, containing 5 and 10 wt. % Cr, was plotted in Figure 8. The wear resistance of the Al-20Zn-Cr alloys were found to be around 5 to 8 fold greater than the Al-20Zn alloys, after 1.76 km sliding distance. The presence of 20 wt. % Zn in the Al-Cr alloys during heat treatment enhanced the inter-diffusion process and formation of homogenous Al-Zn-Cr alloys, as shown in Figure 3e. The Cr particles have almost disappeared although EDX analysis, in Figure 8c,d, showed two phase alloys with different Cr contents. As mentioned above, the addition of Cr to the base alloy increased the hardness compared to the Al-20Zn alloys.

4. Conclusions

Based on the results of the present study, the following conclusions could be reached:

1. Binary Al-Cr alloys from powders, heat-treated at different conditions, could not establish a homogenous microstructure.
2. The presence of 20 wt. % Zn in the Al-Cr alloys enhanced the interdiffusion and densification process during heat treatment, due to the formation of a liquid phase leading to a homogenius microstructure.
3. Both Zn addition and the heat treatment temperature affect the hardness values and the structure of the Al-base matrix
4. Under tension load, improvement in the strength of Al-Cr alloys was obtained after Zn addition and the heat-treatment process.
5. The increase of the Cr and Zn additives in the Al-based alloys would result in a decrease in the corrosion rate. As Al-20Zn-10Cr alloys have corrosion rate of 30% less than pure aluminum, the heat treated Al-based alloys had an approximate corrosion rate of 7% less than as-extruded Al-based alloys.
6. The wear resistance of the ternary Al-Zn-Cr heat treated at 500 °C for one hour was about 5 times higher than that of the binary Al-Zn alloys. The alloy of Al-20Zn-5Cr, heat-treated at 500 °C for one hour followed by water quenching, exhibited the highest wear resistance among the investigated alloys.

Author Contributions: All authors contributed equally in designing and manufacturing of the studied alloys, conceiving, designing and performing the experiments, collecting and analyzing the data, and finally discussion association of the results. I would like to also indicate that the corresponding author was responsible for addressing all reviewers' remarks and journal correspondence.

Conflicts of Interest: The authors also declare no conflict of interest.

References

1. Liu, S.; Zhong, Q.; Zhang, Y.; Liu, W.; Zhang, X.; Deng, D. Investigation of quench sensitivity of high strength Al-Zn-Mg-Cu alloys by time–temperature-properties diagrams. *Mater. Des.* **2010**, *31*, 3116–3120. [CrossRef]
2. Ezuber, H.; El-Houd, A.; El-Shawesh, F. A study on the corrosion behavior of aluminum alloys in seawater. *Mater. Des.* **2008**, *29*, 801–805. [CrossRef]
3. Kalkanl, A.; Yılmaz, S. Synthesis and characterization of aluminum alloy 7075 reinforced with silicon carbide particulates. *Mater. Des.* **2008**, *29*, 775–780. [CrossRef]
4. Nassef, A.; El-Hadek, M. Mechanics of hot pressed aluminum composites. *Int. J. Adv. Manuf. Technol.* **2015**, *76*, 1905–1912. [CrossRef]
5. El-Hadek, M.; Kassem, M. Failure behavior of Cu-Ti-Zr-based bulk metallic glass alloys. *J. Mater. Sci.* **2009**, *44*, 1127–1136. [CrossRef]

6. El-Hadek, M.; Kaytbay, S. Al$_2$O$_3$ Particle Size Effect on Reinforced Copper Alloys: An Experimental Study. *Strain* **2009**, *45*, 506–515. [CrossRef]

7. Nassef, A.; El-Hadek, M. Microstructure and Mechanical Behavior of Hot Pressed Cu-Sn Powder Alloys. *Adv. Mater. Sci. Eng.* **2016**, *2016*, 1–10. [CrossRef]

8. Kurtuldu, G.; Jarry, P.; Rappaz, M. Influence of Cr on the nucleation of primary Al and formation of twinned dendrites in Al-Zn-Cr alloys: Can icosahedral solid clusters play a role? *Acta Mater.* **2013**, *61*, 7098–7108. [CrossRef]

9. Kurtuldu, G.; Jarry, P.; Rappaz, M. Influence of icosahedral short range order on diffusion in liquids: A study on Al-Zn-Cr alloys. *Acta Mater.* **2016**, *115*, 423–433. [CrossRef]

10. Del Arco, M.; Rives, V.; Trujillano, R.; Malet, P. Thermal behavior of Zn-Cr layered double hydroxides with hydrotalcite-like structures containing carbonate or decavanadate. *J. Mater. Chem.* **1996**, *6*, 1419–1428. [CrossRef]

11. Peng, G.; Chen, K.; Fang, H.; Chen, S. Effect of Cr and Yb additions on microstructure and properties of low copper Al-Zn-Mg-Cu-Zr alloy. *Mater. Des.* **2012**, *36*, 279–283. [CrossRef]

12. He, Z.; Su, X.; Peng, H.; Liu, L.; Wu, C.; Wang, J. 600 °C isothermal section of the Al-Cr-Zn ternary phase diagram. *J. Alloys Compd.* **2015**, *649*, 1239–1245. [CrossRef]

13. Li, C.B.; Han, S.Q.; Liu, S.D.; Deng, Y.L.; Zhang, X.M. Grain structure effect on quench sensitivity of Al-Zn-Mg-Cu-Cr alloy. *Trans. Nonferr. Met. Soc. China* **2016**, *26*, 2276–2282. [CrossRef]

14. Fourmentin, R.; Avettand-Fènoël, M.N.; Reumont, G.; Perrot, P. The Fe-Zn-Al-Cr system and its impact on the galvanizing process in chromium-added zinc baths. *J. Mater. Sci.* **2008**, *43*, 6872–6880. [CrossRef]

15. Shaha, S.K.; Czerwinski, F.; Kasprzak, W.; Friedman, J.; Chen, D.L. Ageing characteristics and high-temperature tensile properties of Al-Si-Cu-Mg alloys with micro-additions of Cr, Ti, V and Zr. *Mater. Sci. Eng. A* **2016**, *652*, 353–364. [CrossRef]

16. Shin, S.-S.; Lim, K.-M.; Park, I.-M. Effects of high Zn content on the microstructure and mechanical properties of Al-Zn-Cu gravity-cast alloys. *Mater. Sci. Eng. A* **2017**, *679*, 340–349. [CrossRef]

17. Chinh, N.Q.; Jenei, P.; Gubicza, J.; Bobrukb, E.V.; Valiev, R.Z.; Langdon, T.G. Influence of Zn content on the microstructure and mechanical performance of ultrafine-grained Al-Zn alloys processed by high-pressure torsion. *Mater. Lett.* **2017**, *186*, 334–337. [CrossRef]

18. Staišiūnas, L.; Miečinskas, P.; Leinartas, K.; Selskis, A.; Grigucevičienė, A.; Juzeliūnas, E. Sputter-deposited Mg-Al-Zn-Cr alloys—Electrochemical characterization of single films and multilayer protection of AZ31 magnesium alloy. *J. Corros. Sci.* **2014**, *80*, 487–493. [CrossRef]

19. Wang, H.; Zhang, R.; Hu, X.; Wang, C.A.; Huang, Y. Characterization of a powder metallurgy SiC/Cu-Al composite. *J. Mater. Process. Technol.* **2008**, *197*, 43–48. [CrossRef]

20. Ogel, B.; Gurbuz, R. Microstructural characterization and tensile properties of hot pressed Al-SiC composites prepared from pure Al and Cu powders. *Mater. Sci. Eng. A* **2001**, *301*, 213–220. [CrossRef]

21. Chang, Y.; Sun, W.; Xiong, X.; Chen, Z.; Wang, Y.; Hao, Z.; Xu, Y. A novel design of Al-Cr alloy surface sealing for ablation resistant C/C-ZrC-SiC composite. *J. Eur. Ceram. Soc.* **2017**, *37*, 859–864. [CrossRef]

22. El-Hadek, M.; Kassem, M. Characterization of strengthened rapidly quenched Zr-based alloys. *Int. J. Mech. Mater. Des.* **2008**, *4*, 279–289. [CrossRef]

23. El-Hadek, M.; Kaytbay, S. Fracture properties of SPS tungsten copper powder composites. *Metall. Mater. Trans. A* **2013**, *44*, 544–551. [CrossRef]

24. Ozyılmaz, A.T.; Kardas, G.; Erbil, M.; Yazici, B. The corrosion performance of polyaniline on nickel plated mild steel. *Appl. Surf. Sci.* **2005**, *242*, 97–106. [CrossRef]

25. Kaytbay, S.; El-Hadek, M. Wear resistance and fracture mechanics of WC–Co composites. *Int. J. Mater. Res.* **2014**, *105*, 557–565. [CrossRef]

metals

MDPI

Article

Transparent Conducting Film Fabricated by Metal Mesh Method with Ag and Cu@Ag Mixture Nanoparticle Pastes

Hyun Min Nam [1], Duck Min Seo [2], Hyung Duk Yun [2], Gurunathan Thangavel [2], Lee Soon Park [2] and Su Yong Nam [1,*

[1] Department of Graphic Arts Information Engineering, Pukyong National University, Busan 48547, Korea; skagusals12@naver.com
[2] School of Material Science and Engineering, Ulsan National Institute of Science and Technology (UNIST), Ulsan 44919, Korea; seodm@unist.ac.kr (D.M.S.); yun2985@unist.ac.kr (H.D.Y.); gurunathan.t20@unist.ac.kr (G.T.); parkls@unist.ac.kr (L.S.P.)
*   Correspondence: suynam@pknu.ac.kr; Tel.: +82-51-629-6396

Academic Editor: Manoj Gupta
Received: 7 March 2017; Accepted: 11 May 2017; Published: 16 May 2017

Abstract: Transparent conducting electrode film is highly desirable for application in touch screen panels (TSPs), flexible and wearable displays, sensors, and actuators. A sputtered film of indium tin oxide (ITO) shows high transmittance (90%) at low sheet resistance (50 Ω/cm^2). However, ITO films lack mechanical flexibility, especially under bending stress, and have limitation in application to large-area TSPs (over 15 inches) due to the trade-off in high transmittance and low sheet resistance properties. One promising solution is to use metal mesh-type transparent conducting film, especially for touch panel application. In this work, we investigated such inter-related issues as UV imprinting process to make a trench layer pattern, the synthesis of core-shell-type Ag and Cu@Ag composite nanoparticles and their paste formulation, the filling of Ag and Cu@Ag mixture nanoparticle paste to the trench layer, and touch panel fabrication processes.

Keywords: synthesis of core–shell metal nanoparticles; Cu@Ag composite nanoparticle; metal mesh; screen printing; touch screen panel

1. Introduction

Transparent conducting electrode film is highly desirable for application in touch screen panels, flexible organic light emitting diode (OLED), and wearable displays, sensors, and actuators [1–5]. Metal nanoparticles have been used for conducting electrodes by using various fabrication process. One-step direct nanoimprinting of gold nanoparticles was reported by using hexanethiol self-assembled monolayer (SAM)-protected gold particles and polydimethylsiloxane (PDMS) mold [6]. Gold nanoparticles were also imprinted by PDMS mold on the polyimide film, and an organic field effect transistor device was fabricated [7]. The direct imprinting process using a PDMS mold leaves metal nanoparticles on top of the film substrate so that it has limited application in transparent touch screen panel (TSP) fabrication. Silver conductive ink was patterned on the PDMS stretchable substrate by stencil printing to make organic thin film transistor (OTFT) devices [8]. The OTFT device had excellent stretchability up to 150%, but the pattern width was 50 μm due to the stencil printing process, thus limiting the application in TSPs. The direct imprinting and screen printing processes have the merits of being simple processes operated at ambient condition. However, they result in an embossed pattern on the film substrate, so they may not withstand such hard mechanical stresses as bending and stretching.

In this work, we synthesized both Ag and cost-effective Cu@Ag nanoparticles for application to the large size touch screen panels. Indium tin oxide (ITO) has been widely used in transparent conducting electrodes (TCEs) to make small TSPs used in smartphones. However, ITO conductors cannot be used in large TSPs over 15 inches due to the trade-off between high transparency and low electrical resistance required for the operation of touch screen panels and the requirement of increasing the thickness of ITO thin films to lower the electrical resistance. Of the transparent conducting materials such as carbon nanotubes (CNTs), graphene, conducting polymer (PEDOT: PSS; poly (3,4-ethylenedioxythiophene) polystyrene sulfonate), and metal nanostructures, silver nanoparticles and silver-coated copper (Cu@Ag) nanoparticles have high potential for application in large TSPs, including flexible and stretchable versions. This is due to the suitability of the silver nanoparticle paste to the metal mesh-type transparent conducting electrode. In the metal mesh method, the transparent electrodes are patterned by using the trench filling process. In this method, a narrow engraved trench (~2.5 μm) pattern is formed by coating a ultra violet (UV) curing resin on the substrate film and then pressing with a transparent mold followed by UV exposure and demolding. The gap between the trenches is over 100 μm, so the visible light transmittance of the metal mesh film is over 88% while the resistivity of the Ag nanoparticle paste is less than 10 $\Omega\cdot$cm by the percolation mechanism. Metal mesh TCEs also have the merit of easy hard coating layer formation on top of a trench pattern layer filled with silver particles compared to the imprinting or screen printing processes with embossed silver electrode patterns.

2. Materials and Methods

2.1. Ag and Cu@Ag Composite Nanoparticles and Paste

The synthetic process of silver (Ag) nanoparticles is shown in Figure 1. First, butylamine (160 g) (Sigma-Aldrich, Seoul, Korea) and ethanol (320 g) (Sigma-Aldrich, Seoul, Korea) were mixed at 23–25 °C; then, oleic acid (surfactant, 213 g) (Sigma-Aldrich, Seoul, Korea) was added and the mixture solution was heated to 70 °C. After cooling to 55 °C, silver acetate (106.8 g) (Sigma-Aldrich, Seoul, Korea) was added and the temperature was maintained at 40 °C, while aqueous hydrazine (Sigma-Aldrich, Seoul, Korea) and polyvinylpyrrolidone (Sigma-Aldrich, Seoul, Korea) mixture solution was added at a rate of 4 mL/10 min for 10 min. After 30 min stabilization, the silver nanoparticles were precipitated and separated by repeated washing with ethanol.

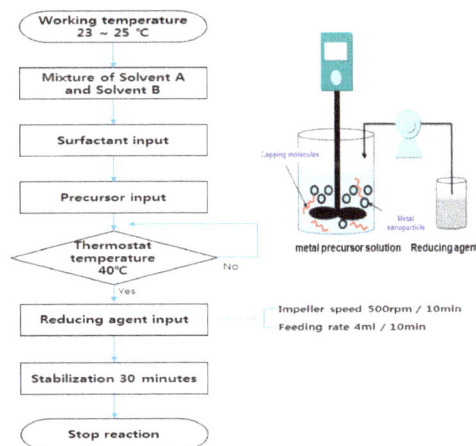

Figure 1. Synthesis of metal nanoparticles.

The Cu@Ag composite nanoparticles were made by electroplating silver onto copper (Cu) nanoparticles [9]. The process is as follows. First, Cu nanoparticles made by a similar method mentioned above for the Ag nanoparticles were washed with HCl aqueous solution and with deionized water. The washed Cu nanoparticles were dispersed in deionized water with polyacrylate surfactant. To this solution were added the $AgNO_3$ and NH_4OH mixture aqueous solution and reacted with stirring followed by separation and drying. The resulting Cu@Ag nanoparticles were recovered by centrifuge and washed with water to remove the capping agent. The Ag and Ag@Cu nanoparticles were mixed with bisphenil—an epoxy acrylate resin dissolved in the diethylene glycol monoethyl ether acetate (ECA) (Sigma-Aldrich, Seoul, Korea) solvent with dispersant BYK-9076 obtained from BYK Additives & Instruments, Wesel, Germany [10]. The resulting Ag/Cu@Ag nanoparticle paste was further mixed by using a three roll-mill (Exakt, Oklahoma City, OK, USA) followed by degassing and rolling.

2.2. Metal Mesh Mold and Trench Layer Patterning

The embossed metal mesh mold was fabricated by using photolithographic and electroplating method. The engraved trench layer pattern was made by first coating a photosensitive UV resin on the optical grade polyethylene terephthalate (PET) and then pressing with the embossed metal mold (less than 10 psi) followed by UV exposure (150 mJ) and demolding. The trench screen panel and engraved pattern of trench layers consisting of sensor and bezel electrodes are shown in Figure 2. As shown in Figure 2, the width of the narrow sensor electrodes was 1–2.5 μm, and the width of the bezel electrodes was 20–50 μm. After filling the Ag/Cu@Ag paste into the trench layers of the sensor and bezel electrodes, the remaining Ag/Cu@Ag pastes on top of the PET film substrate needed to be wiped out. However, during this process, the Ag/Cu@Ag paste in the engraved area of the trench layer is also removed by the wiping process. In order to prevent the wiping-out of filled Ag/Cu@Ag nanoparticle paste from the engraved trench layers, the embossed metal mesh mold was designed to have many small embossed patterns, especially inside the wide bezel electrode area. This design of embossed metal mesh mold was found to be very effective in reducing the wiping-out of the Ag/Cu@Ag paste filled into the engraved trench layer.

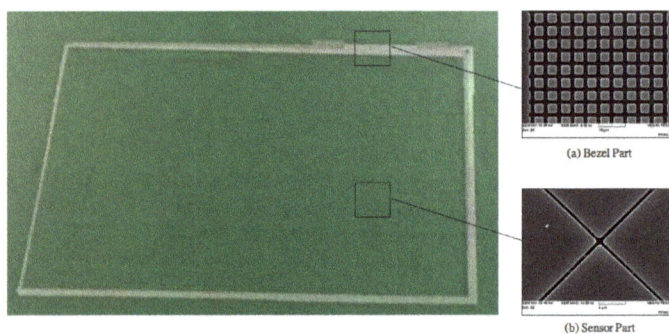

(a) Bezel Part

(b) Sensor Part

Figure 2. Touch sensor panel and scanning electron microscope (SEM) images of engraved (**a**) bezel electrodes and (**b**) sensor electrode patterns made by UV imprinting process with embossed metal mold.

2.3. Filling of Ag/Cu@Ag Nanoparticle Paste and Wiping of Residual Paste

The Ag/Cu@Ag nanoparticle paste was filled into the engraved trench layers consisting of sensor and bezel electrodes on the PET substrate film by using a doctoring machine made by Mino Co. [11] (Gifu, Japan), as shown in Figure 3. The formulations of the photosensitive UV resins are shown in Table 1. Of these UV resins, sample UVT-5-1 was found to give adequate flexibility and demolding property from the metal mold after UV exposure. After filling Ag/Cu@Ag nanoparticle paste into the engraved trench layer, the residual paste was wiped off the embossed part of PET film substrate by a wiping cloth, followed by post-heating of the filled silver paste.

Figure 3. Filling process of Ag/Cu@Ag nanoparticle paste with doctoring machine.

Table 1. Formulation of photosensitive UV resin for imprinting process to make trench layers.

Sample	UV Oligomer	UV Monomer		Photoinitiator
	F-130 (g)	EEEA (g)	HDDA (g)	PI-TPO 25 wt % in EEEA (g)
UVT-1-1	7.0	1.0	1.8	0.2
UVT-2-1	6.0	2.0	1.8	0.2
UVT-3-1	5.0	3.0	1.8	0.2
UVT-3-2	5.0	3.0	1.8	0.4
UVT-4-1	4.0	4.0	1.8	0.2
UVT-4-2	4.0	4.0	1.8	0.4
UVT-5-1	3.0	5.0	1.8	0.2
UVT-5-2	3.0	5.0	1.8	0.4

F-130 ; Ebecryl 8411 (Allnex, Sydney, Australia), EEEA ; Etoxyetoxy ethyl acrylate(Sigma-Aldrich, Seoul, Korea), HDDA ; Hexane diol diacrylate (Sigma-Aldrich, Seoul, Korea), PI-TPO (Ciba, Basel, Switzerland).

2.4. Fabrication of Touch Panel by Using Metal Mesh Films

The structure and fabrication process of the touch panel are shown in Figure 4, utilizing the two transparent metal mesh films made with Ag/Cu@Ag nanoparticle pastes. After lamination of the cover glass and the top/bottom part of the touch sensor with optically clear adhesive (OCA) films, the touch panel circuits and controller were bonded to the flexible printed circuit board (f-PCB) using anisotropic conductive film (ACF).

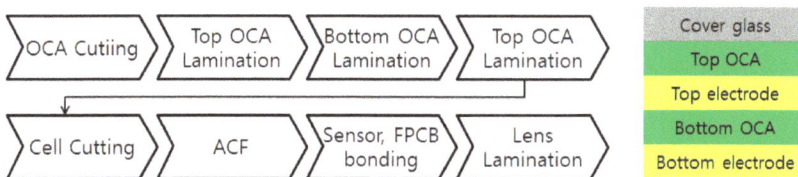

Figure 4. Structure and fabrication process of touch screen panel. ACF: anisotropic conductive film; OCA: optically clear adhesive.

3. Results

3.1. Ag Nanoparticle Synthesis and Paste Properties

Scanning electron microscope (SEM) images of the synthesized Ag nanoparticles are shown in Figure 5. Ag powder in paste (1) exhibited nanoparticles in the 1–25 nm range, which can promote sintering at lower temperature. Ag powder in paste (2) exhibited the aggregation of Ag nanoparticles, and the Ag powder in paste (3) contained large Ag nanoparticles. The three Ag pastes were formulated with different Ag nanoparticles, and their conductivities were checked after coating and thermal curing on PET substrate films. Table 2 indicates that the Ag paste in paste (1) cured at 180 °C for 20 min showed high conductivity. This may be due to the partial sintering of Ag nanoparticles in Ag paste (1) and concurrent high packing density of Ag nanoparticles, thus promoting effective percolation.

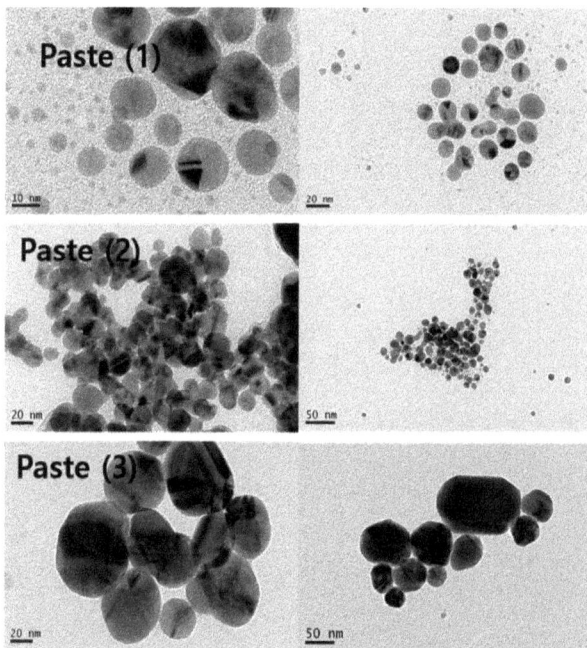

Figure 5. Field emission scanning electron microscopy (FE-SEM) images of synthesized Ag nanoparticles in Ag pastes (**1**) to (**3**).

Table 2. Synthetic condition of Ag nanoparticles and the formulation of Ag nanoparticle pastes. ECA: Diethylene glycol monoethyl ether acetate.

Ag Paste	Ag Powder Synthetic Condition	Ag Powder	Binder Polymer	Solvent	Additive
		88 wt %	4 wt %	7 wt %	1 wt %
Ag Paste (1)	Acid value: 100 / Injection rate: 10 mL/min (40 min)	Ag powder (1)	Bisphenol-A Epoxy acrylate	ECA	BYK-754
Ag Paste (2)	Acid value: 100 / Injection rate: 40 mL/min (10 min)	Ag powder (2)	Bisphenol-A Epoxy acrylate	ECA	BYK-754
Ag Paste (3)	Acid value: 50 / Injection rate: 40 mL/min (10 min)	Ag powder (3)	Bisphenol-A Epoxy acrylate	ECA	BYK-754

Ag Paste	Thermal curing condition and conductivity		
	100 °C, 20 min	130 °C, 20 min	180 °C, 20 min
Ag Paste (1)	-	13–18 mΩ	3–4 mΩ
Ag Paste (2)	-	11–13 mΩ	4–6 mΩ
Ag Paste (3)	-	8–10 mΩ	4–5 mΩ

3.2. Cu@Ag Composite Nanoparticles and Paste Properties

The Cu@Ag composite nanoparticles were synthesized by electroplating method, as described in the experimental part with Ag contents of 20, 30, 40 wt %. The Cu@Ag nanoparticle pastes were formulated with binder polymer content of 4 and 2 wt %, as shown in Table 3. The viscosity of the Cu@Ag nanoparticle paste with 2 wt % of binder polymers was higher than that of 4 wt % paste, while the sheet resistance was lower in the case of metal paste with 2 wt % of binder polymers. However, the conductivity of the Cu@Ag nanoparticle paste was lower than that of the pure Ag nanoparticle paste in Table 2. Therefore, a new metal paste was made by mixing pure Ag and Cu@Ag nanoparticle with 30 wt % Ag in the ratio of 8:2, 5:5, and 2:8 by weight. As shown in Table 4, the Ag/Cu@Ag mixture paste (3) showed low surface resistance for touch panel application along with good filling property into trench layer pattern.

Table 3. Formulation of Cu@Ag nanoparticle pastes and properties.

| Cu@Ag Pastes | Cu@Ag Powder | Cu@Ag Paste Formulation (wt %) | | | | Sheet Resistance (4 Point, Ω/cm^2) |
| | | Cu@Ag Powder | Binder Polymer | Solvent | Additive | |
			Bisphenol-A Epoxy Acrylate	ECA	BYK-9076	
Cu@Ag paste (1)	Cu@Ag (Ag: 20 wt %)	88	4	7	1	500–600
Cu@Ag paste (2)	Cu@Ag (Ag: 30 wt %)	88	4	7	1	380–480
Cu@Ag paste (3)	Cu@Ag (Ag: 40 wt %)	88	4	7	1	200–300
Cu@Ag paste (4)	Cu@Ag (Ag: 20 wt %)	88	2	9	1	450–500
Cu@Ag paste (5)	Cu@Ag (Ag: 30 wt %)	88	2	9	1	250–300
Cu@Ag paste (6)	Cu@Ag (Ag: 40 wt %)	88	2	9	1	150–200

Table 4. Formulation of Ag and Cu@Ag nanoparticle mixture paste and properties.

| Cu@Ag Pastes | Ag and Cu@Ag Paste Mixture Formulation (wt %) | | | | | Sheet Resistance (4 Point, Ω/cm^2) |
| | Cu@Ag Powder | Ag Powder | Binder Polymer | Solvent | Additive | |
			Bisphenol Epoxy Acrylate	ECA	BYK-9706	
Paste (1)	Cu@Ag 70.4 wt %	Ag 17.6 wt %	2	9	1	55–80
Paste (2)	Cu@Ag 44.0 wt %	Ag 44.0 wt %	2	9	1	30–45
Paste (3)	Cu@Ag 17.6 wt %	Ag 70.4 wt %	2	9	1	13–15

3.3. Inlay Filling of Ag and Cu@Ag Mixture Paste and Performance of Touch Screen Panel

The Ag and Cu@Ag mixture paste was filled into the pattered trench layer by using a doctoring machine made by Mino Co., Gifu, Japan. Figure 6 shows that the mixture paste could be filled 80–90% by first filling and almost 100% by second filling with the doctoring machine. After optimizing the metal paste by mixing Ag and Cu@Ag paste at 20:80 wt % ratio and filling process of the mixture paste into the trench layer, the touch panel module was fabricated and the properties were as follows. The optical property of the metal mesh transparent film was L^*: 0.95 and b^*: 0.53, suitable for touch screen panel application. The uniformity of the width of the trench layer with 1 μm design was found to be 0.93 ± 0.007, and the sheet resistance of the metal mesh with 1 μm width were 14.57 ± 0.487 Ω/cm^2), with uniformity of 3.26%. The capacitance change before (C_m = 6.68 pF) and after (C_m = 4.85 pF) 10,000 bending cycles at 60 rpm and radius R = 10 mm also met the specification value of 28.37%, which was within the acceptance range of 30%.

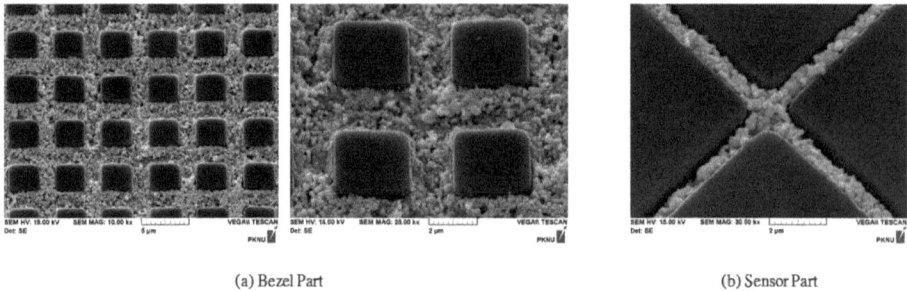

(a) Bezel Part   (b) Sensor Part

Figure 6. SEM images of filled Ag/Cu@Ag mixture paste in (**a**) bezel and (**b**) sensor parts of metal meshes touch screen panel.

4. Conclusions

In this work, we investigated such inter-related issues as the synthesis of core-shell-type Ag and Cu@Ag composite nanoparticles and their paste formulation UV imprinting process to make a trench layer pattern, filling of Ag and Cu@Ag mixture nanoparticle paste to the trench layer, and touch panel fabrication processes. After optimizing the metal paste by mixing Ag and Cu@Ag paste at 20:80 wt % ratio and filling process of the mixture paste into the trench layer, a touch panel module was fabricated and the properties were as follows. The optical property of the metal mesh transparent film was L^*: 0.95 and b^*: 0.53. The uniformity of the width of the trench layer with 1 μm design was found to be 0.93 ± 0.007, and the sheet resistance of the metal mesh with 1 μm width were 14.57 ± 0.487 Ω/cm^2 with uniformity of 3.26%, suitable for touch screen panel application.

Acknowledgments: This work was supported by a Research Grant of Pukyong National University (2016 year).

Author Contributions: Su Yong Nam and Lee Soon Park conceived and designed the experiments; Hyun Min Nam, Duck Min Seo, Hyung Duk Yun and Gurunathan Thangavel performed the experiments.

Conflicts of Interest: The authors declare no conflict of interest.

References

1. Park, S.I.; Xiong, Y.; Kim, R.H.; Elvikis, P.; Meitl, M.; Kim, D.H.; Wu, J.; Yoon, J.; Yu, C.J.; Liu, Z.; et al. Printed assemblies of inorganic light-emitting diodes for deformable and semitransparent displays. *Science* **2009**, *325*, 977–981. [CrossRef] [PubMed]
2. Yu, Z.; Niu, X.; Liu, Z.; Pei, Q. Intrinsically Stretchable Polymer Light-Emitting Devices Using Carbon Nanotube-Polymer Composite Electrodes. *Adv. Mater.* **2011**, *23*, 3989–3994. [CrossRef] [PubMed]
3. Sekitani, T.; Nakajima, H.; Maeda, H.; Fukushima, T.; Aida, T.; Hata, K.; Someya, T. Stretchable active-matrix organic light-emitting diode display using printable elastic conductors. *Nat. Mater.* **2009**, *8*, 494–499. [CrossRef] [PubMed]
4. Han, T.H.; Jeong, S.H.; Lee, Y.; Seo, H.K.; Kwon, S.J.; Park, M.H.; Lee, T.W. Flexible transparent electrodes for organic light emitting diodes. *J. Inf. Disp.* **2015**, *16*, 71–84. [CrossRef]
5. Choi, Y.M.; Kim, K.W.; Lee, E.; Jo, J.; Lee, T.M. Fabrication of a single layer metal-mesh touchscreen sensor using reverse-offset printing. *J. Inf. Disp.* **2015**, *16*, 37–41. [CrossRef]
6. Ko, S.H.; Park, I.; Pan, H.; Grigoropoulos, C.P.; Pisano, A.P.; Luscombe, C.K.; Fréchet, J.M. Direct nanoimprinting of metal nanoparticles for nanoscale electronics fabrication. *Nano Lett.* **2007**, *7*, 1869–1877. [CrossRef] [PubMed]
7. Park, I.; Ko, S.H.; Pan, H.; Grigoropoulos, C.P.; Pisano, A.P.; Fréchet, J.M.; Lee, E.S.; Jeong, J.H. Nanoscale patterning and electronics on flexible substrate by direct nanoimprinting of metallic nanoparticles. *Adv. Mater.* **2008**, *20*, 489–496. [CrossRef]

8. Suh, Y.D.; Jung, J.; Lee, H.; Yeo, J.; Hong, S.; Lee, P.; Lee, D.; Ko, S.H. Nanowire reinforced nanoparticle nanocomposite for highly flexible transparent electrodes: borrowing ideas from macrocomposites in steel-wire reinforced concrete. *J. Mater. Chem. C* **2017**, *5*, 791–798. [CrossRef]
9. Park, M.; Im, J.; Shin, M.; Min, Y.; Park, J.; Cho, H.; Park, S.; Shim, M.B.; Jeon, S.; Chung, D.Y.; et al. Highly stretchable electric circuits from a composite material of silver nanoparticles and elastomeric fibers. *Nat. Nanotechnol.* **2012**, *7*, 803–809. [CrossRef] [PubMed]
10. BYK Additives & Instruments. Available online: http://BYK.com/kr (accessed on 7 March 2017).
11. Mino International Ltd. Available online: http://mino.co.jp (accessed on 7 March 2017).

metals

MDPI

Article

Comparative Investigation of Tungsten Fibre Nets Reinforced Tungsten Composite Fabricated by Three Different Methods

Linhui Zhang [1,2], Yan Jiang [1], Qianfeng Fang [1,2,*], Rui Liu [1], Zhuoming Xie [1,2], Tao Zhang [1], Xianping Wang [1,*] and Changsong Liu [1]

[1] Key Laboratory of Materials Physics, Institute of Solid State Physics, Chinese Academy of Sciences, Hefei 230031, China; zhanglinhuia@126.com (L.Z.); Jiangyanzky@163.com (Y.J.); liurui@issp.ac.cn (R.L.); ZMXie1989@163.com (Z.X.); zhangtao@issp.ac.cn (T.Z.); csliu@issp.ac.cn (C.L.)

[2] Department of Materials Science and Engineering, University of Science and Technology of China, Hefei 230026, China

* Correspondence: qffang@issp.ac.cn (Q.F.); xpwang@issp.ac.cn (X.W.); Tel.: +86-551-6559-1459 (Q.F.); +86-551-6559-1125 (X.W.)

Received: 1 March 2017; Accepted: 20 June 2017; Published: 4 July 2017

Abstract: Tungsten fibre nets reinforced tungsten composites (W_f/W) containing four net layers were fabricated by spark plasma sintering (SPS), hot pressing (HP) and cold rolling after HP (HPCR), with the weight fraction of fibres being 17.4%, 10.5% and 10.5%, respectively. The relative density of the HPCRed samples is the highest (99.8%) while that of the HPed composites is the lowest (95.1%). Optical and scanning electron microscopy and electron back scattering diffraction were exploited to characterize the microstructure, while tensile and hardness tests were used to evaluate the mechanical properties of the samples. It was found that partial recrystallization of fibres occurred after the sintering at 1800 °C. The SPSed and HPed W_f/W composites begin to exhibit plastic deformation at 600 °C with tensile strength (TS) of 536 and 425 MPa and total elongation at break (TE) of 11.6% and 23.0%, respectively, while the HPCRed W_f/W composites exhibit plastic deformation at around 400 °C. The TS and TE of the HPCRed W_f/W composites at 400 °C are 784 MPa and 8.4%, respectively. The enhanced mechanical performance of the W_f/W composites over the pure tungsten can be attributed to the necking, cracking, and debonding of the tungsten fibres.

Keywords: tungsten composites; tungsten-fibre-net reinforcement; powder metallurgy; tensile strength

1. Introduction

Tungsten (W) with a high melting temperature, high erosion resistance, excellent thermal conductivity and low tritium retention has been applied in the manufacturing and electronic industries, lighting engineering, medical, aerospace and military fields, and more recently in the nuclear fields served as the plasma-facing materials [1–3]. However the inherent low-temperature brittleness and recrystallization or irradiation induced embrittlement of tungsten severely limit its application in the extreme conditions of simultaneous high temperature and high flux neutron irradiation [4–6]. So it is important to decrease the embrittlement of tungsten materials.

Two ways are usually employed to enhance the toughness of tungsten: one is by particle dispersion strengthening, and the other is via fibre reinforcement. Different from the dispersion strengthening by adding oxides like La_2O_3 [7] and Y_2O_3 [8] or carbides like TiC [9] and ZrC [10] particles, the fibre strengthening by adding fibres such as carbon fibres [11], SiC fibres [12,13], short fibres of tantalum [14] and tungsten fibres [15] is considered to be an effective approach to improve the toughness via micro-cracking, twinning, bridging, debonding and pullout or a mixture of them [15].

Among the different fibres, tungsten fibres possess high strength, fine ductility, high recrystallization temperature (>1800 °C) and plastic rupture [16], which could contribute substantially to the toughness of tungsten matrix. Riesch and Du et al. carried out extensive studies on small-size fabrication of tungsten fibre reinforced tungsten composites by chemical vapor infiltration or chemical vapor deposition [17–21]. Three-point bending and fibre pullout tests of the samples indicated that tungsten fibres could enhance the toughness of tungsten or tungsten alloy. However, on the economic aspect for low utilization rate of raw materials and harsh reaction conditions, chemical deposition method seems unfavorable. Composites of W_f/W have been prepared by powder metallurgical method, but the density of the materials needs to be improved [22,23] and other preparation methods need to be exploited.

In our previous work [16,24], the spark plasma sintering (SPS) method was exploited to prepare tungsten-fibre reinforced tungsten composites (W_f/W), and compact composites were obtained because of the unique features of SPS (large pulsed DC current provides fast heating/cooling rate and short consolidation time, which helps obtain high density and fine grains). However, the non-uniform temperature distribution limits the large-scale production of W_f/W composites. Except for SPS, the hot pressing (HP) method is also used to consolidate composite materials, and is able to provide more uniform heat distribution and enables preparation of large-size samples. In addition, cold rolling is a promising method to enhance material properties due to the porosity reduction and texture evolution in the rolling process [25], thus the cold rolling after HP (HPCR) may further improve the performance of the composites. In this work, different preparation methods based on SPS, HP and HPCR were exploited to prepare tungsten-fibre-net reinforced W_f/W samples. The influences of preparation method on the microstructure, Vickers hardness and tensile properties of the W_f/W composites were comparatively investigated.

2. Experimental Section

2.1. Preparation and Characterization

Commercial K-doped tungsten fibres with diameter of 150 μm were purchased from Honglu Molybdenum Company, Xiamen, China. The chemical composition of the fibres is given in our previous work [16,24]. The theoretical density of the W fibres can be considered as 19.3 g/cm^3, as same as the value of pure tungsten. Long tungsten fibres were woven into a net-like shape using a braider, as presented in Figure 1. The fibres along the YO direction act as supporting effect of the nets, while the fibres along the XO direction (which is also the tensile and rolling direction) will play the major role in improving the toughness. The diameter of the nets is about 20 mm for SPS as shown in Figure 1a and about 60 mm for HP and HPCR as shown in Figure 1b. The enlarged picture at the part I in Figure 1b is shown in Figure 1I for clarity. In this work, each sample contained four tungsten-fibre-net layers according to the optimization of fibre contents in our earlier work [24].

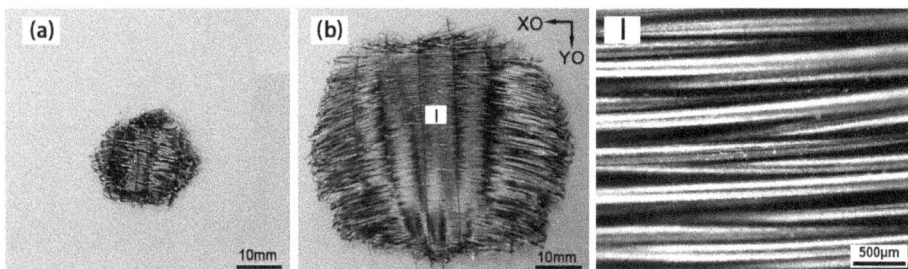

Figure 1. Photograph of a tungsten fibre net for SPSed samples (**a**), HPed and HPCRed samples (**b**), and the enlarged picture at the part I in b (**I**).

Pure tungsten powders (purity > 99.9%, average particle size 600 nm) were ball-milled for 4 h in a planetary ball mill in argon atmosphere to increase the activity with a powder-to-ball weight ratio of 1:8 and rotation speed of 240 rpm. During ball milling, tungsten carbide balls and mortars were used to minimize the possible contamination by impurities. Tungsten fibre nets were buried in the tungsten powders. To keep a roughly equal distance between the net layers (about 750 μm), equal amount of W powder was put into the space between the neighboring net layers. The fibre mass fraction in the composite was calculated just by weighting the fibres and matrix powders respectively before mixing.

The samples with the diameter of 20 mm were sintered by spark plasma sintering (SPS, FCT systeme GmbH, Rauenstein, Germany) according to the procedure as described in [26], which is shown in Figure 2a again for clarity, where the temperature was 1800 °C and the pressure was 47 MPa. The samples with the diameter of 60 mm were sintered in a hot-pressing vacuum furnace (HP, Shanghai Chen Xin Electric Furnace Co., Ltd., CXZT-60-23Y, Shanghai, China) with a heating or cooling rate of about 10 °C/min. The detailed temperature and pressure profile of the HP sintering program were illustrated in Figure 2b: (i) heated from room temperature to 700 °C and held for 60 min; (ii) heated to 1300 °C and held for 180 min; (iii) heated to the sintering temperature 1800 °C and held for 90 min; and (iv) cooled down to room temperature (RT). At the same time, the pressure in the graphite die was increased linearly to 56 MPa with a changing rate of about 0.28 MPa/min, held for 300 min and then decreased to zero. After that, The HPed samples were preheated in a high temperature furnace (KSL-1100X, Hefei Kejing Materials Technology Co., Ltd., Hefei, China) at 1100 °C for 5 min, and then rolled by a rolling machine (Wuxi Guancheng Machinery Co., Ltd., Wuxi, China) with the roller radius of 100 mm, the roller length of 200 mm and the rotating speed of 500 r/min [27]. The samples were also preheated between the successive rolling. The HPed plates were rolled three times with a total thickness reduction of 38%. Hereafter, the cold rolled HPed samples were abbreviated as HPCRed ones.

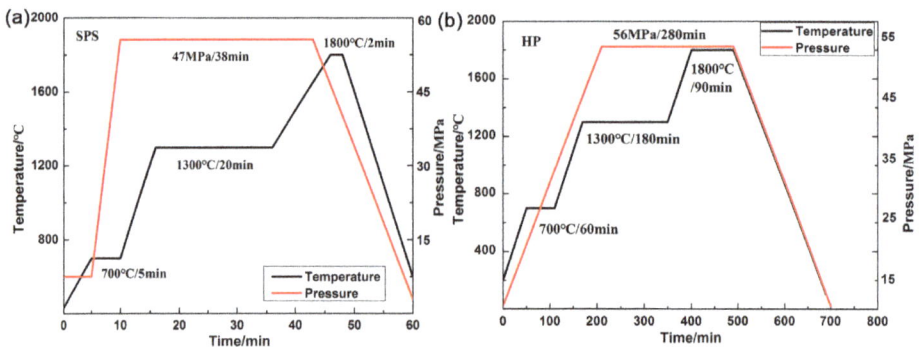

Figure 2. Temperature and pressure profile of SPS and HP procedure: (**a**) SPS and (**b**) HP.

2.2. Testing Procedure

The density of samples was determined by Archimedes principle. The sintered samples and the inlaid original wire were polished and then chemically etched with a 10% aqueous solution of $K_3Fe(CN)_6$ and NaOH. The distribution of the tungsten fibres in the composites was investigated by optical microscopy (ZEIZZ-Axio Scope.Al, Carl Zeiss AG, Oberkochen, Germany). The Vickers micro-hardness of the sample was tested by the Vickers Indenter (HV-1000 A, Laizhou ITC Test Instrument Co., Ltd, Laizhou, China) at room temperature with a load of 200 g and a dwell time of 10 s. The hardness of matrix was tested along the direction either perpendicular or parallel to the pressing direction, either close to or away from the fibre. Each sample was indented for 8 times in different locations and the average value was adopted.

In order to obtain information about the grain size, aspect ratio of grains, angles of grain boundary and anisotropy or isotropy of fibre and matrix in samples, the Electron Back Scattering Diffraction (EBSD) measurements were carried out. The samples for EBSD were mechanically polished and then electro polished by NaOH (2%) aqueous solution. The EBSD data was collected and analyzed using the software of HKL Tango (Version-2010, Oxford Instruments, Oxford, UK) with the resolution rate of 80–90% and a step from 0.25 to 0.4 μm depending on the grain size, and the raw data was presented. Microstructure of the samples and the fracture surfaces of the tensile tested samples were characterized with a field-emission scanning electron microscope (FESEM Sirion 200, FEI, Hillsboro, OR, USA).

The dog-bone-shaped tensile samples were cut along the XO direction with a working length of 5 mm as in [9] and a cross-section of 1.4×1.5, 1.4×2.2, 1.4×1.8 mm^2 for the SPS, HP, and HPCR samples, respectively. Tensile experiments were carried out at various temperatures using an Instron-5967 machine (Instron Corporation, Boston, MA, USA) with a constant displacement rate of 0.06 mm/min. In this work, the loading direction is parallel to the XO direction due to high mechanical performance [28], the number of tested samples was three to five.

3. Results and Discussion

3.1. Density and Vickers Hardness

The mass fraction of W fibres, relative density and Vickers hardness of the W_f/W composites are listed in Table 1. The estimated mass fraction of fibres in SPSed, HPed and HPCRed samples is 17.4%, 10.5% and 10.5%, respectively. As listed in Table 1, the relative densities of the SPSed (97.5%) and HPed composites (95.1%) are a little lower than that of the HPCRed samples (99.8%), which is close to 100% of pure dense tungsten. The overall average hardness of fibres in composites is approximately 537 ± 7 HV$_{0.2}$ for all three kinds of samples, which is about 100 HV$_{0.2}$ smaller than that of the original tungsten fibre (623.6 ± 11 HV$_{0.2}$) due to the grain growth of fibre in the sintering process [29], and more details will be analyzed in the following sections. Indicated by the lower density, the hardness of the matrix in the HPed samples (331.3 ± 8 HV$_{0.2}$) is lower than that of the SPSed (431.3 ± 8 HV$_{0.2}$) and HPCRed samples (488.2 ± 810 HV$_{0.2}$).

Table 1. Density, Mass fraction of fibres and Vickers micro-hardness of the fibre and matrix in the SPSed, HPed and HPCRed samples.

Different Samples	Mass Fraction of Fibre	Relative Density (%)	Vickers Hardness/HV$_{0.2}$	
			Fibre	Matrix
Original fibre	100%	100%	623.6 ± 11	-
SPSed	$17.4 \pm 0.1\%$	$97.5 \pm 0.3\%$	537.4 ± 9	431.3 ± 8
HPed	$10.5 \pm 0.1\%$	$95.1 \pm 0.2\%$	538.5 ± 7	331.9 ± 6
HPCRed	$10.5 \pm 0.1\%$	$99.8 \pm 0.1\%$	536.4 ± 5	476.4 ± 2

3.2. Distribution of Fibres

To investigate the distribution of the tungsten-fibre-nets in the SPSed, HPed and HPCRed samples, cross-sectional optical micrographs of the samples with a low magnification are shown in Figure 3. Due to different grain size and grain orientation, the fibres and matrix show different corrosion extent by the etching solution. All the darker areas indicate the tungsten fibres, where the wave strips are the fibres in YO orientation while the round circles represent fibres in XO orientation. From Figure 3a it can be seen that the fibres distribute neatly in the net layer of the SPSed samples, and the net layers are separated one by one with a nearly equal distance. In the HPed samples (Figure 3b), the fibre layers are also evenly and regularly aligned. After cold rolling, the HPCRed samples are much thinner so that a slight variation in the fibre distribution and orientations can be obtained from the tomographic visualization as shown in Figure 3c. Before the rolling phase of preparation, the fibres have a circular shape with an interlayer spacing of about 560 μm. After rolling, the fibres present elliptical shape with an interlayer spacing of about 350 μm. The 38% reduction in spacing is consistent with the thickness reduction of the whole sample.

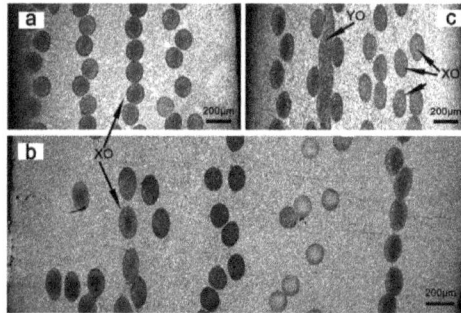

Figure 3. Optical micrographs of the W$_f$/W composites: (**a**) SPSed, (**b**) HPed, and (**c**) HPCRed.

The SEM images are presented in Figure 4 for (a) SPSed, (b) HPed and (c) HPCRed samples fractured at RT. It can be seen that there is no debonding seen in any of the samples between the tungsten fibre and the matrix. In addition, the cross section of fibre in the SPSed and HPed samples is almost circular, while that in the HPCRed samples is elliptical with an aspect ratio of about 5:2 owing to the rolling.

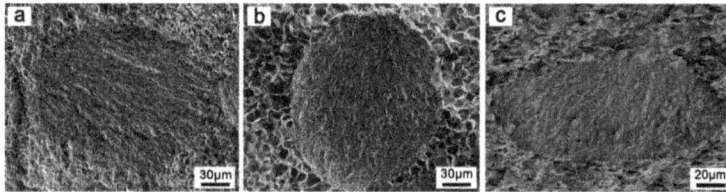

Figure 4. The Scanning Electron Microscope (SEM) images of the W$_f$/W composites: (**a**) SPSed, (**b**) HPed, and (**c**) HPCRed.

3.3. Microstructure of the Original and Sintered Fibres

Figure 5a shows the EBSD results of the original and sintered W fibres, where the black lines mean the high-angle grain boundaries ($\theta > 10°$) while the gray lines mean the low-angle grain boundaries ($\theta < 10°$). Figure 5(a1) shows the EBSD results of the original W fibres. The ZO direction is perpendicular to the XO and YO directions. The texture of the original tungsten fibre is <101> as shown in Figure 5(a2). Figure 5(a1) also indicates that the grains of the original tungsten fibre are tens of microns long in the XO direction and several microns wide in the YO and ZO directions, corresponding to a length/width aspect ratio of about 5.2:1 (as shown in Figure 5(a3)). It is interesting to note that the elongated mother-grains ($\theta > 10°$) are composed of fine equiaxed sub-grains ($\theta < 10°$) with the grain size less than 1 μm. In addition, the ratio of small angle grain boundaries in the original fibres is as high as 60%.

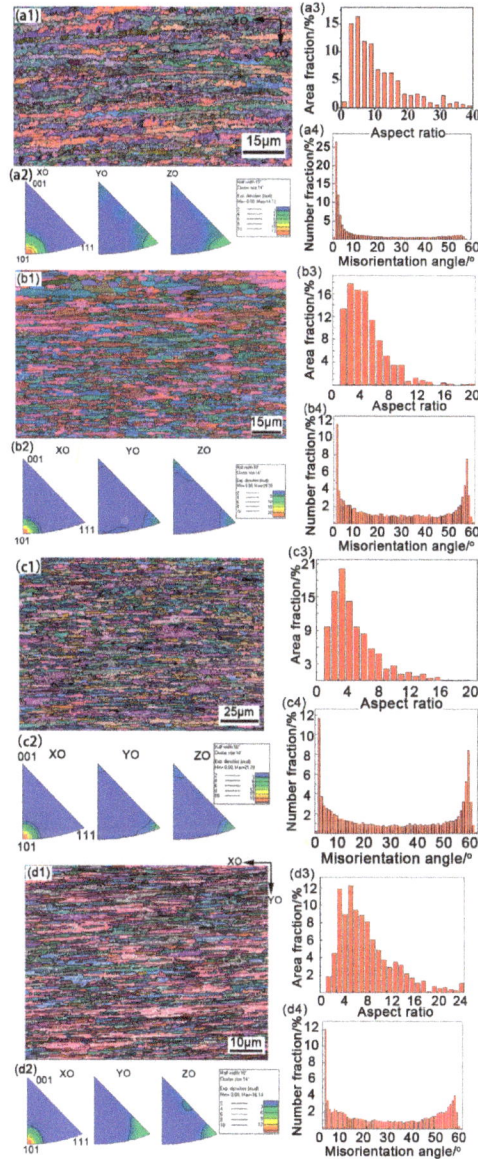

Figure 5. The Electron Backscattered Diffraction (EBSD) images of a fibre in original W fibres-**a**, SPSed-**b**, HPed-**c**, HPCRed-**d** samples: (**1**) grain boundary and Euler angle map, (**2**) the inverse pole figures, (**3**) length-to-diameter ratio, and (**4**) grain boundary misorientation map.

It is well known that the mechanical properties are determined by the microstructure of materials, so it is necessary to investigate the microstructure of a fibre and matrix in the sintered samples. The microstructure of fibres in the SPSed, HPed and HPCRed samples is clearly demonstrated by the EBSD results as shown in Figure 5b–d, respectively. According to the deformation texture analysis, the grain width at the edge of the fibre in SPSed samples is a little larger than that in the central, as shown in Figure 5(b1). However, the texture of fibre in SPSed samples is <101>, which is consistent with the original fibre, as shown in Figure 5(b2). On the other hand, the average aspect ratio of grains with misorientation angles $\theta > 10°$ is about 3.7:1 (Figure 5(b3)). From the grain boundary misorientation map shown in Figure 5(b4), it can be noted that the percentage of grains with misorientation angle <10° decreases from the value of 60% in the original fibre to 31.5%. The elimination of some small angle grain boundaries indicates that partial recrystallization occurred in the SPS sintering process. For the HPed samples the texture of fibre is also <101>, as shown in Figure 5(c1,c3). However, the average aspect ratio of grains with misorientation angles $\theta > 10°$ is about 3.2:1, as shown in Figure 5(c3), which is a little smaller than that in the SPSed samples. The percentage of grains with misorientation angles $\theta < 10°$ decreases to 26.8%, indicating that partial recrystallization occurred in the HP sintering process more severely than in the SPS process. For the HPCRed samples, however, the texture density of fibres in the XO direction decreases, and the orientation disperses along <111> in both the YO and ZO directions as shown in Figure 5(d1,d2). According to Figure 5(d3), the grains become longer and the average aspect ratio is about 5.8:1. Furthermore, the proportion of grains with $\theta < 10°$ is increased to 31.6% (Figure 5(d4)). All the variation of the grains in the fibres could be attributed to the rolling process.

As well known, high temperature annealing can result in reduction of dislocation density and elimination of some low-angle grain boundaries [30]. Therefore, during the high temperature sintering, stress in the grain of fibre is gradually released and the number of small angle grain boundaries decreases due to grain growth and coalesce. As a result, the percentage of grains with misorientation angle <10° decreases from the value of 60% in the original fibre to 31.5% in SPSed and 26.8% in the HPed samples. However the grains are still filiform and the texture of fibre in both SPSed and HPed samples is almost same. On the other hand, the cold rolling exhibits the opposite effect to sintering, and after rolling the percentage of grains with misorientation angle <10° increases from 26.8% in the HPed samples to 31.6% in the HPCRed samples. Meanwhile, pressure exerted in the ZO direction could result in grain elongation along the XO direction when sintering. This is why the average length-to-diameter ratio of the grain increased from 5.2:1 in the original fibres to 3.7:1, 3.2:1, and 5.8:1 in the fibres of the SPSed, HPed, and HPCRed samples, respectively.

3.4. Grains of Matrix

In order to research the influence of tungsten fibres on the size and misorientation angle of grains in matrix, the EBSD results of the matrix area near and far away from the tungsten fibres in the SPSed, HPed and HPCRed samples were studied. Figure 6 shows the EBSD results of the tungsten matrix near and far away from a fibre in the SPSed samples. There is no preferred orientation of tungsten grains as shown in the grain boundary—inverse pole figure map (Figure 6(a1)). The average size of grains with misorientation angles $\theta > 10°$ is 4.3 μm which is calculated from the distribution range from about 2 to 10 μm (as shown in Figure 6(a2)), while the grains with small misorientation angle ($\theta < 10°$) take a very small proportion (6.8%). As a contrast, Figure 6(b1) shows the EBSD results of the W matrix far away from the fibre in the SPSed samples. There is also no preferred orientation of tungsten grain. However, the average size of grains with misorientation angles $\theta > 10°$ and the percentage of grains with a small misorientation angle ($\theta < 10°$) decreases to 3.6 μm and 4.2%, respectively.

For the HPed samples, the EBSD results of W matrix near and far away from tungsten fibre were shown in Figure 7. Comparing Figure 7(a1) with Figure 7(b1), there is no obvious texture, but the grain size is much different. The average size of grains with large misorientation angle ($\theta > 10°$) is 8.1 μm near the tungsten fibre and 11.4 μm far away from the fibre. In addition, the percentage of grains with small misorientation angle ($\theta < 10°$) near and far away from tungsten fibre in the HPed samples is 7.8% and 12.0%, respectively.

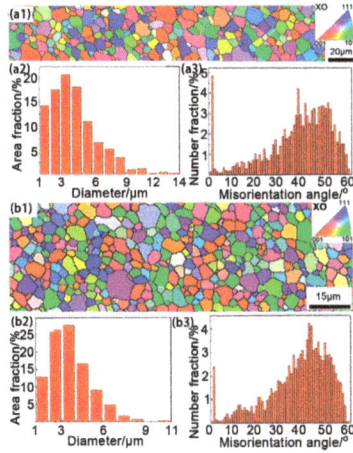

Figure 6. EBSD images of tungsten matrix near the fibre-**a**, far away from the fibre-**b** in SPSed samples: (**1**) grain boundary and IPF map (inset), (**2**) grain size, and (**3**) grain boundary misorientation map.

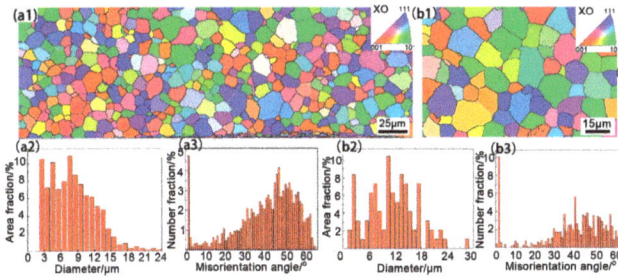

Figure 7. EBSD images of tungsten matrix near the fibre-**a**, far away from the fibre-**b** in HPed samples: (**1**) grain boundary and IPF map (inset), (**2**) grain size, and (**3**) grain boundary misorientation map.

Figure 8 shows the EBSD results of tungsten matrix near and far away from tungsten fibre in the HPCRed samples, which is indicated by white (high-angle grain boundaries) and black lines (low-angle grain boundaries). Intuitively, tungsten grains in the matrix are elongated along the rolling direction in the HPCRed samples and the average grain length/width ratio is about 2.6:1 near the fibre (see Figure 8(a2)) and 2.1:1 away from the fibre (see Figure 8(b2)). There is also a large number of small angle grain boundaries and the proportion of grains with small misorientation angles ($\theta < 10°$) is about 72.7% near the fibre (see Figure 8(a3)) and 78.3% away from the fibre (see Figure 8(b3)).

To sum up, the EBSD results of all samples were shown in Table 2. It can be seen that the W fibres embedded in the W matrix have great influence on the microstructure of the tungsten matrix during different sample preparation. In the SPS sintering process, because of the non-uniform current distribution and the higher density of fibre than matrix, the temperature is higher in and near the fibre, which can lead to grain growth in the fibre as shown in Figure 6(a1), and result in the lower proportion of small angle grain boundaries ($\theta < 10°$) and the larger grain size near the fibre. However, the exposure time to high temperature is not long enough to make the grains change from the strip-like to the equiaxed in tungsten fibres. In the HP sintering process, although the temperature is more uniform, the long exposure time to high temperature can result in a lower percentage of small angle grain boundaries in the fibre and the larger grain size in the matrix owing to the recrystallization.

In the HPCRed samples the plenty of small angle boundaries in the matrix both near and far away from the fibre and the small grain size of W matrix can be attributed to the severe plastic deformation and dynamic recrystallization during rolling, as in the cold rolled pure W samples [31].

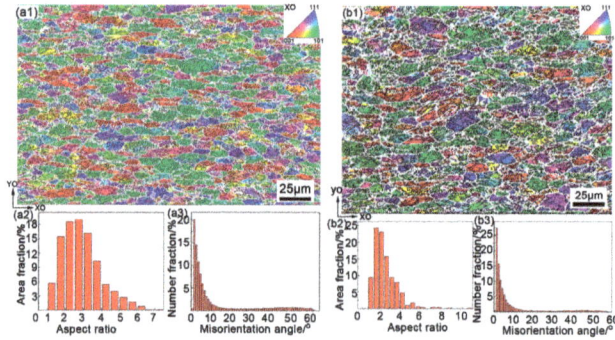

Figure 8. EBSD images of tungsten matrix near the fibre-**a**, far away from the fibre-**b** in HPCRed samples: (**1**) grain boundary and IPF map, (**2**) length-diameter ratio, and (**3**) grain boundary misorientation map.

Table 2. Small angle grain boundaries ($\theta < 10°$) and grain size of the original fibre, the fibre and matrix in SPSed, HPed and HPCRed samples.

Fibre	The Share of Small Angle Grain Boundaries $\theta < 10°$/%			Average Length-Diameter Ratio		
	60.1 ± 0.2			5.2 ± 0.1		
	Fibre	Near the Fibre	Away from the Fibre	Fibre	Near the Fibre	Away from the Fibre
SPSed	31.5 ± 0.1	6.8 ± 0.2	4.2 ± 0.1	3.7 ± 0.1	4.3 ± 0.3 μm (grain size)	3.6 ± 0.2 μm (grain size)
HPed	26.8 ± 0.2	7.8 ± 0.1	12.0 ± 0.2	3.2 ± 0.1	8.1 ± 0.2 μm (grain size)	11.4 ± 0.2 μm (grain size)
HPCRed	31.6 ± 0.3	72.7 ± 0.3	78.3 ± 0.2	5.8 ± 0.2	2.6 ± 0.1	2.1 ± 0.2

3.5. Tensile Properties and Fracture Microstructure

Tensile tests were performed at different temperatures for the SPSed, HPed and HPCRed samples, and the curves of engineering stress versus strain were shown in Figure 9. From such curves the average tensile strength (TS) and total elongation at break (TE) at different temperatures can be obtained. For the SPSed and HPed samples, there is almost no plastic deformation at 500 °C. When tested at 600 °C, the TS values of SPSed and HPed samples are 536 and 425 MPa, which are higher than that of pure tungsten [32], while the TE values are 11.6% and 23.0%, respectively. Because the recrystallization occurred in the HP sintering process is more severe than in the SPS process as shown in Figures 6 and 7, the smaller grain size in the matrix can enhance the TS at comparative temperature, and the SPSed samples with the smaller grain size exhibit higher TS. As for the HPCRed samples, it exhibits almost no plastic deformation at 300 °C, but the TS at 300 °C is as high as 816 MPa. When tested at 400 °C, the HPCRed samples undergo observable plastic deformation with a TE of 8.4% and a TS of 784 MPa. At 500 °C, the TE further increase up to 17.5% and the TS reduces to 750 MPa. The fracture energy of the HPCRed samples is obviously larger than that of the SPSed and HPed samples, because the proportion of small angle grain boundaries is higher than others in both the fibres and matrix. The mechanical properties of the composites were studied preliminary, and other methods will be used to improve that.

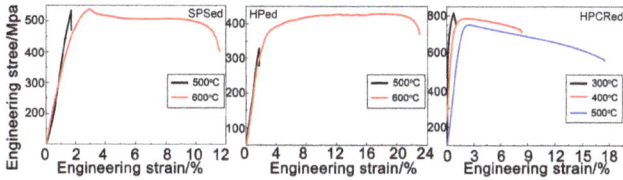

Figure 9. Tensile behavior of the W_f/W composites prepared with different methods between 300 °C and 600 °C.

The ductile fracture surfaces of the SPSed, HPed and HPCRed tensile tested specimens were shown in Figure 10. The SPSed (Figure 10a) and HPed samples (Figure 10b) show fibre cracking and necking at 600 °C, which can be seen more clearly from the insets. In the HPCRed samples, fibre crack can be identified in Figure 10(c2) at 400 °C and Figure 10(d2) at 500 °C, whereas the matrix shows intergranular fracture behavior in both cases. This suggests that the enhanced tensile strength of the HPCRed samples over the SPSed and HPed samples can be mainly attributed to the elongated grains. However, a necking behavior instead of a plastic plateau after reaching the ultimate stress makes the HPCR method a bit defective, and whether a modified HPCR method is satisfied to obtain compact and tough W_f/W composites needs more research.

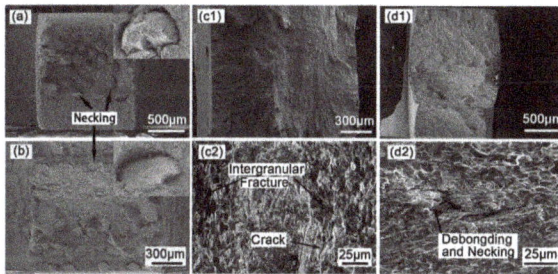

Figure 10. SEM micrographs of fracture surface in W_f/W composites tensile-tested at different temperatures: (**a**) SPS-600 °C, (**b**) HP-600 °C, (**c**) HPCR-400 °C, and (**d**) HPCR-500 °C. The insets on the top right corner of (**a**,**b**) are the enlarged views over a single fibre.

4. Conclusions

Tungsten-fibre-net-reinforced tungsten composites (W_f/W) containing four layers of nets with enhanced fracture energy were synthesized by three different methods including spark plasma sintering (SPS), hot pressing (HP), and cold rolling after hot pressing (HPCR). The total thickness of the SPSed, HPed and HPCRed samples were 2.18, 3.52 and 2.08 mm, respectively, with corresponding fibre mass fraction of 17.4%, 10.5% and 10.5%, respectively. The microstructure, tensile property and fibre texture of these samples were investigated. The main results derived from such investigations can be concluded as follows:

The relative density of all samples was above 95.10%, while the highest relative density of 99.80% is reached in the HPCRed samples. The hardness of the sintered fibres in all samples is around 537 $HV_{0.2}$ which is smaller than the value of the original fibres (about 624 $HV_{0.2}$). The proportion of grains with low misorientation angles <10° decreased from the value of 60% in the original fibre to 27–32%. However, the texture of fibres is not significantly affected by the high temperature (1800 °C) sintering, which is <101> in fibre direction. These results indicate that partial recrystallization of fibres occurred at the 1800 °C sintering processes. To avoid such partial recrystallization of fibres lower sintering temperature should be adopted.

The SPSed and HPed W_f/W composites begin to exhibit ductile behavior at 600 °C. The tensile strength of the SPSed and HPed samples at 600 °C is 536 and 425 MPa, and the total elongation at break is 11.6% and 23.0%, respectively. The HPCRed W_f/W composites however, begin to exhibit plastic deformation at 400 °C, and the TS and TE are 784 MPa and 8.4%, respectively. The enhanced mechanical performance of the W_f/W composites can be attributed to the necking, cracking, and debonding of fibres.

Acknowledgments: This work was supported by the National Magnetic Confinement Fusion Program (Grant No. 2015GB112000), and the National Natural Science Foundation of China (Grant Nos. 51301164, 11274305, 11475216, 11674319, 51601189), and the Anhui Provincial Natural Science Foundation of China (1408085QE77).

Author Contributions: L.Z. and Y.J. carried out all the experiments and prepared the draft of the manuscript. Q.F. and X.W. suggested the original ideas and modified the manuscript. R.L., Z.X., T.Z. and C.L. gave suggestions in the discussion.

Conflicts of Interest: The authors declare no conflict of interest.

References

1. Chong, F. Feasibility of employing tungsten for plasma facing material in fusion devices. *Mater. Rev.* **2016**, *30*, 39–41.
2. Hao, J.; Shu, X.; Jin, S.; Zhang, X.; Zhang, Y.; Lu, G.-H. A comparison of interatomic potentials for modeling tungsten nanocluster structures. *Nucl. Instrum. Methods Phys. Res. Sect. B Beam Interact. Mater. Atoms* **2017**, *393*, 180–185. [CrossRef]
3. Wang, K.; Bannister, M.E.; Meyer, F.W.; Parish, C.M. Effect of starting microstructure on helium plasma-materials interaction in tungsten. *Acta Mater.* **2017**, *124*, 556–567. [CrossRef]
4. Hu, X.; Koyanagi, T.; Fukuda, M.; Kumar, N.A.P.K.; Snead, L.L.; Wirth, B.D.; Katoh, Y. Irradiation hardening of pure tungsten exposed to neutron irradiation. *J. Nucl. Mater.* **2016**, *480*, 235–243. [CrossRef]
5. Khan, A.; Elliman, R.; Corr, C.; Lim, J.J.H.; Forrest, A.; Mummery, P.; Evans, L.M. Effect of rhenium irradiations on the mechanical properties of tungsten for nuclear fusion applications. *J. Nucl. Mater.* **2016**, *477*, 42–49. [CrossRef]
6. Van Renterghem, W.; Uytdenhouwen, I. Investigation of the combined effect of neutron irradiation and electron beam exposure on pure tungsten. *J. Nucl. Mater.* **2016**, *477*, 77–84. [CrossRef]
7. Cui, K.; Shen, Y.Z.; Yu, J.; Ji, B. Microstructural characteristics of commercial purity W and W-1% La_2O_3 alloy. *Int. J. Refract. Met. Hard Mater.* **2013**, *41*, 143–151. [CrossRef]
8. Liu, R.; Xie, Z.M.; Hao, T.; Zhou, Y.; Wang, X.P.; Fang, Q.F.; Liu, C.S. Fabricating high performance tungsten alloys through zirconium micro-alloying and nano-sized yttria dispersion strengthening. *J. Nucl. Mater.* **2014**, *451*, 35–39. [CrossRef]
9. Miao, S.; Xie, Z.M.; Yang, X.D.; Liu, R.; Gao, R.; Zhang, T.; Wang, X.P.; Fang, Q.F.; Liu, C.S.; Luo, G.N.; et al. Effect of hot rolling and annealing on the mechanical properties and thermal conductivity of W-0.5 wt. % TaC alloys. *Int. J. Refract. Met. Hard Mater.* **2016**, *56*, 8–17. [CrossRef]
10. Xie, Z.M.; Liu, R.; Miao, S.; Yang, X.D.; Zhang, T.; Fang, Q.F.; Wang, X.P.; Liu, C.S.; Lian, Y.Y.; Liu, X.; et al. High thermal shock resistance of the hot rolled and swaged bulk W-ZrC alloys. *J. Nucl. Mater.* **2016**, *469*, 209–216. [CrossRef]
11. Kanerva, M.; Johansson, L.S.; Campbell, J.M.; Revitzer, H.; Sarlin, E.; Brander, T.; Saarela, O. Hydrofluoric-nitric-sulphuric-acid surface treatment of tungsten for carbon fibre-reinforced composite hybrids in space applications. *Appl. Surf. Sci.* **2015**, *328*, 418–427. [CrossRef]
12. Shi, X.L.; Wang, M.; Zhang, S.; Zhang, Q.X. Fabrication and properties of W-20Cu alloy reinforced by titanium nitride coated SiC fibers. *Int. J. Refract. Met Hard Mater.* **2013**, *41*, 60–65. [CrossRef]
13. Guo, C.Y.; Zhang, C.B.; He, L.L.; Jin, B.H.; Shi, N.L. Microstructure characterization of long W core SiC fiber. *J. Mater. Sci. Technol.* **2007**, *23*, 677–684.
14. Livramento, V.; Nunes, D.; Correia, J.; Carvalho, P.; Mateus, R.; Hanada, K.; Shohoji, N.; Fernandes, H.; Silva, C.; Alves, E. Tungsten-tantalum composites for plasma facing components. *Mater. Energy* **2010**, 4–8.

15. Riesch, J.; Buffiere, J.Y.; Hoschen, T.; di Michiel, M.; Scheel, M.; Linsmeier, C.; You, J.H. In situ synchrotron tomography estimation of toughening effect by semi-ductile fibre reinforcement in a tungsten-fibre-reinforced tungsten composite system. *Acta Mater.* **2013**, *61*, 7060–7071. [CrossRef]

16. Jiang, Y.; Zhang, L.H.; Fang, Q.F.; Zhang, T.; Wang, X.P.; Hao, T.; Liu, C.S. Toughness enhancement of tungsten reinforced with short tungsten fibres. *Mater. Sci. Eng. A* **2017**, *690*, 208–213. [CrossRef]

17. Riesch, J.; Hoschen, T.; Linsmeier, C.; Wurster, S.; You, J.H. Enhanced toughness and stable crack propagation in a novel tungsten fibre-reinforced tungsten composite produced by chemical vapor infiltration. *Phys. Scr.* **2014**, *T159*, 1–8. [CrossRef]

18. Riesch, J.; Aumann, M.; Coenen, J.W.; Gietl, H.; Holzner, G.; Höschen, T.; Huber, P.; Li, M.; Linsmeier, C.; Neu, R. Chemically deposited tungsten fibre-reinforced tungsten—The way to a mock-up for divertor applications. *Nucl. Mater. Energy* **2016**, *9*, 75–83. [CrossRef]

19. Du, J.; Hoschen, T.; Rasinski, M.; You, J.H. Interfacial fracture behavior of tungsten wire/tungsten matrix composites with copper-coated interfaces. *Mater. Sci. Eng. A Struct. Mater. Prop. Microstruct. Process.* **2010**, *527*, 1623–1629. [CrossRef]

20. Du, J.; Hoschen, T.; Rasinski, M.; Wurster, S.; Grosinger, W.; You, J.H. Feasibility study of a tungsten wire-reinforced tungsten matrix composite with ZrO_x interfacial coatings. *Compos. Sci. Technol.* **2010**, *70*, 1482–1489. [CrossRef]

21. Du, J.; Hoeschen, T.; Rasinski, M.; You, J.H. Shear debonding behavior of a carbon-coated interface in a tungsten fiber-reinforced tungsten matrix composite. *J. Nucl. Mater.* **2011**, *417*, 472–476. [CrossRef]

22. Jasper, B.; Coenen, J.W.; Riesch, J.; Höschen, T.; Bram, M.; Linsmeier, C. Powder metallurgical tungsten fiber-reinforced tungsten. *Mater. Sci. Forum* **2015**, *825–826*, 125–133. [CrossRef]

23. Coenen, J.W.; Mao, Y.; Almanstötter, J.; Calvo, A.; Sistla, S.; Gietl, H.; Jasper, B.; Riesch, J.; Rieth, M.; Pintsuk, G.; et al. Advanced materials for a damage resilient divertor concept for DEMO: Powder-metallurgical tungsten-fibre reinforced tungsten. *Fusion Eng. Des.* **2016**. [CrossRef]

24. Zhang, L.H.; Jiang, Y.; Fang, Q.F.; Zhang, T.; Wang, X.P.; Liu, C.S. Toughness and microstructure of tungsten fibre net-reinforced tungsten composite produced by spark plasma sintering. *Mater. Sci. Eng. A Struct. Mater. Prop. Microstruct. Process.* **2016**, *659*, 29–36. [CrossRef]

25. Huang, Q.Y.; Tang, A.T.; Ma, S.D.; Pan, H.C.; Song, B.; Gao, Z.Y.; Rashad, M.; Pan, F.S. Enhancing thermal conductivity of Mg-Sn alloy sheet by cold rolling and aging. *J. Mater. Eng. Perform.* **2016**, *25*, 2356–2363. [CrossRef]

26. Xie, Z.M.; Liu, R.; Fang, Q.F.; Zhou, Y.; Wang, X.P.; Liu, C.S. Spark plasma sintering and mechanical properties of zirconium micro-alloyed tungsten. *J. Nucl. Mater.* **2014**, *444*, 175–180. [CrossRef]

27. Zeng, L.F.; Gao, R.; Fang, Q.F.; Wang, X.P.; Xie, Z.M.; Miao, S.; Hao, T.; Zhang, T. High strength and thermal stability of bulk Cu/Ta nanolamellar multilayers fabricated by cross accumulative roll bonding. *Acta Mater.* **2016**, *110*, 341–351. [CrossRef]

28. Zhang, B.; Fu, H.M.; Sha, P.F.; Zhu, Z.W.; Dong, C.; Zhang, H.F.; Hu, Z.Q. Anisotropic compressive deformation behaviors of tungsten fiber reinforced Zr-based metallic glass composites. *Mater. Sci. Eng. A Struct. Mater. Prop. Microstruct. Process.* **2013**, *566*, 16–21. [CrossRef]

29. Chen, Z.C.; Zuo, T.Y.; Zhou, M.L. The recrystallization mechanism of doped tungsten wire. *J. Mater. Sci. Lett.* **1990**, *9*, 782–784.

30. Xia, F.; Wang, M.; Li, Z.; Wei, H.; Chen, C.; Jia, Y.; Lei, Q. Recrystallization and texture of tungsten strip. *J. Cent. South Univ. Sci. Technol.* **2013**, *44*, 1359–1366.

31. Reiser, J.; Hoffmann, J.; Jantsch, U.; Klimenkov, M.; Bonk, S.; Bonnekoh, C.; Rieth, M.; Hoffmann, A.; Mrotzek, T. Ductilisation of tungsten (W): On the shift of the brittle-to-ductile transition (BDT) to lower temperatures through cold rolling. *Int. J. Refract. Met. Hard Mater.* **2016**, *54*, 351–369. [CrossRef]

32. Xie, Z.M.; Zhang, T.; Liu, R.; Fang, Q.F.; Miao, S.; Wang, X.P.; Liu, C.S. Grain growth behavior and mechanical properties of zirconium microalloyed and nano-size zirconium carbide dispersion strengthened tungsten alloys. *Int. J. Refract. Met. Hard Mater.* **2015**, *51*, 180–187. [CrossRef]

metals

MDPI

Article

Aluminum and Nickel Matrix Composites Reinforced by CNTs: Dispersion/Mixture by Ultrasonication

Sónia Simões [1,*], Filomena Viana [1], Marcos A. L. Reis [2] and Manuel F. Vieira [1]

[1] CEMMPRE, Department of Metallurgical and Materials Engineering, University of Porto, R. Dr. Roberto Frias, 4200-465 Porto, Portugal; fviana@fe.up.pt (F.V.); mvieira@fe.up.pt (M.F.V.)
[2] Faculdade de Ciências Exatas e Tecnologia, Universidade Federal do Pará, Abaetetuba, PA 68440-000, Brazil; marcosallan@ufpa.br
* Correspondence: ssimoes@fe.up.pt; Tel.: +35-122-041-3113

Received: 21 June 2017; Accepted: 18 July 2017; Published: 22 July 2017

Abstract: The main challenge in the production of metal matrix composites reinforced by carbon nanotubes (CNTs) is the development of a manufacturing process ensuring the dispersion of nanoparticles without damaging them, and the formation of a strong bond with the metallic matrix to achieve an effective load transfer, so that the maximum reinforcement effect of CNTs will be accomplished. This research focuses on the production by powder metallurgy of aluminum and nickel matrix composites reinforced by CNTs, using ultrasonication as the dispersion and mixture process. Microstructural characterization of nanocomposites was performed by optical microscopy (OM), scanning and transmission electron microscopy (SEM and TEM), electron backscattered diffraction (EBSD) and high-resolution transmission electron microscopy (HRTEM). Microstructural characterization revealed that the use of ultrasonication as the dispersion and mixture process in the production of Al/CNT and Ni/CNT nanocomposites promoted the dispersion and embedding of individual CNT in the metallic matrices. CNT clusters at grain boundary junctions were also observed. The strengthening effect of the CNTs is shown by the increase in hardness for all nanocomposites. The highest hardness values were observed for Al/CNT and Ni/CNT nanocomposites, with a 1.00 vol % CNTs.

Keywords: metal matrix composites; nickel; aluminum; carbon nanotubes; powder metallurgy; ultrasonication; microstructural characterization

1. Introduction

Metal matrix composites demonstrate a range of fascinating properties, especially high mechanical ones, including high strength and stiffness, a desirable coefficient of thermal expansion and good damping properties [1,2]. Several investigations have been conducted into the development of metal matrix composites with different types of nanometric reinforcing materials. Ceramic nanoparticles, carbon nanotubes (CNTs) and graphene have been reported as the most promising reinforcements [3–5].

Among these, the CNTs stand out for their extraordinary properties, making them highly attractive for use as reinforcements. CNTs have been used to reinforce several metallic matrices, such as aluminum [6,7], nickel [8,9], copper [10,11], titanium [12] and magnesium [13,14], due to their extraordinary mechanical properties and excellent electrical conductivity [15,16].

The successful development of a production process that promotes a uniform and dense dispersion of CNTs in the matrix, without damaging them, is essential for obtaining the expected strengthening of nanocomposites. An effective load transfer from the metallic matrices to the CNTs is the key for the successful production of these nanocomposites. Several techniques have been suggested as adequate for this: conventional sintering [12,17], hot pressing [18,19], hot extrusion [20,21], thermal spraying [22,23] and electrodeposition [24,25], which appear to be the most promising. The possibility of producing these nanocomposites by powder metallurgy routes is quite interesting, since it allows the production of dense, near net shape components with complex geometries. However, the sintering temperatures for some of these matrices, such as nickel, copper and titanium, can induce high CNT damage and thus compromise the mechanical properties.

Several methods have been reported to obtain a uniform dispersion of CNTs through the matrix, with the mechanical ones, such as ball milling or sonication, being the most common [26–31]. In the ball milling method, a large amount of energy is involved, since the dispersion is achieved by collisions of dense and rigid balls with the CNTs. Due to this high-energy milling, CNTs suffer damage, which causes a degradation of their properties. However, this method performed with lower energy and shorter times can be used with the aim of reducing the length of the nanotubes [26]. The ultrasonication method involves the dispersion of the CNTs in a liquid, using ultrasound energy. It is a very efficient method for obtaining untangled CNTs dispersed in liquids, such as ethanol, isopropanol, ethylene glycol or acids [27]. The efficiency of this dispersion technique depends on the liquid, ultrasound energy, time and type of CNTs; the time required for dispersion is a crucial factor, since a considerable length of time leads to CNT damage [29,30].

In previous studies [29–31], ultrasonication was used to disperse and mix the metallic powders with CNTs, which is an effective dispersion/mixing process causing only a small amount of damage to the CNT structure. However, the studies on dispersion methods already published focus on a specific matrix, and it is difficult to generalize to matrices with a different composition or particle size. In this context, the main aim of this research is to extend the dispersion/mixture process already used on the Al/CNT nanocomposites to nickel matrix nanocomposites. Nickel is a high sintering temperature metal with mechanical strength higher than aluminum, which will allow testing the effect of the manufacture conditions on CNTs' damage. Also, nickel does not form carbides by reacting with CNT as it was observed in Al/CNT nanocomposites.

For the evaluation of the effects of this process on the production of the nanocomposites, it is essential to understand the relationship between the microstructure and the mechanical properties. Microstructural characterization was performed by optical microscopy (OM), scanning and transmission electron microscopies (SEM and TEM), electron backscattered diffraction (EBSD) and high-resolution transmission electron microscopy (HRTEM); mechanical properties evaluation was undertaken by Vickers microhardness tests.

2. Materials and Methods

2.1. Materials

CNTs used in this investigation (from Fibermax Nanocomposites) are multi-walled carbon nanotubes (MWCNTs). Figure 1 shows the morphology and structure of the as-received CNTs. The MWCNTs exhibit inner and outer diameters of 5 ± 1 and 19 ± 6 nm respectively. As-received CNTs were entangled and had a large aspect ratio (Figure 1a). In order to achieve good dispersion in the production of the nanocomposites, the untangling of CNTs is essential before they are mixed with the metallic powders.

A more detailed observation of MWCNT morphology is presented in the HRTEM image of Figure 1b. The image shows an MWCNT with 20 walls. These walls show few defects; however, it is possible to observe the bamboo-type structure characteristic of MWCNTs produced by chemical vapor deposition.

The aluminum and nickel powders (from Goodfellow) have a maximum particle size of 65 μm and 60 μm respectively, and a purity of 99.5%. Figure 2 shows the SEM images of as-received aluminum and nickel powders and the distributions of the particle sizes.

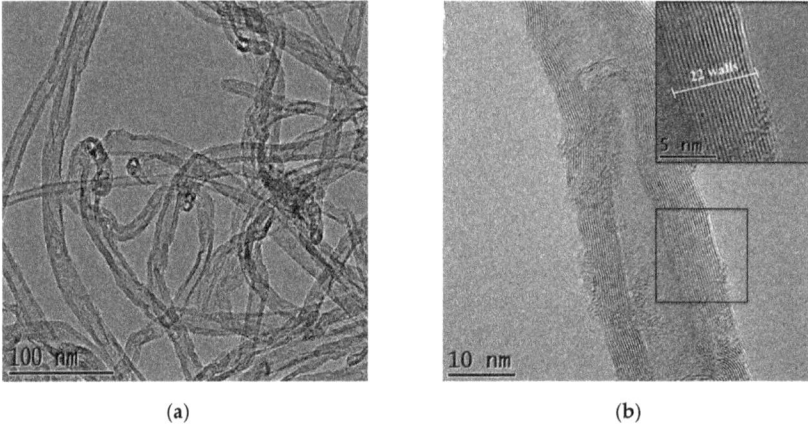

(a)   (b)

Figure 1. (**a**) TEM image of as-received carbon nanotubes (CNTs), and (**b**) high-resolution transmission electron microscopy (HRTEM) image of a CNT showing the morphology and structure of the multi-walled carbon nanotubes (MWCNTs) as well as detail showing the number of walls measured for this CNT.

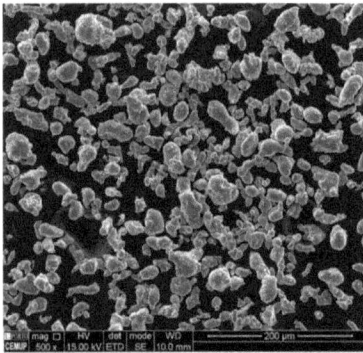

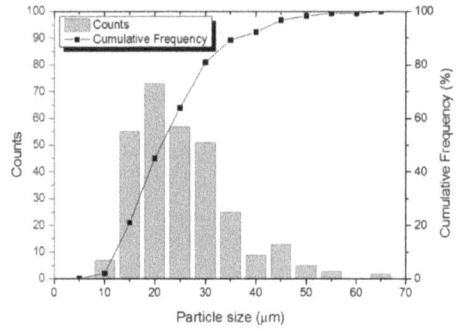

(a)   (b)

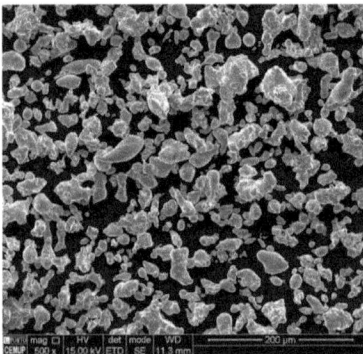

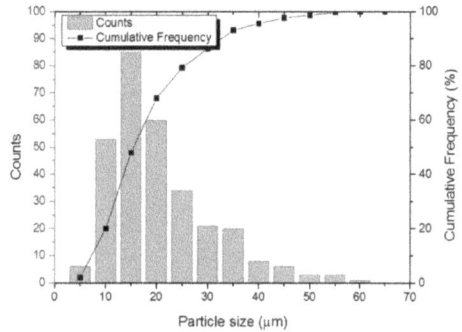

(c)   (d)

Figure 2. SEM images and particle size distributions of: (**a**) and (**b**) aluminum and (**c**) and (**d**) nickel powders.

2.2. Production of Nanocomposites

To produce the nanocomposites, the MWCNTs were dispersed and mixed with metallic powders in isopropanol using an ultrasonicator for 15 min, as described in [29,30]. The mixtures were dried and uniaxially pressed with 300 MPa for Al/CNTs and 900 MPa for Ni/CNTs. Compacts of 6 mm diameter and 2 mm thickness were produced with different content of CNTs (ranging from 0.50 to 2.00 vol %) and pressureless sintered under a vacuum better than 10^{-2} Pa. The sintering was performed at 640 °C with a dwell time of 90 min for Al/CNTs and at 950 °C with a dwell time of 120 min for Ni/CNT nanocomposites.

2.3. Nanocomposites Characterization

2.3.1. Microstructural Characterization

Microstructural characterization of cross-sections perpendicular to the compaction direction of the nanocomposites was performed by optical microscopy (OM) (DM4000, Leica Microsystems, Wetzlar, Germany), scanning and transmission electron microscopies (SEM and TEM), selected area electron diffraction (SAED), electron backscattered diffraction (EBSD), high-resolution transmission electron microscopy (HRTEM) and fast Fourier transform (FFT) analysis. A high-resolution FEI QUANTA 400 FEG SEM (FEI Company, Hillsboro, OR, USA), a JEOL JEM 2010F (JEOL Ltd., Tokyo, Japan) and an FEI Tecnai G2 (FEI Company, Hillsboro, OR, USA) were used for this purpose. Electron transparent cross sections of nanocomposites were prepared by the lift-out technique at 5–30 keV, using the focused ion beam (FIB) (FEI FIB200, FEI Company, Hillsboro, OR, USA).

The distribution and size of CNT clusters through the matrices and the composite average grain size were evaluated using an OM and Leica Application Suite software (Leica Microsystems, Wetzlar, Germany). The dispersion of the CNT clusters was assessed by measuring the number and average diameter of the clusters in five fields of 489×653 μm^2.

2.3.2. Mechanical Characterization

Mechanical characterization was performed by microhardness tests. The hardness was evaluated by Vickers microhardness using a 98 mN load (Duramin-1, Struers A/S, Ballerup, Denmark); ten tests were performed on cross-sections of each sample.

Al/CNT and Ni/CNT nanocomposites and aluminum and nickel samples produced under the same processing conditions were also tested.

3. Results and Discussion

3.1. Microstructural Characterization of Nanocomposites

Al/CNT and Ni/CNT nanocomposites were produced by powder metallurgy using ultrasonication as the dispersion/mixture process. Different CNT contents were used to evaluate the conditions leading to a higher strengthening by CNT reinforcement. Figure 3 shows the microstructures of the nanocomposites produced with 1.00 vol % of CNTs. From the OM images of Figure 3, it is observed that the microstructures of the Al/CNT and Ni/CNT nanocomposites are very similar. In these OM images, it is possible to observe that equiaxied grains characterize the nanocomposites; at some grain boundary junctions, dark areas are observed that are larger than typical grain boundaries. By SEM analysis (Figure 4) it is clear that these zones correspond to CNT clusters.

TEM and HRTEM observations revealed the presence of individual CNTs well-dispersed and embedded in the metallic matrices (Figure 5). In Figure 5a,b some examples of CNTs (marked with black arrows) inside the grains of the matrix can be observed. HRTEM images of Figure 5c revealed a CNT well-bonded to the metallic matrix. In a previous work [29], the formation of Al_4C_3 by reaction of the CNTs with the aluminum matrix was observed by HRTEM in a nanocomposite produced by this method. The reaction of the CNTs with the aluminum matrix was also reported by other researchers [32,33]. In contrast, there was no evidence of any type of reaction between nickel and CNTs.

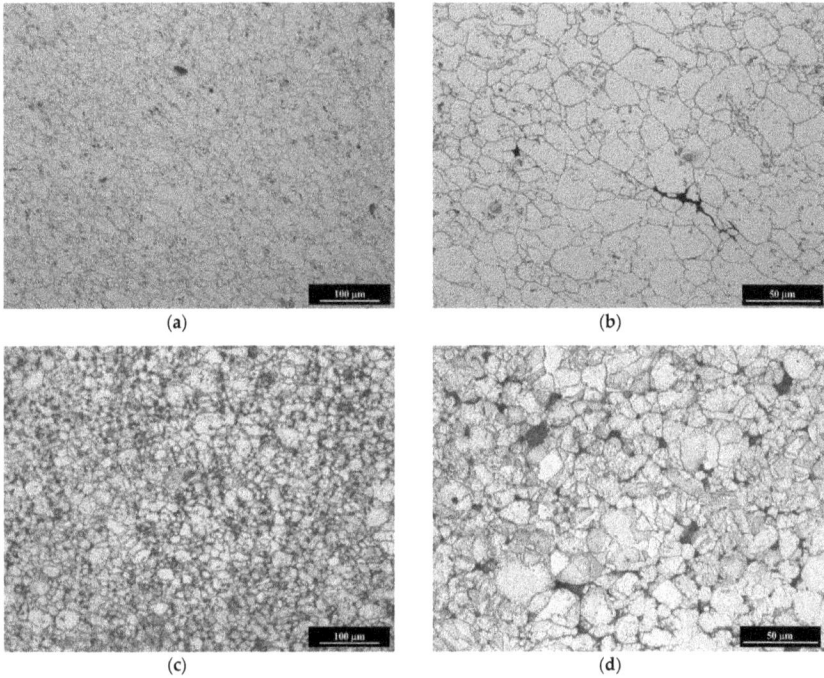

Figure 3. Optical microscopy (OM) images of the nanocomposites produced with 1.00 vol % of CNTs: (a) and (b) Al/CNTs and (c) and (d) Ni/CNTs.

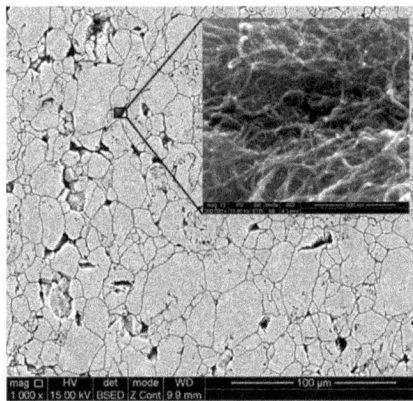

Figure 4. SEM image of the Al/CNT nanocomposites produced with 1.00 vol % of CNTs .

The number and size of the CNT clusters depends on the content of reinforcement. For both matrices, it is observed that a better dispersion, of smaller cluster size, is obtained for the nanocomposites produced with 1.00 vol % of CNTs. The increase in the content of CNTs promotes the increase in the size of the clusters and a non-uniform dispersion, measured by OM (Table 1). These two parameters are essential in order to understand the effect of the dispersion process and of the content of CNTs in the production of the nanocomposites.

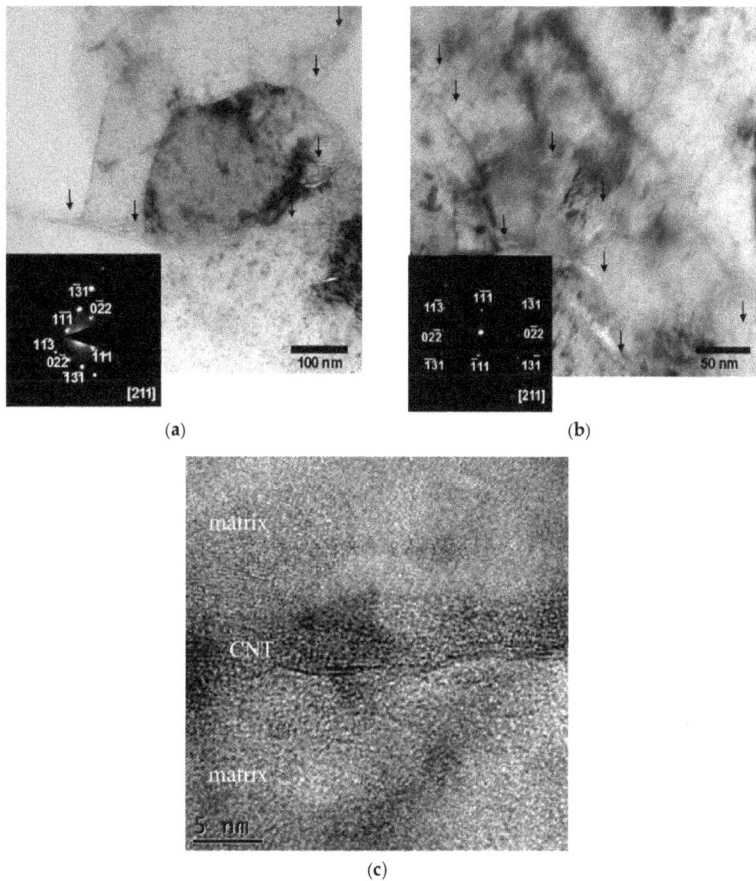

Figure 5. TEM images of the (**a**) Al/CNTs and (**b**) Ni/CNTs with the selected area electron diffraction (SAED) of the matrix and (**c**) HRTEM image showing a CNT.

The powder metallurgy process is characterized by the production of samples with some porosity. The aluminum and nickel samples exhibit a porosity of around 1% (Table 1). This porosity hinders the identification of clusters in nanocomposites. The SEM observations of the microstructures reveal that the pores are filled by CNTs, though it is not guaranteed that all pores will have been filled. In this context, the percentage of pores and CNT clusters was measured jointly. Table 1 shows the values as well as the maximum size of the CNT clusters. The effect of the CNT content on the percentage of pores, plus CNT clusters for the Al/CNT and Ni/CNT nanocomposites is evident. As expected, the increase in the content of CNTs leads to an increase in the number of pores plus CNT clusters. The best dispersion is obtained for the nanocomposites produced with 1.00 vol % of CNTs as for this composition the nanocomposites revealed the smaller size of the CNT clusters.

In order to evaluate the influence of the introduction of the CNTs on the microstructure of aluminum and nickel matrices, the grain size and crystallographic orientation were investigated by

EBSD. Figure 6 shows the grain size maps and distributions of samples produced with 0.00 vol % and 1.00 vol % of CNTs.

Table 1. Maximum size of CNT clusters, percentage of porosity and CNT clusters and average grain size as a function of the CNT content.

Composites	CNT Content (vol %)	Porosity and CNT Clusters (vol %)	Maximum Size of CNT Clusters (μm)	Average Grain Size (μm)
	0	1.00 [1]	-	17
	0.50	4.88	89	16
Al/CNT	0.75	6.53	102	15
	1.00	5.95	78	16
	1.50	7.38	177	16
	0	1.05 [1]	-	15
	0.50	3.22	83	14
Ni/CNT	0.75	3.96	155	13
	1.00	6.67	137	14
	1.50	12.47	310	16

[1] This value only represents the vol % of pores.

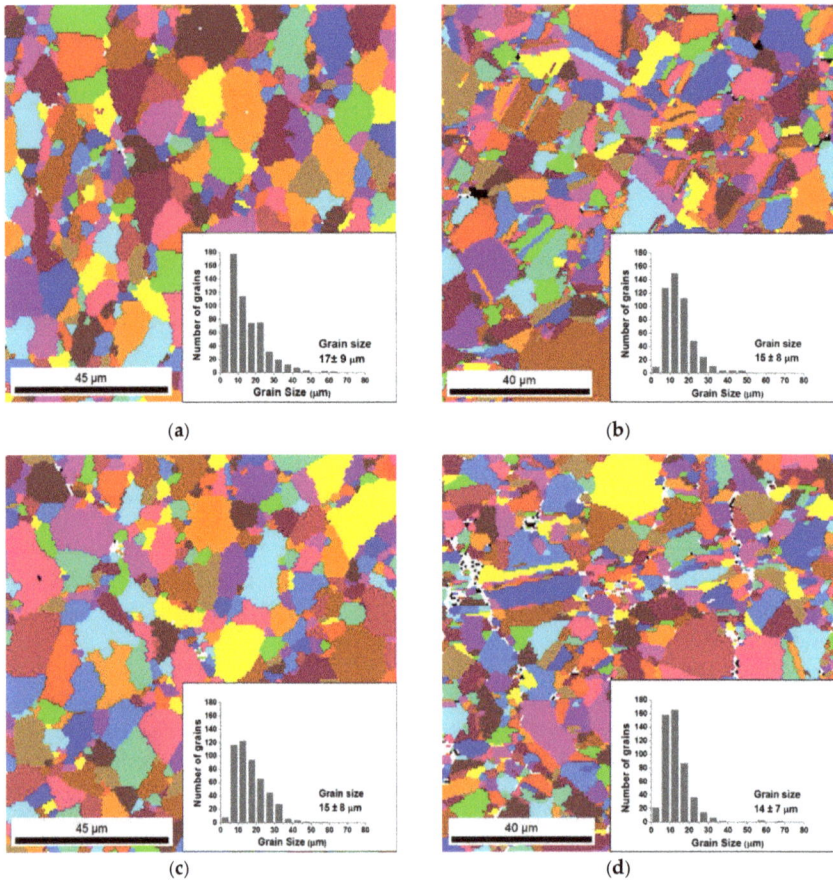

Figure 6. Grain size maps and distributions of the (**a**) aluminum and (**b**) nickel matrices and (**c**) Al/CNTs and (**d**) Ni/CNTs nanocomposites.

All nanocomposites had an average grain size similar to the sample produced without CNTs (pure metallic matrix subject to the same production conditions). Table 1 presents the average grain sizes for the pure metal and nanocomposites. Contrary to what has been reported by other authors [34], grain refinement is not clearly observed for Al/CNT and Ni/CNT nanocomposites.

The inverse pole figures of the two metallic matrices and nanocomposites can be seen in Figure 7. The analysis of this figure showed that the addition of CNTs affects the crystallographic orientation of the grains. However, the pure matrices and the Al/CNT and Ni/CNT nanocomposites do not have a strong preferred crystallographic orientation (texture); i.e., the incorporation of CNTs in the metallic matrices does not promote the formation of a strong texture. This aspect is very important, since a strong texture can significantly affect the mechanical properties.

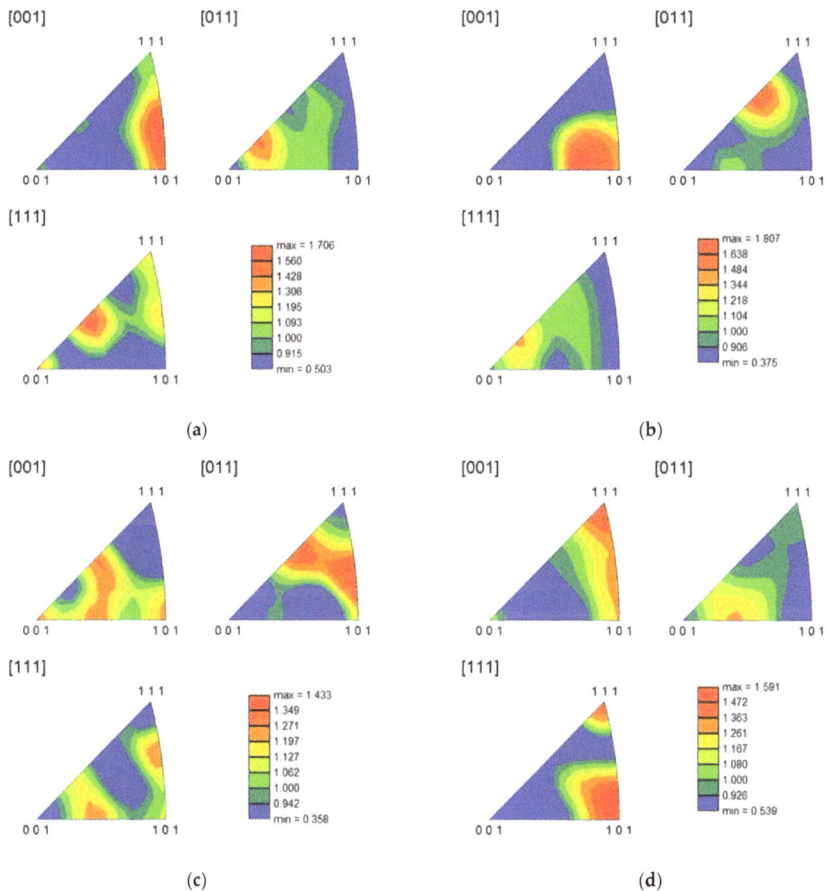

Figure 7. Inverse pole figures of the (a) aluminum and (b) nickel matrices and (c) Al/CNTs and (d) Ni/CNTs nanocomposites.

Microstructural characterization has shown that it is possible to produce nanocomposites of different matrices (aluminum and nickel) by powder metallurgy, using ultrasonication as a process for dispersing and mixing the powders and CNTs in a single step. The behavior of the two matrices is quite similar. In both cases, the content of CNTs seems to significantly affect their dispersion. Summing up, the results indicate that to obtain the best CNT dispersion, 1.00 vol % of CNTs must be used for both matrices. Higher contents of CNTs only promote the formation of larger clusters and a less uniform dispersion. The CNTs introduction does not promote significant microstructural changes, either in grain size or in the formation of textures.

3.2. Mechanical Characterization of Nanocomposites

The mechanical characterization of the nanocomposites was performed by Vickers hardness tests. Figure 8 shows the results of the hardness for the Al/CNT and Ni/CNT nanocomposites reinforced with different contents of CNTs; the bar chart shows the increase in the hardness of the nanocomposites relative to the samples without CNTs produced for comparison purposes. For both metal matrix composites, the highest hardness values were observed for 1.00 vol % of CNTs; these values correspond to an increase in 50% when compared to the hardness of pure metallic matrices.

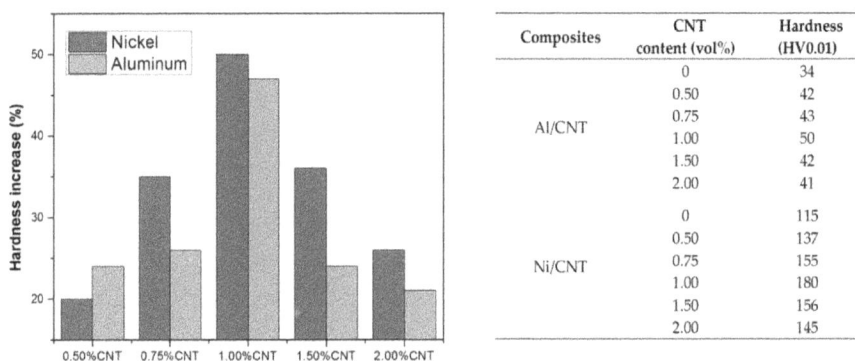

Composites	CNT content (vol%)	Hardness (HV0.01)
	0	34
	0.50	42
	0.75	43
Al/CNT	1.00	50
	1.50	42
	2.00	41
	0	115
	0.50	137
	0.75	155
Ni/CNT	1.00	180
	1.50	156
	2.00	145

Figure 8. The increase in the hardness (HV 0.01) of the nanocomposites relative to pure metallic matrices. The hardness values of all samples are listed in the table.

The observed strengthening cannot be explained by a Hall–Petch or texture effect, since nanocomposites and metallic samples produced by the same method have similar grain sizes and weak textures. The CNTs embedded in the matrices form continuous and bonded interfaces with them (as can be observed in the HRTEM image presented in Figure 5c), thus assuring the effective load transfer, reinforcing the nanocomposites. The formation of carbides can also contribute to the strengthening; as previously mentioned, this was only observed in Al/CNT nanocomposites [29]. However, not all CNTs contribute to the strengthening of nanocomposites; the CNTs of the clusters observed in the grain junctions are not bonded to the matrix and cannot assure the load transfer. The increase of CNT content above 1 vol % leads to the formation of larger clusters, increasing the heterogeneous distribution of CNTs and thus leading to a softening.

4. Conclusions

Aluminum and nickel nanocomposites reinforced by CNTs were successfully produced via a classical powder metallurgy route using ultrasonication as the dispersion/mixture process.

Individual CNTs dispersed and embedded in the matrix were observed through HRTEM, which is an essential factor for reinforcement. This observation confirms that the use of the ultrasonication method to disperse and mix is effective in the production of CNTs reinforced nanocomposites. However, the CNTs are also observed in clusters, mainly at grain boundary junctions. The better dispersion results were observed for the nanocomposites produced with 1.00 vol % of CNTs leading to the highest hardness values. The 50% hardness increase demonstrated the strengthening effect of the CNTs. Above this, CNT content larger clusters are formed, which decreases the strengthening effect.

Acknowledgments: This research is sponsored by FEDER funds through the program COMPETE-Programa Operacional Factores de Competitividade and by national funds through FCT-Fundação para a Ciência e a Tecnologia-under the project PEst-C/EME/UI0285/2013. The authors are grateful to CEMUP-Centro de Materiais da Universidade do Porto for expert assistance with SEM and EBSD.

Author Contributions: Sónia Simões produced and characterized the nanocomposites; Sónia Simões, Filomena Viana and Manuel F. Vieira discussed the microstructural and mechanical results. Marcos A. L. Reis and Manuel F. Vieira performed the mechanical tests. All the authors participated in the design of the experiments and cooperated in writing this paper.

Conflicts of Interest: The authors declare no conflict of interest.

References

1. Kainer, K.U. Custom-made materials for automative and aerospace engineering. In *Metal Matrix Nanocomposites*; WILLEY-VCH Verlag GmbH & Co. KGaA: Weinheim, Germany, 2006; pp. 1–48.
2. Shirvanimoghaddam, K.; Hamim, S.U.; Akbari, M.K.; Fakhrhoseini, S.M.; Pakseresht, A.H.; Ghasali, E.; Zabet, M.; Munir, K.S.; Jia, S.; Davim, J.P.; et al. Carbon fiber reinforced metal matrix composites: Fabrication processes and properties. *Compos. Part A* **2017**, *92*, 70–96. [CrossRef]
3. Casati, R.; Vedani, M. Metal matrix composites reinforced by nano-particles-A review. *Metals* **2014**, *4*, 65–83. [CrossRef]
4. Zhang, Z.; Chen, D.L. Contribution of Orowan strengthening effect in particulate-reinforced metal matrix nanocomposites. *Mater. Sci. Eng. A* **2008**, *483–484*, 148–152. [CrossRef]
5. Zhang, Z.; Chen, D.L. Consideration of Orowan strengthening effect in particulate-reinforced metal matrix nanocomposites: A model for predicting their yield strength. *Scr. Mater.* **2006**, *54*, 1321–1326. [CrossRef]
6. Mansoor, M.; Shahid, M. Carbon-nanotube-reinforced aluminum composite by induction melting. *J. Appl. Res. Technol.* **2016**, *14*, 215–225. [CrossRef]
7. Ostovan, F.; Matori, K.A.; Toozandehjani, M.; Oskoueian, A.; Yusoff, H.M.; Yunjus, R.; Ariff, A.H.M. Nanomechanical behavior of multi-walled carbon nanotubes particulate reinforced aluminum nanocomposites prepared by ball milling. *Materials* **2016**, *9*, 140. [CrossRef]
8. Jeon, Y.S.; Byun, J.Y.; Oh, T.S. Electrodeposition and mechanical properties of Ni-carbon nanotube nanocomposite coatings. *J. Phys. Chem. Solids* **2008**, *69*, 1391–1394. [CrossRef]
9. Suárez, S.; Soldera, F.; Oliver, C.G.; Acevedo, D.; Mucklich, F. Thermomechanical behavior of bulk Ni/MWNT composites produced via powder metallurgy. *Adv. Eng. Mater.* **2012**, *14*, 499–502. [CrossRef]
10. Kim, B.J.; Oh, S.Y.; Yun, H.S.; Ki, J.H.; Kim, C.J.; Baik, S.; Lim, B.S. Synthesis of Cu-CNT nanocomposite powder by ball milling. *J. Nanosci. Nanotechnol.* **2009**, *9*, 7393–7397. [CrossRef] [PubMed]
11. Bor, A.; Ichinkhorloo, B.; Uyanga, B.; Lee, J.; Choi, H. Cu/CNT nanocomposite fabrication with different raw material properties using a planetary ball milling process. *Powder Technol.* **2016**. In Press. [CrossRef]
12. Li, S.; Sun, B.; Imai, H.; Mimoto, T.; Kondoh, K. Powder metallurgy titanium metal matrix composites reinforced with carbon nanotubes and graphite. *Compos. Part A* **2013**, *48*, 57–66. [CrossRef]
13. Goh, C.S.; Wei, J.; Lee, L.C.; Gupta, M. Ductility improvement and fatigue studies in Mg-CNT nanocomposites. *Compos. Sci. Technol.* **2008**, *68*, 1432–1439. [CrossRef]
14. Han, G.Q.; Du, W.B.; Wang, Z.H.; Liu, K.; Li, S.B.; Du, X. Effective dispersion of CNTs to fabricate CNT/Mg nanocomposites. *Mater. Sci. Forum* **2015**, *816*, 470–475. [CrossRef]
15. Lei, X.; Natsuki, T.; Shi, J.; Ni, Q.Q. Analysis of carbon nanotubes on the mechanical properties at atomic scale. *J. Nanomater.* **2011**, *2011*, 805313. [CrossRef]
16. Demczyk, B.G.; Wang, Y.M.; Cumings, J.; Hetman, M.; Han, W.; Zettl, A.; Ritchie, R.O. Direct mechanical measurement of the tensile strength and elastic modulus of multiwalled carbon nanotubes. *Mater. Sci. Eng. A Struct.* **2002**, *334*, 173–178. [CrossRef]
17. Pérez-Bustamante, R.; Estrada-Guel, I.; Antúnez-Flores, L.; Miki-Yoshida, M.; Ferreira, P.J.; Martínez-Sánchez, R. Novel Al-matrix nanocomposites reinforced with multi-walled carbon nanotubes. *J. Alloys Compd.* **2008**, *450*, 323–326. [CrossRef]
18. Kuzumaki, T.; Miyazawa, K.; Ichinose, H.; Ito, K. Processing of carbon nanotube reinforced aluminum composite. *J. Mater. Res.* **1998**, *13*, 2445–2449. [CrossRef]
19. Xu, C.L.; Wei, B.Q.; Ma, R.Z.; Liang, J.; Ma, X.K.; Wu, D.H. Fabrication of aluminum–carbon nanotube composites and their electrical properties. *Carbon* **1999**, *37*, 855–858. [CrossRef]

20. Deng, C.; Zhang, X.; Ma, Y.; Wang, D. Fabrication of aluminum matrix composite reinforced with carbon nanotubes. *Rare Metals* **2007**, *26*, 450–455. [CrossRef]
21. Kwon, H.; Leparoux, M. Hot extruded carbon nanotube reinforced aluminum matrix composite materials. *Nanotechnology* **2012**, *23*, 415701. [CrossRef] [PubMed]
22. Kwon, H.; Park Dae, H.; Silvain Jean, F.; Kawasak, A. Investigation of carbon nanotube reinforced aluminum matrix composite materials. *Compos. Sci. Technol.* **2010**, *70*, 546–550. [CrossRef]
23. Laha, T.; Chen, Y.; Lahiri, D.; Agarwal, A. Tensile properties of carbon nanotube reinforced aluminum nanocomposite fabricated by plasma spray forming. *Compos. Part A* **2009**, *40*, 589–594. [CrossRef]
24. Nguyen, J.; Holland, T.B.; Wen, H.; Fraga, M.; Mukherjee, A.; Lavernia, E. Mechanical behavior of ultrafine-grained Ni–carbon nanotube composite. *J. Mater. Sci.* **2014**, *49*, 2070–2077. [CrossRef]
25. Kim, S.K.; Oh, T.S. Electrodeposition behavior and characteristics of Ni-carbon nanotube composite coatings. *Trans. Nonferr. Metal Soc.* **2011**, *21*, s68–s72. [CrossRef]
26. Arai, S.; Saito, T.; Endo, M. Cu-MWCNT composite films fabricated by electrodeposition. *J. Electrochem. Soc.* **2010**, *157*, D147–D153. [CrossRef]
27. Smart, S.K.; Ren, W.C.; Cheng, H.M.; Lu, G.Q.; Martin, D.J. Shortened double-walled carbon nanotubes by high-energy ball milling. *Int. J. Nanotechnol.* **2007**, *4*, 618–633. [CrossRef]
28. Simões, S.; Viana, F.; Vieira, M.F. Carbon Nanotubes and their Nanocomposites. In *Nanomaterials and Nanocomposites-Zero- to Three-Dimensional Materials and Their Composites*; Wiley-VCH Verlag GmbH: Weinheim, Germany, 2016; pp. 75–101.
29. Simões, S.; Viana, F.; Reis, M.A.L.; Vieira, M.F. Improved dispersion of carbon nanotubes in aluminum composites. *Compos. Struct.* **2014**, *108*, 992–1000. [CrossRef]
30. Simões, S.; Viana, F.; Reis, M.A.L.; Vieira, M.F. Influence of dispersion/mixture time on mechanical properties of Al-CNTs nanocomposites. *Compos. Struct.* **2015**, *126*, 114–122. [CrossRef]
31. Simões, S.; Viana, F.; Reis, M.A.L.; Vieira, M.F. Microstructural characterization of aluminum-carbon nanotube nanocomposites produced using different dispersion methods. *Microsc. Microanal.* **2016**, *22*, 725–732. [CrossRef] [PubMed]
32. Chen, B.; Li, S.; Imai, H.; Jia, L.; Umeda, J.; Takahashi, M.; Kondoh, K. An approach for homogeneous carbon nanotube dispersion in Al matrix composites. *Mater. Des.* **2015**, *72*, 1–8. [CrossRef]
33. Chen, B.; Shen, J.; Ye, X.; Imai, H.; Umeda, J.; Takahashi, M.; Kondoh, K. Solid-state interfacial reaction and load transfer efficiency in carbon nanotubes (CNTs)-reinforced aluminum matrix composites. *Carbon* **2017**, *114*, 198–208. [CrossRef]
34. Suárez, S.; Lasserre, F.; Mucklich, F. Mechanical properties of MWNT/Ni bulk composites: Influence of the microstructural refinement on the hardness. *Mater. Sci. Eng. A Struct.* **2013**, *587*, 381–386. [CrossRef]

metals

MDPI

Article

Significantly Enhancing the Ignition/Compression/Damping Response of Monolithic Magnesium by Addition of Sm_2O_3 Nanoparticles

Milli Suchita Kujur [1,2], Ashis Mallick [1], Vyasaraj Manakari [2], Gururaj Parande [2], Khin Sandar Tun [2] and Manoj Gupta [2,*]

[1] Department of Mechanical Engineering, Indian Institute of Technology (ISM), Dhanbad,
     Jharkhand 826004, India; millisuchitakujur@gmail.com (M.S.K.); mal123_us@yahoo.com (A.M.)
[2] Department of Mechanical Engineering, National University of Singapore, 9 Engineering Drive 1,
     Singapore 117576, Singapore; mbvyasaraj@u.nus.edu (V.M.); gururaj.parande@u.nus.edu (G.P.);
     mpekhst@nus.edu.sg (K.S.T.)
*   Correspondence: mpegm@nus.edu.sg; Tel.: +65-6516-6358

Received: 7 July 2017; Accepted: 6 September 2017; Published: 9 September 2017

Abstract: The present study reports the development of Mg–Sm_2O_3 nanocomposites as light-weight materials for weight critical applications targeted to reduce CO_2 emissions, particularly in the transportation sector. Mg-0.5, 1.0, and 1.5 vol % Sm_2O_3 nanocomposites are synthesized using a powder metallurgy method incorporating hybrid microwave sintering and hot extrusion. The microstructural studies showed dispersed Sm_2O_3 nanoparticles (NPs), refinement of grain size due to the presence of Sm_2O_3 NPs, and presence of limited porosity. Microhardness and dimensional stability of pure Mg increased with the progressive addition of Sm_2O_3 NPs. The addition of 1.5 vol % of Sm_2O_3 NPs to the Mg matrix enhanced the ignition temperature by ~69 °C. The ability of pure Mg to absorb vibration also progressively enhanced with the addition of Sm_2O_3 NPs. The room temperature compressive strengths (CYS and UCS) of Mg–Sm_2O_3 nanocomposites were found to be higher without having any adverse effect on ductility, leading to a significant increase in energy absorbed prior to compressive failure. Further, microstructural characteristics are correlated with the enhancement of various properties exhibited by nanocomposites.

Keywords: Magnesium; Sm_2O_3 nanoparticles; compression properties; damping; microstructure; ignition

1. Introduction

Magnesium (Mg) is the lightest structural metal and the third most abundant element in the Earth's hydrosphere and sixth most abundant in the Earth's crust, making it readily available [1]. Mg has a density of 1.74 g/cc which is ~35.56%, ~61.39%, and ~77.89% lower than that of aluminium (2.7 g/cc), titanium (4.506 g/cc), and iron (7.87 g/cc), respectively [2]. In addition to this, Mg exhibits good mechanical and thermal properties, damping capacity, excellent castability, and machinability [3]. Mg-based materials have always been used in various industrial sectors such as aerospace, energy, construction, automotive, security, and defense, all of which are crucial to the sustainability and growth of the global economy. Increased demand for light-weighting drives the interest in Mg-based materials to be used in above-mentioned sectors striving for weight reduction, higher fuel efficiency, and payload capacity leading to reduced CO_2 emissions. However, one of the major challenges is the necessity to reduce environmental impact both in their production, end-use, and recyclability. Adoption of fabrication techniques with reduced processing time and cost such as hybrid microwave sintering adopted in this study can result in significant energy savings which is economically viable for industries and environmentally friendly in the reduction of CO_2 emissions [4].

Currently, Mg is also finding application in biomedical engineering owing to its superior biocompatibility [5]. Mg, being an important electrolyte for human metabolism, is the fourth most abundant cation present in the body and is a cofactor in more than 300 enzyme systems that help in regulating diverse biochemical reactions in the body [6]. Overall, 99% of total Mg present in our body is in bone, muscles, and soft-muscular tissues [7]. Mg exhibits elastic modulus (41–45 GPa) closer to that of human bone (3–20 GPa) in comparison to other materials such as titanium (100–110 GPa) and stainless steel (189–205 GPa) showing, in addition, no indication of local or systemic toxicity and hence is therefore being encouraged as a biomaterial by the scientific community [8]. It is biocompatible as well as biodegradable [9] which further helps in eliminating corrective surgery and patient trauma. However, its extensive use is limited owing to its limited room temperature ductility, creep, corrosion resistance, and performance at elevated temperature [2]. Hence, researchers are actively investigating Mg-based nanocomposites as lightweight structural materials having low density, high strength, stiffness, and durability with improved corrosion resistance and acceptable performance at elevated temperatures [10]. Nanocomposites form an emerging class of materials with extremely good mechanical properties coupled with thermal integrity [11] owing to the presence of dimensionally stable ceramic or metallic reinforcements, which provide high mechanical strength as well as ductility [4].

As per the literature, Mg and its alloys have already been successfully incorporated with ceramic oxide nanoparticles (NPs) such as ZnO, Al_2O_3, ZrO_2, TiO_2, and Y_2O_3 and characterized for various mechanical properties [3]. Results obtained so far are promising, warranting further exploration of new systems. Rare earth elements (REEs) have been recently used to alloy pure Mg with encouraging results [12–17]. Rare earth oxides are natural choices as the addition of rare earth elements (REEs) as alloying elements enhances various properties of magnesium. However rare earth elements are toxic in nature and the use of rare earth oxides in nano-length scale is a viable option as they can be used in lower amounts to realize the improvement in properties (typically less than 2% by volume). There are seventeen rare earth oxides (REOs) and they have similar chemical properties by nature [6]. REOs, owing to their strong rare-earth–oxygen interactions have been reported to significantly reduce the grain size which leads to an increase in strength owing to a refined microstructure [18]. Further, they help in gathering of segregated solute impurities at the grain boundaries, alleviating the concentration of such deleterious solutes in the lattice and leading to improved ductility [19]. Hence, the addition of REOs may help in realizing a good strength-ductility combination. Also, incorporation of REOs at the nanoscale and spreading them into the interior of the grains encourages dislocation trapping by the REO particles in the interior of the grains, helping sustain work hardening and resulting in uniform elongation [19]. For example, Mg has been incorporated with yttrium oxide NPs and the tensile properties improved both at room and elevated temperature [11]. However, the main concern that arises with the use of REOs as a part of biomedical applications is their relatively unknown effects on the physiological system and research needs to be carried out if these oxides are too toxic for use as biomaterials [20].

Sm_2O_3 is one such REO having density 8.347 g/cc with high hardness Vickers (438 HV), high melting temperature (2335 °C), elastic modulus (183 GPa), Gibbs free energy (-1734.9 KJ·Mol^{-1}) [21,22]. Sm_2O_3 is an important rare earth oxide and its current scope lies in the field of solar cells, semiconductor gas, biochemical sensors, laser and photonic devices, precision guided weapons, and is also an active catalyst for CO hydrogenation [23]. Sm_2O_3 is also used as a bone-seeking radiopharmaceutical providing therapeutic irradiation to osteoblastic bone metastases [24]. The in-vitro analysis of Sm_2O_3 was found to be excellent with an appropriate cell response for a bone-contacting material and could support the initial stages of osteogenesis [24]. The response of osteoblast-like cells to Sm_2O_3 assure the non-cytotoxicity of the material and biofunctionality making it potentially useful in the field of biomaterials [24].

Also, with the ban on Mg being lifted by Federal Aviation Administration (FAA) in 2015, there is a renewed interest in replacing Al alloy based materials in the aircraft seat components, and until now, Elektron®WE43, Elektron®21 and Elektron®675 alloys have complied with new FAA regulations [25]. Incorporation of rare earth elements have resulted in a better ignition performance of Mg wherein Mg-10.6Y alloy and Mg3.5Y0.8Ca alloys have reported no ignition until ~1000 °C [16]. Although the research on the ignition resistance of Mg-based nanocomposites is still in its early stages, promising results for Mg-alloys can be of great encouragement to develop materials with superior overall properties. Accordingly, in the present work, the microstructural, ignition, compression, and damping properties are analyzed for the addition of 0.5, 1.0, and 1.5 vol % of Sm_2O_3 NPs in pure Mg.

2. Materials and Methods

2.1. Materials

The choice of material for metal matrix and reinforcement was Mg and Sm_2O_3, respectively. Magnesium powder of 98.5% purity with a size range of 60–300 μm was supplied by Merck, Germany. Sm_2O_3 powder with a size range of 20–30 nm was supplied by US Research Nanomaterials, Inc., Houston, TX, USA.

2.2. Synthesis

Powder Metallurgy technique assisted by hybrid microwave sintering was used for synthesis of pure Mg and (0.5, 1 and 1.5 vol %) Sm_2O_3 nanocomposites. Blending was done for the carefully weighed samples of pure Mg powder and Sm_2O_3 NPs in a RETSCH PM-400 mechanical alloying machine (Haan, Germany). Cold compaction was done post-blending at a uniaxial pressure of 1000 psi. The obtained billets (35-mm diameter and 40-mm height) were sintered using a hybrid microwave sintering technique at 630 °C in a 2.45 GHz, 900 W Sharp microwave oven. The benefits of hybrid microwave sintering over conventional sintering has been previously reported [10]. The billets were soaked at 450 °C for 2 h prior to extrusion. A 150 T hydraulic press was used to extrude the billets at a die temperature of 400 °C at an extrusion ratio of 20.25:1 to get cylindrical rods of 8 mm diameter. The extruded samples were characterized for microstructural, physical, and mechanical properties.

2.3. Microstructural Characterization

The metallographic polished samples in the extruded state were studied to investigate the grain size, reinforcement distribution, and interfacial integrity between the Mg matrix and Sm_2O_3 reinforcement. The OLYMPUS metallographic microscope (Leica Microsystems (SEA) Pte Ltd., Singapore, Singapore) and JEOL JSM-5800 LV Scanning Electron Microscope (SEM) coupled with Energy Dispersive Spectroscopy (EDS) (Jeol USA Inc., Peabody, MA, USA) were used for the microstructural characterization studies. The grain size was estimated by using a mathematical code developed in-house.

2.4. Physical Properties

2.4.1. Density Measurement

The density of three polished extruded samples for each composition was measured using the Archimedes principle. An A&D HM-202 electronic balance (Bradford, MA, USA) with an accuracy of 0.0001 g was used for measuring accurately the weights of the polished samples separately both in air and when immersed in distilled water. The theoretical densities of the samples were calculated assuming they are dense and there is no Mg–Sm_2O_3 interfacial reaction. The volume percentage of porosity in each case was computed using the theoretical and experimental density values.

2.4.2. Microhardness

Flat and polished specimens were used for micro hardness testing. A Shimadzu HMV automatic digital micro hardness tester (Kyoto, Japan) with a Vickers indenter with a phase angle of 136° was utilized for this purpose. The samples were subjected to a micro indentation load of 245 gf for a dwell time of 15 s. The tests were performed as per ASTM E384-16 standard. 15 readings were taken for each sample to ensure repeatability and accuracy of the results.

2.4.3. Coefficient of Thermal Expansion (CTE)

The CTE values of pure Mg and Mg–Sm_2O_3 nanocomposites were analyzed using a LINSEIS TMA PT 1000LT thermo-mechanical analyzer (Tokyo, Japan). The argon flow was maintained at 0.1 litres per min (lpm) while heating rate was set at 5 °C/min. The displacement of the samples was measured using an alumina probe in a temperature range of 50–400 °C. Three samples were tested to check the response.

2.4.4. Ignition Temperature

The ignition temperature of the extruded pure Mg and Mg–Sm_2O_3 nanocomposite samples (2 mm × 2 mm × 1 mm) was determined using a Shimadzu DTG-60H Thermo Gravimetric Analyser (Kyoto, Japan). The samples were heated from 30 to 750 °C at a heating rate of 10 °C/min in purified air with a flow rate of 50 mL/min. Three samples were tested to check the response.

2.5. Mechanical Properties

2.5.1. Compression Testing and Fracture Behavior

The compressive properties of the extruded pure Mg and Mg–Sm_2O_3 nanocomposite samples were determined in accordance with ASTM test method E9-09 using an MTS-810 testing machine (Eden Prairie, MN, USA) with a strain rate set at 8.334×10^{-5} s^{-1} on test specimens of 8 mm diameter and 8 mm length. Five samples were tested for each composition to ensure the repeatability of the test. Fractography was done for the compressive fractured samples using JEOL JSM-5800 LV Scanning Electron Microscope (SEM) to get a better understanding of the possible modes of the failure.

2.5.2. Elastic Modulus and Damping Behavior

A damping analyzer (IMCE, Genk, Belgium) was used for analyzing the elastic modulus and damping response of pure Mg and Mg–Sm_2O_3 nanocomposites using a resonant frequency as per ASTM E1876-09. Two samples were used for the analysis were of 60 mm in length and 8 mm in diameter.

3. Results and Discussion

3.1. Microstructural Characterization

The grain size measurements for pure Mg and the nanocomposite samples are shown in Table 1. The size of the grains for pure Mg was found to decrease with the increasing presence of Sm_2O_3 NPs. Mg-0.5 vol % Sm_2O_3, Mg-1.0 vol % Sm_2O_3, Mg-1.5 vol % Sm_2O_3 exhibited grain size of 21.4 µm, 17.6 µm, and 12.9 µm which is ~16.73%, ~31.51%, and ~49.80% lower than that of pure Mg (25.7 µm), respectively. The reduction in grain size for the nanocomposite can be attributed to the grain boundary pinning mechanism where the Sm_2O_3 NPs pinned the recrystallized grains of Mg, hence restraining its growth, and not to particle stimulated nucleation [26]. Dispersed Sm_2O_3 NPs within the Mg matrix can be seen in Figure 1a–c. This can be attributed to the proper selection of blending, compaction, and sintering parameters. Further, it can also be stated that the extrusion process having high extrusion ratio (20.25:1) is able to break the agglomeration and disperse the NPs uniformly as can be seen in Figure 1. However, some agglomerated sites may still be present after the extrusion process. This dispersed

Sm_2O_3 NPs promotes more uniform heating through microwaves and demonstrates the effectiveness of using hybrid microwave sintering for the synthesis of $Mg-Sm_2O_3$ nanocomposites.

Table 1. Results of microstructure, coefficient of thermal expansion (CTE), and microhardness studies.

Material	Grain Size (μm)	Aspect Ratio	Hardness (Hv)	CTE ($\times 10^{-6}$/K)
Pure Mg	25.7 ± 2.6	1.39 ± 0.41	54 ± 2	26.27
Mg-0.5 Sm_2O_3	21.4 ± 2.3 (↓16.73%)	1.36 ± 0.20	64 ± 3 (↑18.52%)	25.82 (↓1.71%)
Mg-1.0 Sm_2O_3	17.6 ± 1.8 (↓31.51%)	1.33 ± 0.16	71 ± 2 (↑31.48%)	24.99 (↓4.87%)
Mg-1.5 Sm_2O_3	12.9 ± 2.2 (↓49.80%)	1.52 ± 0.30	73 ± 1 (↑35.18%)	24.11 (↓8.22%)

(↑x%) and (↓x%) indicates the percentage increase and decrease in the property with respect to pure Mg by x%, respectively.

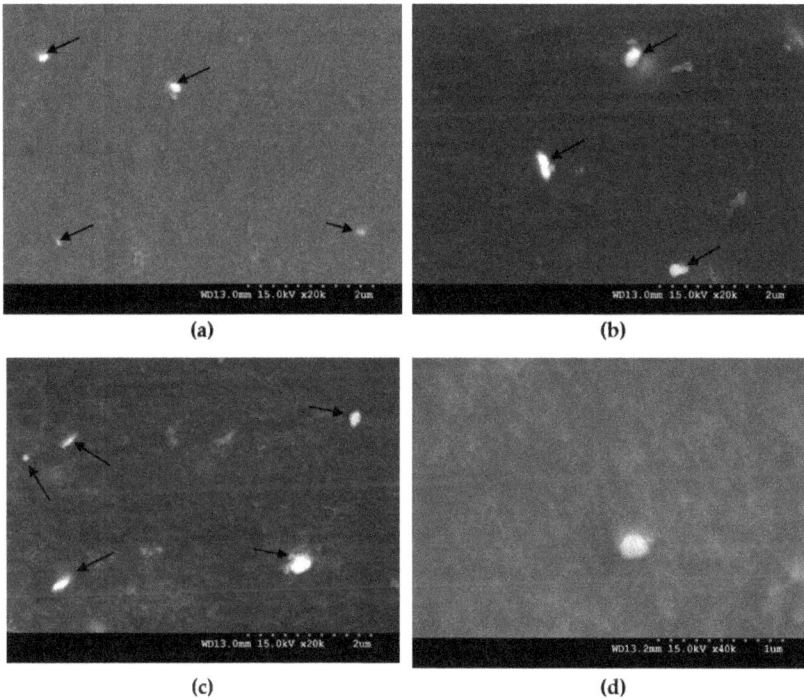

Figure 1. Scanning electron microscopy (SEM) images showing the distribution of Sm_2O_3 nanoparticles (indicated by the arrows) in $Mg-Sm_2O_3$ nanocomposites: (**a**) Mg-0.5 vol % Sm_2O_3 (**b**) Mg-1 vol % Sm_2O_3 (**c**) Mg-1.5 vol % Sm_2O_3 (**d**) interfacial integrity of Mg-1.0 vol % Sm_2O_3 nanocomposite.

3.2. Physical Properties

3.2.1. Density

The theoretical and experimental densities of the pure Mg and $Mg-Sm_2O_3$ nanocomposite samples are shown in Table 2. The experimental density of the nanocomposite samples was found to be increasing with the progressive addition of Sm_2O_3 NPs and can be attributed to the difference in density between Sm_2O_3 NPs (8.34 g/cc) and pure Mg (1.74 g/cc). As seen from Table 2, the porosity values are also found to be increasing with increasing density and the Mg-1.5 vol % Sm_2O_3 nanocomposite exhibited highest porosity value. The low wettability, agglomeration rate, and pore nucleation at the $Mg-Sm_2O_3$ interface could be the main reasons behind the increase in porosity with the increase in vol % of Sm_2O_3 NPs in Mg [27]. Further, no observed macrostructural defects were observed on sintered samples and extruded rods, demonstrating the efficiency of hybrid microwave sintering and hot extrusion to synthesize near-dense magnesium-based nanocomposites [28].

Table 2. Density measurements of pure magnesium and Mg-Sm_2O_3 nanocomposites.

Material	Theoretical Density (g/cc)	Experimental Density (g/cc)	Porosity (%)
Pure Mg	1.74	1.7265 ± 0.0142	0.78
Mg-0.5 Sm_2O_3	1.773	1.7287 ± 0.0147	2.50
Mg-1.0 Sm_2O_3	1.806	1.7531 ± 0.0047	2.92
Mg-1.5 Sm_2O_3	1.839	1.7838 ± 0.0085	3.00

3.2.2. Microhardness

The microhardness measurements for pure Mg and the nanocomposite samples are shown in Table 1. The progressive addition of Sm_2O_3 NPs resulted in a steady increase in the hardness values of pure Mg. With the addition of 1.5 vol % Sm_2O_3, a maximum average value of 73 Hv was observed which is ~35.18% greater than that of pure Mg (54 Hv). The presence of high hardness Sm_2O_3 NPs (430 Hv), reduced grain size, and constraint to localized deformation during indentation due to the presence of harder Sm_2O_3 NPs are the main inferences for the increase in the microhardness of the nanocomposite samples.

3.2.3. Coefficient of Thermal Expansion (CTE)

The coefficient of thermal expansion values for pure Mg and Mg-Sm_2O_3 nanocomposites samples are shown in Table 1. The CTE values for Mg-0.5 vol % Sm_2O_3 (25.82×10^{-6}/K), Mg-1.0 vol % Sm_2O_3 (24.99×10^{-6}/K), and Mg-1.5 vol % Sm_2O_3 (24.11×10^{-6}/K) were found to be ~1.71%, ~4.87, and ~8.22% lower than that of pure Mg (26.27×10^{-6}/K). The CTE values of pure Mg follows a linear decreasing trend with the increasing addition of Sm_2O_3 NPs, which is found to be in accordance with the theory that the thermal expansion of composites is governed by the competing interactions of expansion of the Mg matrix and the constraint of reinforcement particles through their interfaces. This behavior can be attributed to the lower CTE value of Sm_2O_3 (8.5×10^{-6}/K) reinforcement as compared to that of pure Mg (26.27×10^{-6}/K) and the presence of ceramic reinforcements in the matrix, hence maintaining the dimensional stability of pure Mg.

For comparison purposes, theoretical CTE values of Mg–Sm_2O_3 nanocomposites were determined by using the Rule of Mixture (ROM) model [29], which is expressed as:

$$\alpha_c = \alpha_m \cdot v_m + \alpha_p \cdot v_p \tag{1}$$

where α is the CTE, 10^{-6}/K; v is the volume fraction and subscripts c, m, p refers to the composite, matrix, and reinforcement phase, respectively. It was observed that the experimental values of CTE obtained were lower than the theoretical values of 27.1, ~27.01, ~26.91, and ~26.82 × 10^{-6}/K for pure Mg, Mg-0.5 vol % Sm_2O_3, Mg-1 vol % Sm_2O_3, and Mg-1.5 vol % Sm_2O_3, respectively. This can be attributed to the overriding effect of presence and increasing amount of Sm_2O_3 over that of increasing level of porosity of Mg–Sm_2O_3 nanocomposites (Table 2).

3.2.4. Ignition Properties

The onset of ignition occurs only when the stable surface oxide of Mg-based materials tends to lose its protective properties [30]. Pure Mg can auto-ignite in solid state during heating due to the rapid increase in localized heat that causes melting and evaporation of the metal locally. When the Mg vapor is in contact with air at the gas/metal interface, the metal ignites. However, with the modification of the chemistry of the material by the addition of thermally stable alloying elements and reinforcements, the mechanism changes and can delay the onset of ignition [16]. Therefore, an attempt has been made to study and analyze the effect of thermally stable rare earth oxides (Sm_2O_3 in the current system) on the ignition temperature of monolithic magnesium.

The ignition results using thermogravimetric analysis (TGA) are presented in Table 3. The ignition point is 610 and 613 °C for 0.5 and 1.0 vol % Sm_2O_3 and then escalates to 650 °C for 1.5 vol % of Sm_2O_3 which is ~69 °C more than pure Mg (581 °C). The ignition temperature for 1.5 vol % of Sm_2O_3 is found to be higher than most of the commercially available magnesium alloys such as AZ31, AZ61, AZ63, AZ91, AM50, AM60, ZK40A, ZK51A, and ZK60A [16] (Table 3). The ascending behavior of the point of ignition and the enhanced resistance to ignition with increasing volume fraction of Sm_2O_3 NPs may be attributed to the lower CTE values, thus maintaining the thermal and dimensional stability of the nanocomposites [26]. In addition, the presence of Sm_2O_3 NPs in the Mg matrix assists in reducing specific areas of oxidation and hence assists in delaying the onset of ignition in the Mg-Sm_2O_3 nanocomposites. It has been previously reported that Sm_2O_3 is a stable protective oxide in air at temperatures up to at least 593 °C [31]. Further, the molecular volume of oxide ($\varnothing^a$) is 1.21 for Sm_2O_3 and both its oxidation states (+2 and +3) form stable protective oxides [32]. Since Sm_2O_3 has a high affinity to oxygen owing to strong rare-earth–oxygen interactions, a protective barrier layer is formed on the surface of Mg, restraining its reaction with oxygen. The results in Table 3 indicated that 1.5 vol % of Sm_2O_3 is most effective and the variation in ignition temperature between 0.5 and 1 vol % Sm_2O_3 was negligible, suggesting the requirement of a certain critical threshold of Sm_2O_3 to significantly enhance the ignition temperature. Further work is continuing in this area.

Table 3. Results of ignition temperatures characterization.

Material	Ignition Temperature (°C)	Thermal Conductivity (W/m·K)
Pure Mg	581	135
Mg-0.5 Sm_2O_3	610	134.32
Mg-1.0 Sm_2O_3	613	133.65
Mg-1.5 Sm_2O_3	650	132.97
AZ31	628	
AZ61	559	
AZ63	573	
AZ91	600	
AM50	585	-
AM60	525	
ZK40A	500	
ZK51A	552	
ZK60A	499	

The ignition temperature values of commercially available magnesium alloys such as AZ31, AZ61, AZ63, AZ91, AM50, AM60, ZK40A, ZK51A, and ZK60A are compiled from references given in [16].

The thermal conductivity (W/m·K) at 400 °C calculated theoretically by rule of mixtures and ignition temperature can be correlated for the Mg-Sm_2O_3 nanocomposites to understand the ignition behavior [26]. The ignition temperature ascends with the decrease in thermal conductivity with the increasing content in the volume fraction of Sm_2O_3, as seen from Table 3. It can be inferred that that the addition of Sm_2O_3 NPs helped in increasing the insulating property of pure Mg [33]. Further, the thermal conductivity of the composites is directly related to the amount of reinforcement added to the matrix, and in view of the ability of reinforcement to reduce the availability of metallic matrix for ignition, also leads to increased ignition performance with progressive addition of Sm_2O_3 NPs [34].

However, the underlying dominating mechanisms determining the ignition temperatures of Mg-based materials are complex and are not very well known, especially of the nanocomposites. Therefore, further study on the mechanisms of increase/decrease in the ignition temperature with the presence of Sm_2O_3 NPs and the dependence of the ignition characteristics on the size of the NPs will be an interesting direction of research.

3.3.Mechanical Properties

3.2.5. Compression Properties and Fracture Behavior

The room temperature compressive properties and their stress–strain relationship is shown in Table 4 and Figure 2, respectively. As it can be seen from Table 4, the compressive yield strength (0.2 CYS) of Mg increased from 74 to 87 MPa, 118, and 128 MPa with the addition of 0.5, 1.0, and 1.5 vol % of Sm_2O_3 NPs, respectively. The ultimate compressive strength (UCS) values for the Mg–Sm_2O_3 nanocomposites also increased with the progressive addition of Sm_2O_3 NPs with Mg-1.5 vol % Sm_2O_3 exhibiting the maximum UCS of 395 MPa, which is ~58.63% greater than pure Mg. The compressive fracture strain values of Mg–Sm_2O_3 nanocomposites also increased up to 1 vol % addition of Sm_2O_3 NPs, and maximum fracture strain was exhibited by Mg-1 vol % Sm_2O_3 with 20.1% (~15.51% greater than pure Mg). With the further addition of Sm_2O_3 NPs (1.5 vol %), a decrease in fracture strain value of 17.2% was observed. This reduction in fracture strain for Mg-1.5 vol % Sm_2O_3 (~1.14% lesser than pure Mg) is very marginal and the addition of Sm_2O_3 NPs helps in maintaining the ductility while increasing the strengths significantly. The increase in the 0.2 CYS and UCS of Mg-Sm_2O_3 nanocomposites can be attributed to (a) presence of fairly dispersed, hard Sm_2O_3 NPs [35]; (b) significant grain refinement (Table 1) [33]; (c) effective transfer of load from the Mg matrix to Sm_2O_3 NPs [36]; (d) mismatch of the elastic modulus and coefficient of thermal expansion values leading to generation of dislocations [37], and (e) Orowan strengthening due to the presence of Sm_2O_3 NPs [37]. The energy absorption (EA) during the process of compressive loading until failure also increased with the progressive addition of Sm_2O_3 NPs. Mg-1 vol % Sm_2O_3 nanocomposite exhibited the maximum EA value of 42.9 MJ/m^3 which is ~60.07% greater than that of pure Mg. The enhanced EA of the nanocomposites with respect to pure Mg shows its potential to be used in damage tolerant designs.

Table 4. Results of room temperature compression testing.

Material	0.2 CYS (MPa)	UCS (MPa)	Fracture Strain (%)	Energy Absorbed (MJ/m^3)
Pure Mg	74 ± 3	249 ± 6	17.4 ± 0.3	26.8 ± 0.7
Mg-0.5 Sm_2O_3	87 ± 1 (↑17.56%)	285 ± 6 (↑14.45%)	19.8 ± 0.6 (↑13.79%)	33.0 ± 1.3 (↑23.13%)
Mg-1.0 Sm_2O_3	118 ± 2 (↑59.45%)	331 ± 7 (↑32.93%)	20.1 ± 0.7 (↑15.51%)	42.9 ± 2.9 (↑60.07%)
Mg-1.5 Sm_2O_3	128 ± 5 (↑72.97%)	395 ± 7 (↑58.63%)	17.2 ± 0.5 (↓1.14%)	41.2 ± 2.4 (↑53.73%)

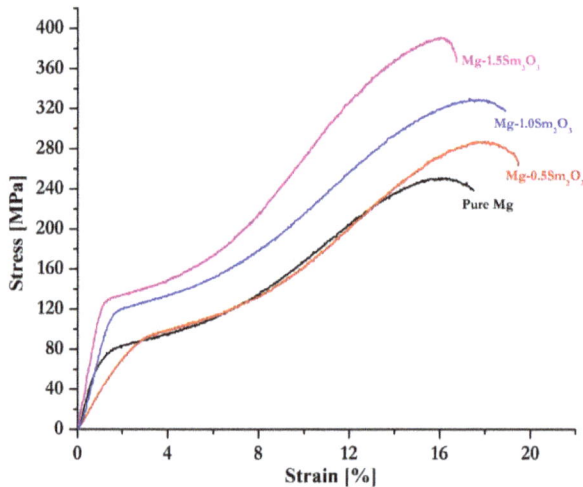

Figure 2. Stress–strain curves of pure Mg and the synthesized Mg-Sm_2O_3 nanocomposites during compression loading.

Under compression along the extrusion direction of Mg based materials, deformation occurs by twin, followed by slip [38]. The upward concave (sigmoidal) nature of compressive flow curves (Figure 2) with high work hardening further affirms that the compressive deformation occurs by a twinning process. Though the grain refinement is believed to suppress both twin and slip [39], in turn enhancing the strength of nanocomposites, this effect comes at the expense of compressive ductility. Conversely, the results show that compressive failure strain of the developed Mg-Sm_2O_3 nanocomposites were superior when compared to pure Mg upto 1 vol % addition of Sm_2O_3 NPs and was comparable to pure Mg for Mg-1.5 vol % Sm_2O_3. This can be explained by the fact that, in addition to the role of grain size on the activation energy of twinning, texture also influences the deformation twinning. Additions of rare earth oxides, such as Sm_2O_3, can facilitate grain alignment, favoring both slip and deformation twinning [40]. Thus, the presence of Sm_2O_3 NPs would help in delaying the twinning by activation of additional competing slip/twin deformation modes by the change in crystallographic orientation, thereby contributing to increased failure strain. Further studies, such as electron back-scattered diffraction (EBSD), are required to confirm the evolution of crystallographic texture in pure Mg due to incorporation of Sm_2O_3 NPs. To confirm the mode of failure under compression, fracture studies were performed. Under compressive loading, fracture surfaces are at about 45 degrees with respect to the compression testing direction. Shear bands were observed (see Figure 3), which is an indication of shear mode of failure.

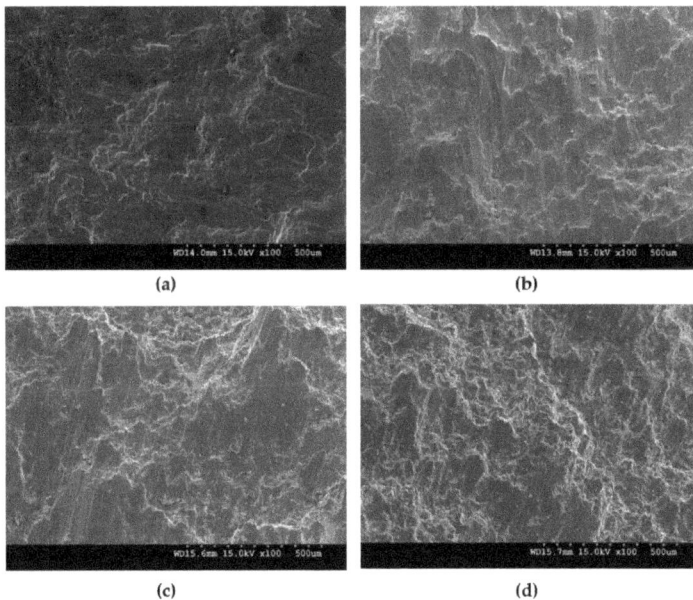

(a) (b) (c) (d)

Figure 3. Fractographs after compressive loading of: (**a**) Pure Mg; (**b**) Mg-0.5 vol % Sm_2O_3; (**c**) Mg-1 vol % Sm_2O_3, and (**d**) Mg-1.5 vol % Sm_2O_3.

Figure 4 provides an overview of the improvements observed in UCS and fracture strain by the addition of Sm_2O_3 NPs to pure Mg in comparison to other ceramic, hybrid (ceramic + metallic), and amorphous reinforcements synthesized by powder metallurgy. It can be seen from Figure 4 that, Mg–Sm_2O_3 nanocomposites exhibited the highest UCS with better fracture strain in comparison to other ceramic reinforcements (represented by green bubble). Also, the strength values were comparable to those of hybrid (represented by blue bubble) and amorphous (represented by red bubble) particle-reinforced Mg composites. Further, it should be noted that the room temperature compressive properties of the synthesized nanocomposites are superior or comparable to the properties of commercial magnesium alloys such as AZ91, WE43, WE54, and ME21 [1,41]. The combined enhancement in both strength-ductility properties which is better than most powder metallurgy synthesized nanocomposites justifies the use of Sm_2O_3 as a reinforcement, and would encourage researchers in the Mg community to further study its behavior and suitability for various commercial applications.

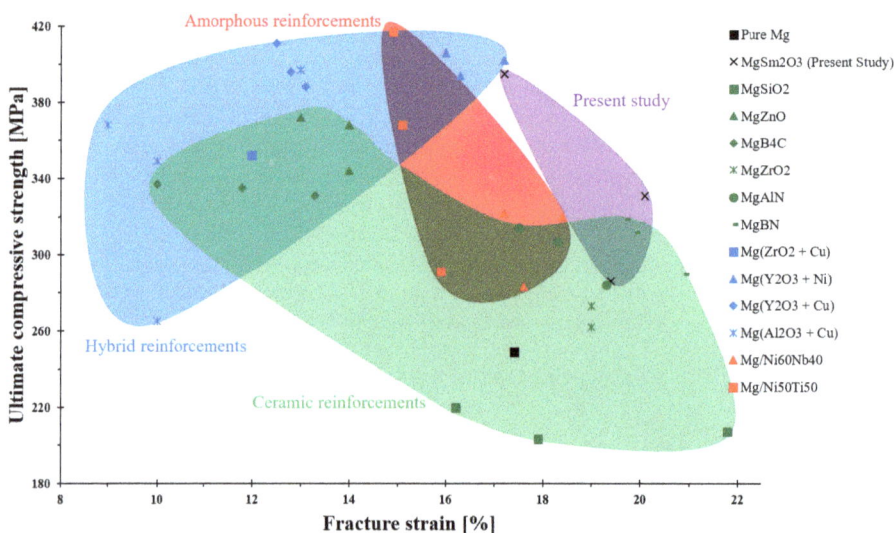

Figure 4. Bubble chart illustrating the superior compressive behaviour of Mg-Sm$_2$O$_3$ nanocomposites compiled from references [3,10].

3.2.6. Elastic Modulus and Damping Characteristics

Figure 5 shows a set of amplitude–time plots of representative samples and Table 5 lists damping loss rate, damping capacity, and elastic moduli of pure Mg and their nanocomposite samples. The vibration signal from each sample is recorded in terms of amplitude vs. time in free vibration mode. The results clearly indicate that the amplitude and time taken to stop the vibration are different for each material and the addition of Sm_2O_3 NPs significantly enhances the damping characteristics of pure Mg. It can be seen from Figure 5 that the amplitude decreases gradually for pure Mg as against steeper fall in Mg-Sm$_2$O$_3$ nanocomposites. With the addition of 0.5 and 1 vol % Sm_2O_3 in Mg matrix, the time taken to damp the vibrations is reduced significantly from 0.65 to 0.37 and 0.29 s, respectively. Most of the vibrations are ceased in less than ~0.2 s with the addition of 1.5 vol % Sm_2O_3.

Table 5. Elastic modulus and damping characteristics of Mg-Sm$_2$O$_3$ nanocomposites.

Material	Damping Loss Rate	Damping Capacity	Elastic Modulus (GPa)
Pure Mg	8.2 ± 0.2	0.000394 ± 0.000021	42.3 ± 0.14
Mg-0.5 Sm$_2$O$_3$	20.2 ± 0.4 (* 2.46)	0.000719 ± 0.000017 (↑82.48%)	43.7 ± 0.1 (↑3.30%)
Mg-1 Sm$_2$O$_3$	29.35 ± 1.2 (* 3.57)	0.001049 ± 0.00058 (↑166.24%)	45.4 ± 0.08 (↑7.32%)
Mg-1.5 Sm$_2$O$_3$	36.65 ± 0.9 (* 4.47)	0.0011395 ± 0.0008 (↑189.21%)	44.9 ± 0.2 (↑6.14%)

(* x) indicates the increase in the property with respect to pure Mg by x times; (↑x%) and (↓x%) indicates the percentage increase and decrease in the property with respect to pure Mg by x%, respectively.

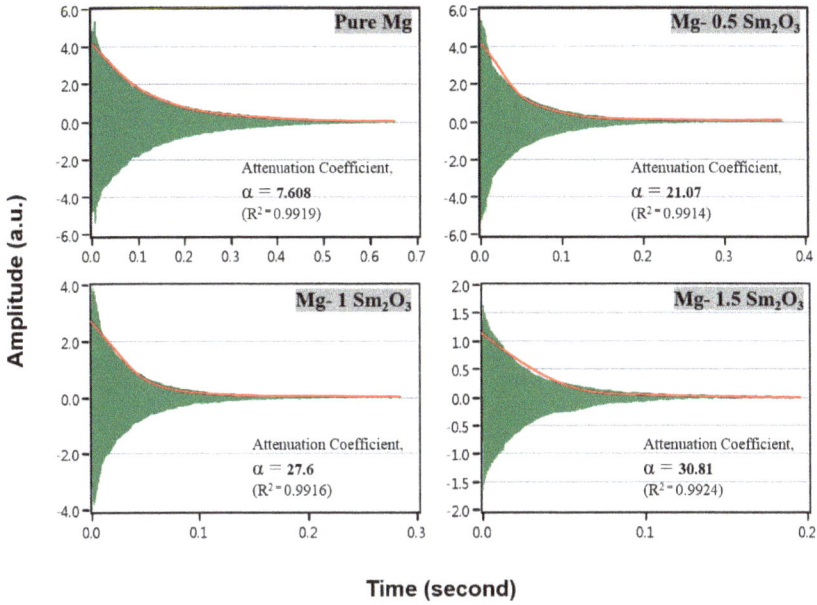

Figure 5. Damping characteristics of Mg and its nanocomposite samples.

The damping loss rate (L) which is the ability of a material to absorb vibration [42], showed an increase with the addition of Sm_2O_3 NPs (Table 5) and Mg-1.5 vol % Sm_2O_3 exhibited the maximum value of ~36.65 (~4.47 times greater than that of pure Mg). Damping loss rate as a function of volume percent of the reinforcement follows a linear fit and can be expressed using Equation (2) as,

$$L = 9.425 + 18.9 \times X \text{ (vol \%)}, (R^2 = 0.9877) \tag{2}$$

Further, the damping capacity (Q^{-1}) of Mg is also seen to be increasing with the addition of Sm_2O_3 NPs. The highest damping capacity of 0.0011395 (~189.21% rise as compared to pure Mg) is shown by Mg-1.5 vol % Sm_2O_3 nanocomposite. This overall enhancement of the damping properties of Mg-Sm_2O_3 nanocomposites might be due to the presence of a plastic zone around reinforcement, increase in dislocation density, and due to other damping sources, such as grain boundary sliding mechanisms, defects, and porosities which are analyzed for their validity in the following paragraphs.

A more qualitative evaluation of damping would be by considering the attenuation coefficients. In magnesium and its alloys, the amplitude of free vibration gradually decreases with increase in time and the difference lies in the steepness of the curve which is quantified by the attenuation coefficient. In this study, the material vibrates at a resonant frequency when excited by cyclic external force and when this external force is removed, the resonant-vibration dampens gradually. Then, the amplitude of a damping vibration, $A(t)$, can be expressed as

$$A(t) = A_o \, \exp\left[-\alpha t + i\frac{\pi}{f_r}t\right] \tag{3}$$

where, "t" is the time after removal of the external force, "A_0" denotes the amplitude at $t = 0$, "α" represents the attenuation (damping) coefficient (which depends on the damping capacity of materials), and f_r denotes the resonant frequency. Here, in the case of the Mg-Sm_2O_3 nanocomposites, various vibration modes are excited, and therefore we cannot obtain the attenuation coefficient by fitting Equation (3) to the vibration–damping curve. In order to determine α in such a case, the maximum value of positive amplitude of the damping curve has been picked up and fitted according to the following equation:

$$A(t) = A_o \, e^{(-\alpha)t} + C \tag{4}$$

The amplitude and the apparent attenuation coefficients of Mg-Sm_2O_3 nanocomposites have been obtained qualitatively, where C denotes the fitting coefficient. An increasing trend in α is clearly evident from Figure 5 with the increasing amount of Sm_2O_3 NPs. A notable enhancement in the value of α from 7.608 to 30.81 is observed in case of Mg-1.5 vol % Sm_2O_3 nanocomposite as compared to pure Mg, demonstrating significant rise in damping capability.

The CTE of Sm_2O_3 and Mg is 8.5 [43] and 27.1 $\times$ 10^{-6}/K, respectively. This difference in thermal expansion coefficient between Mg and Sm_2O_3 might induce high residual stresses around the particulates in the Mg matrix, resulting in the formation of plastic deformation zone at the particle/matrix interface. According to the plastic zone damping model proposed by Carreno-Morelli et al. [44], the damping capacity of a material depends directly on the volume fraction of plastic zone. Therefore, progressive increase in the energy dissipation of pure Mg matrix can be attributed to the higher amount of plastic zone around Sm_2O_3 NPs and further, at higher volume fractions effects are multifold resulting in such a rise in the damping capacity of nanocomposites. Further, significantly higher damping capacities realized for Mg-1 vol % Sm_2O_3 and Mg-1.5 vol % Sm_2O_3 nanocomposites (~166.24% and 189.21% rise as compared to pure Mg, respectively) can be due to overlapping of plastic zones, caused when the plastic zone is larger compared to smaller inter-particulate distances as the volume fraction of NPs increases. This increase in the presence of plastic zones due to the presence of Sm_2O_3 NPs leading to an increase in the hardness of the nanocomposite samples (Table 1) is found to be substantially high when compared to pure Mg.

Also, thermal mismatch between the constituents leads to higher dislocation density in the matrix. The increase in dislocation density is given as follows [45],

$$\rho = \frac{9.6 \, \Delta\alpha\Delta T V_p}{bd} \tag{5}$$

where, "$\Delta\alpha$" is the thermal expansion coefficient mismatch between the matrix alloy and the filler ($\times 10^{-6}$/K), "ΔT" is the difference between working and final temperatures (°C), "b" is denoted by the Burger vector, "V_p" represents volume fraction (%), and "d" is the diameter of reinforcement (m). The CTE difference between Sm_2O_3 and Mg is around 18.6 $\times$ 10^{-6}/K. Dislocation density can be quite significant at the interface and it increases with increasing Sm_2O_3 content. For magnesium-based materials, increased dislocations are favorable for the damping enhancement as dislocation pinning contributes to the damping behavior of magnesium nanocomposites [46]. The increase in dislocation density can also be attributed to the presence of hard Sm_2O_3 NPs in the magnesium matrix [47]. In addition, the crystal structure will be distorted locally at the matrix/reinforcement interface due to the presence of two-dimensional defects at the interface. Thereby, atoms may slip up at the interface, resulting in flexible dislocation movement and leading to higher damping response [48].

Further, it has also been observed that defects play an important role in tailoring damping properties. Chung [49] suggested that defects may shift the locations during vibration, acting as internal friction resources leading to higher damping capacities. The presence of porosity further augments the damping capacity due to the heterogeneous stress–strain distribution, causing stress concentrations which results in higher dislocation movements [50]. From Table 1, an increase in the addition of Sm_2O_3 NPs increased the porosity levels of Mg-Sm_2O_3 nanocomposites. The highest damping capacity is observed for Mg-1.5 vol % Sm_2O_3 nanocomposite which has maximum matrix porosity levels as seen from Table 1. Based on the aforementioned reasons, the damping is likely to be dominated by the presence of porosity and the microstructural variations due to Sm_2O_3 NPs in Mg-Sm_2O_3 nanocomposites [35]. Elasto-thermodynamic damping and grain boundary damping are not expected to be significant in this study due to room temperature operation conditions, sample dimensions, and frequency magnitude.

Along with the compression and hardness properties, elastic modulus and damping capacity are the two important properties to investigate for applications targeting orthopedic implants. High elastic modulus (such as exhibited by steel and titanium) as compared to the natural bone results in stress-shielding effects and decreases the stimulation of new bone growth, leading to implant failure [5]. The high damping capacity of a metallic implant helps in mitigating the vibrations caused when the patient moves and suppresses the stresses developed at the bone/implant interface to achieve better osseointegration [51]. The addition of Sm_2O_3 NPs enhanced the damping characteristics of pure Mg with marginal increase in the elastic modulus (Table 5). This marginal increase in the elastic modulus with increasing amount of Sm_2O_3 NPs can be attributed to the presence of Sm_2O_3 which exhibits a higher elastic modulus of about 183 GPa [22]. However, it was observed that elastic modulus of all samples remained lower than theoretical values, which can be attributed to the presence of porosity. The effect of presence of high modulus Sm_2O_3 NPs was largely negated by the presence of relatively higher amounts of porosity in the composites.

4. Biomechanical Properties

The compressive and elastic modulus properties can be collectively termed as "biomechanical properties". The biomechanical properties of Mg-Sm_2O_3 nanocomposites are compared with natural bone, cortical bone tissue, Ti-6Al-4V alloy, 316L stainless steel, and Co-Cr alloy in Table 6. The compressive strength is necessary for development and growth of bone and is responsible for deposition of bone material and if the compressive stress exceeds the UCS of bone it will eventually fracture. To avoid stress shielding effects, it is very important that the compressive strength of the material should not exceed the strength of the surrounding bone [52]. Natural bone exhibits certain hierarchical structures of nanometer dimensions within bone matrices and so implants are of ample concern for bone repair and regeneration in biomedical applications [53]. From the results shown in Table 6 compiled from references [8,35,54–56], the 0.2 CYS, UCS, and elastic modulus for the nanocomposite samples are closer to that of bone and bone tissues and could improve the interface between the nanocomposite and bone cells. The favored biomaterials, titanium alloys (Ti-6Al-4V), 316L stainless steel, and Co-Cr alloys exhibit significantly higher elastic modulus and are preferred only as permanent fixtures [13]. Further, in these materials there is a possibility of leaching of ions by corrosion or wear, thus decreasing their biocompatibility and causing tissue loss [57]. Also, as stated in [53], mismatch in elastic modulus of bone and steels/titanium/Co-Cr alloys is likely to result in bone resorption and loosening of implants. Hence, Mg-Sm_2O_3 nanocomposites may effectively increase the stimulation of new bone growth and re-modelling which increases the implant stability, making it favorable for applications in temporary implants avoiding stress-morbidity to the patient. The results of this study suggest that the biomechanical properties of the nanocomposites have an advantage over titanium alloys, 316L stainless steel, and Co-Cr alloys.

Table 6. Comparison between biomechanical properties of Mg-Sm_2O_3 nanocomposites with Ti-6Al-4V alloy, 316L stainless steel, Co-Cr alloy, and hard and soft tissues of the human body.

Material	Density (g/cc)	0.2% CYS (MPa)	UCS (MPa)	Fracture Strain (%)	Elastic Modulus (GPa)
Natural Bone	1.8–2.1 [a]	130–180 [a]	-	-	3–20 [a]
Cortical Bone	-	-	131–224 [b]	2–12 [b]	15–30 [b]
Ti-6Al-4V alloy	4.43 [c]	970 [c]	-	-	113.8 [c]
316L Stainless Steel	8.0 [d]	170–310 [a]	-	-	193 [d]
Co-Cr alloy	9.12–9.24 [e]	-	283–313 [e]	-	222–240 [e]
Pure Mg	1.7265	74	249	17.4	42.3
Mg-0.5 Sm_2O_3	1.7287	87	285	19.8	43.7
Mg-1 Sm_2O_3	1.7531	118	331	20.1	45.4
Mg-1.5 Sm_2O_3	1.7838	128	395	17.2	44.9

Compiled from reference: [a]—[8]; [b]—[35]; [c]—[54]; [d]—[56]; [e]—[55].

5. Conclusions

Mg-Sm$_2$O$_3$ nanocomposites were successfully synthesized using the powder metallurgy method including hybrid microwave sintering. Simultaneously, mechanical, microstructural, and damping properties were determined and analyzed. The following conclusions can be made from this study:

1. The grain size reduced with the progressive incorporation of Sm$_2$O$_3$ NPs to pure Mg, with Mg-1.5 vol % Sm$_2$O$_3$ exhibiting a maximum of 46.7% reduction in grain size with respect to pure Mg.
2. The hardness of pure Mg increased with the increasing amount of Sm$_2$O$_3$ with Mg-1.5 vol % Sm$_2$O$_3$ showing a maximum increase of ~37%.
3. The CTE values reduced with the incorporation of Sm$_2$O$_3$ NPs in pure Mg with Mg-1.5 vol % Sm$_2$O$_3$ showing a reduction of ~8.22% and the ignition temperature of Mg-1.5 vol % Sm$_2$O$_3$ showed the highest resistance to ignition (enhancement by ~69 °C), indicating superior thermal and dimensional stability.
4. The damping loss rate and damping capacity of pure Mg enhanced with the increasing amount of Sm$_2$O$_3$ NPs, with the Mg-1.5 vol % Sm$_2$O$_3$ nanocomposite displaying the best damping response (~4.5 times better than pure Mg).
5. The best compressive strength was exhibited by the Mg-1.5 vol % Sm$_2$O$_3$ nanocomposite with 0.2 CYS and UCS values increasing by ~56% and 53% when compared to pure Mg. The ductility values of Mg-Sm$_2$O$_3$ composites were either better or similar to pure Mg.
6. The superior compressive and damping properties with elastic modulus closer to natural bone makes Mg-Sm$_2$O$_3$ composites a potential choice as implant materials.

This study introduces a lightweight Mg-Sm$_2$O$_3$ nanocomposite with an excellent combination of strength, ductility, ignition resistance, and damping behavior. The superior-performance nanocomposite presented in this study has great potential in automobile and aerospace applications, and can be extended to others including defense, sport, electronic, and biomedical sectors. However, a considerable amount research is still necessary to validate these materials for their tensile, dynamic, high temperature, corrosion, fatigue, and wear properties before seeing their widespread use in industrial applications. Also, further study into the ignition mechanisms and detailed analysis is necessary to further exploit Mg nanocomposites and validate them for aerospace and defense applications. The promising results obtained in this study by the addition of rare earth oxides (REO) such as Sm$_2$O$_3$ to Mg presents a potential for the progress in research towards development of other REO-reinforced Mg nanocomposites to further ascertain the viability and the usefulness of such nanocomposites.

Author Contributions: Ashis Mallick and Manoj Gupta and Khin Sandar Tun proposed the original project and supervised the investigation. Milli Suchita Kujur performed the experiments. Milli Suchita Kujur, Vyasaraj Manakari and Gururaj Parande analyzed the data, and wrote the paper with assistance from all authors. All authors contributed to the discussions in the manuscript.

Conflicts of Interest: The authors declare no conflict of interest.

References

1. Gupta, M.; Sharon, N.M.L. *Magnesium, Magnesium Alloys, and Magnesium Composites*; John Wiley & Sons: New York, NY, USA, 2011.
2. Mordike, B.; Ebert, T. Magnesium: Properties—Applications—Potential. *Mater. Sci. Eng. A* **2001**, *302*, 37–45. [CrossRef]
3. Gupta, M.; Wong, W. Magnesium-based nanocomposites: Lightweight materials of the future. *Mater. Charact.* **2015**, *105*, 30–46. [CrossRef]
4. Gupta, M.; Parande, G.; Manakari, V. An insight into high performance magnesium alloy/nano-metastable-syntactic composites. In *Proceedings of the 17th Australian International Aerospace Congress: AIAC 2017*, Melbourne, Australia, 26 February–2 March 2017; p. 270.

5. Manakari, V.; Parande, G.; Gupta, M. Selective laser melting of magnesium and magnesium alloy powders: A review. *Metals* **2016**, *7*, 2. [CrossRef]

6. Gröber, U.; Schmidt, J.; Kisters, K. Magnesium in prevention and therapy. *Nutrients* **2015**, *7*, 8199–8226. [CrossRef] [PubMed]

7. Classen, H.; Nowitzki, S. The clinical importance of magnesium. 2. The indications for supplementation and therapy. *Fortschr. Med.* **1990**, *108*, 198–200. [PubMed]

8. Staiger, M.P.; Pietak, A.M.; Huadmai, J.; Dias, G. Magnesium and its alloys as orthopedic biomaterials: A review. *Biomaterials* **2006**, *27*, 1728–1734. [CrossRef] [PubMed]

9. Sietsema, W. Animal models of cortical porosity. *Bone* **1995**, *17*, S297–S305. [CrossRef]

10. Sankaranarayanan, S.; Gupta, M. Review on mechanical properties of magnesium (nano) composites developed using energy efficient microwaves. *Powder Metall.* **2015**, *58*, 183–192. [CrossRef]

11. Ubaid, F.; Matli, P.R.; Shakoor, R.A.; Parande, G.; Manakari, V.; Mohamed, A.M.A.; Gupta, M. Using B_4C nanoparticles to enhance thermal and mechanical response of aluminum. *Materials* **2017**, *10*, 621. [CrossRef] [PubMed]

12. Barta, C.A.; Sachs-Barrable, K.; Jia, J.; Thompson, K.H.; Wasan, K.M.; Orvig, C. Lanthanide containing compounds for therapeutic care in bone resorption disorders. *Dalton Trans.* **2007**, 5019–5030. [CrossRef] [PubMed]

13. Bayani, H.; Saebnoori, E. Effect of rare earth elements addition on thermal fatigue behaviors of AZ91 magnesium alloy. *J. Rare Earths* **2009**, *27*, 255–258. [CrossRef]

14. Haque, N.; Hughes, A.; Lim, S.; Vernon, C. Rare earth elements: Overview of mining, mineralogy, uses, sustainability and environmental impact. *Resources* **2014**, *3*, 614–635. [CrossRef]

15. Kang, Y.-B.; Jin, L.; Chartrand, P.; Gheribi, A.E.; Bai, K.; Wu, P. Thermodynamic evaluations and optimizations of binary mg-light rare earth (La, Ce, Pr, Nd, Sm) systems. *Calphad* **2012**, *38*, 100–116. [CrossRef]

16. Tekumalla, S.; Gupta, M. An insight into ignition factors and mechanisms of magnesium based materials: A review. *Mater. Des.* **2017**, *113*, 84–98. [CrossRef]

17. Walker, J.; Shadanbaz, S.; Woodfield, T.B.; Staiger, M.P.; Dias, G.J. Magnesium biomaterials for orthopedic application: A review from a biological perspective. *J. Biomed. Mater. Res. Part B Appl. Biomater.* **2014**, *102*, 1316–1331. [CrossRef] [PubMed]

18. Cockeram, B. The fracture toughness and toughening mechanism of commercially available unalloyed molybdenum and oxide dispersion strengthened molybdenum with an equiaxed, large grain structure. *Metall. Mater. Trans. A* **2009**, *40*, 2843–2860. [CrossRef]

19. Liu, G.; Zhang, G.; Jiang, F.; Ding, X.; Sun, Y.; Sun, J.; Ma, E. Nanostructured high-strength molybdenum alloys with unprecedented tensile ductility. *Nat. Mater.* **2013**, *12*, 344–350. [CrossRef] [PubMed]

20. Kirkland, N.T. Magnesium biomaterials: Past, present and future. *Corros. Eng. Sci. Technol.* **2012**, *47*, 322–328. [CrossRef]

21. Navrotsky, A.; Lee, W.; Mielewczyk-Gryn, A.; Ushakov, S.V.; Anderko, A.; Wu, H.; Riman, R.E. Thermodynamics of solid phases containing rare earth oxides. *J. Chem. Thermodyn.* **2015**, *88*, 126–141. [CrossRef]

22. Martienssen, W.; Warlimont, H. Springer handbook of condensed matter and materials data. Springer Science & Business Media: Berlin, Germany, 2006.

23. Muneer, I.; Farrukh, M.A.; Javaid, S.; Shahid, M.; Khaleeq-ur-Rahman, M. Synthesis of Gd_2O_3/Sm_2O_3 nanocomposite via sonication and hydrothermal methods and its optical properties. *Superlattices Microstruct.* **2015**, *77*, 256–266. [CrossRef]

24. Herath, H.; Di Silvio, L.; Evans, J. In vitro evaluation of samarium (III) oxide as a bone substituting material. *J. Biomed. Mater. Res. Part A* **2010**, *94*, 130–136. [CrossRef] [PubMed]

25. Marker, T. *Development of a Laboratory-Scale Flammability Test for Magnesium Alloys Used in Aircraft Seat Construction*; Federal Aviation Administration William J. Hughes Technical Center: Egg Harbor Township, NJ, USA, 2014.

26. Parande, G.; Manakari, V.; Meenashisundaram, G.K.; Gupta, M. Enhancing the tensile and ignition response of monolithic magnesium by reinforcing with silica nanoparticulates. *J. Mater. Res.* **2017**, *32*, 1–10. [CrossRef]

27. Reddy, M.P.; Ubaid, F.; Shakoor, R.; Parande, G.; Manakari, V.; Mohamed, A.; Gupta, M. Effect of reinforcement concentration on the properties of hot extruded $Al-Al_2O_3$ composites synthesized through microwave sintering process. *Mater. Sci. Eng. A* **2017**, *696*, 60–69. [CrossRef]

28. Fida Hassan, S.; Al-Aqeeli, N.; Gasem, Z.; Tun, K.; Gupta, M. Magnesium nanocomposite: Increasing copperisation effect on high temperature tensile properties. *Powder Metall.* **2016**, *59*, 66–72. [CrossRef]

29. Vaidya, R.U.; Chawla, K. Thermal expansion of metal-matrix composites. *Compos. Sci. Technol.* **1994**, *50*, 13–22. [CrossRef]

30. Czerwinski, F. Overcoming barriers of magnesium ignition and flammability. *Adv. Mater. Process.* **2014**, *172*, 28–31.

31. Schwartz, M. *Encyclopedia and Handbook of Materials, Parts and Finishes*; Crc Press: Boca Raton, FL, USA, 2016.

32. Phillips, W. Oxidation of several lanthanide elements. *J. Less Common Met.* **1964**, *7*, 139–143. [CrossRef]

33. Nguyen, Q.; Gupta, M. Microstructure and mechanical characteristics of AZ31B/Al$_2$O$_3$ nanocomposite with addition of Ca. *J. Compos. Mater.* **2009**, *43*, 5–17. [CrossRef]

34. Manakari, V.; Parande, G.; Doddamani, M.; Gupta, M. Enhancing the ignition, hardness and compressive response of magnesium by reinforcing with hollow glass microballoons. *Materials* **2017**, *10*, 997. [CrossRef] [PubMed]

35. Parande, G.; Manakari, V.; Meenashisundaram, G.K.; Gupta, M. Enhancing the hardness/compression/damping response of magnesium by reinforcing with biocompatible silica nanoparticulates. *Int. J. Mater. Res.* **2016**, *107*, 1091–1099. [CrossRef]

36. Nguyen, Q.; Gupta, M. Enhancing compressive response of AZ31B magnesium alloy using alumina nanoparticulates. *Compos. Sci. Technol.* **2008**, *68*, 2185–2192. [CrossRef]

37. Hassan, S.; Gupta, M. Development of high strength magnesium copper based hybrid composites with enhanced tensile properties. *Mater. Sci. Technol.* **2003**, *19*, 253–259. [CrossRef]

38. Zhang, D.; Jiang, L.; Zheng, B.; Schoenung, J.; Mahajan, S.; Lavernia, E.; Beyerlein, I. Deformation twinning (update). *Ref. Modul. Mater. Sci. Eng.* **2015**, 1–24. Available online: https://doi.org/10.1016/B978-0-12-803581-8.02878-2 (accessed on 31 October 2016).

39. Zhang, D.; Wen, H.; Kumar, M.A.; Chen, F.; Zhang, L.; Beyerlein, I.J.; Schoenung, J.M.; Mahajan, S.; Lavernia, E.J. Yield symmetry and reduced strength differential in Mg-2.5 Y alloy. *Acta Mater.* **2016**, *120*, 75–85. [CrossRef]

40. Stanford, N.; Barnett, M. The origin of "rare earth" texture development in extruded Mg-based alloys and its effect on tensile ductility. *Mater. Sci. Eng. A* **2008**, *496*, 399–408. [CrossRef]

41. Chen, Y.; Tekumalla, S.; Guo, Y.; Gupta, M. Introducing Mg-4Zn-3Gd-1Ca/ZnO nanocomposite with compressive strengths matching/exceeding that of mild steel. *Sci. Rep.* **2016**, *6*. [CrossRef] [PubMed]

42. Kumar, A.; Meenashisundaram, G.K.; Manakari, V.; Parande, G.; Gupta, M. Lanthanum effect on improving cte, damping, hardness and tensile response of Mg-3Al alloy. *J. Alloy. Compd.* **2017**, *695*, 3612–3620. [CrossRef]

43. Stecura, S.; Campbell, W.J. *Thermal Expansion and Phase Inversion of Rare-Earth Oxides*; US Dept. of the Interior, Bureau of Mines: Washington, DC, USA, 1961; Volume 5847.

44. Carreno-Morelli, E.; Urreta, S.; Schaller, R. Mechanical spectroscopy of thermal stress relaxation at metal–ceramic interfaces in aluminium-based composites. *Acta Mater.* **2000**, *48*, 4725–4733. [CrossRef]

45. Yi, H.-K.; Liu, Z.-T.; Li, F.-H. Investigation on room temperature damping vs strain amplitude behaviors of hypereutectic Al-17Si-xLa alloys. *J. Funct. Mater.* **2003**, *34*, 525–527.

46. Anilchandra, A.; Surappa, M. Microstructure and damping behaviour of consolidated magnesium chips. *Mater. Sci. Eng. A* **2012**, *542*, 94–103. [CrossRef]

47. Srikanth, N.; Zhong, X.; Gupta, M. Enhancing damping of pure magnesium using nano-size alumina particulates. *Mater. Lett.* **2005**, *59*, 3851–3855. [CrossRef]

48. Xiuqing, Z.; Haowei, W.; Lihua, L.; Naiheng, M. In situ synthesis method and damping characterization of magnesium matrix composites. *Compos. Sci. Technol.* **2007**, *67*, 720–727. [CrossRef]

49. Chung, D. Review: Materials for vibration damping. *J. Mater. Sci.* **2001**, *36*, 5733–5737. [CrossRef]

50. Zhang, J.; Gungor, M.; Lavernia, E. The effect of porosity on the microstructural damping response of 6061 aluminium alloy. *J. Mater. Sci.* **1993**, *28*, 1515–1524. [CrossRef]

51. Tsai, M.-H.; Chen, M.-S.; Lin, L.-H.; Lin, M.-H.; Wu, C.-Z.; Ou, K.-L.; Yu, C.-H. Effect of heat treatment on the microstructures and damping properties of biomedical Mg–Zr alloy. *J. Alloy. Compd.* **2011**, *509*, 813–819. [CrossRef]

52. Mallick, K. *Bone Substitute Biomaterials*; Elsevier: Amsterdam, The Netherlands, 2014.

53. Xu, T.; Zhang, N.; Nichols, H.L.; Shi, D.; Wen, X. Modification of nanostructured materials for biomedical applications. *Mater. Sci. Eng. C* **2007**, *27*, 579–594. [CrossRef]

54. Cho, K.; Niinomi, M.; Nakai, M.; Liu, H.; Santos, P.F.; Itoh, Y.; Ikeda, M. *Tensile and Compressive Properties of Low-Cost High-Strength β-Type Ti-Mn Alloys Fabricated by Metal Injection Molding*; Wiley Online Library: San Diego, CA, USA, 2015; pp. 499–503.

55. Henry, D. *Materials and Coatings for Medical Devices: Cardiovascular*; ASM International: Geauga County, OH, USA, 2009; pp. 151–186.

56. Handbook, M. *Properties and Selection: Nonferrous Alloys and Special-Purpose Materials*; ASM International: Geauga County, OH, USA, 1990; Volume 2, p. 102.

57. Puleo, D.A.; Huh, W.W. Acute toxicity of metal ions in cultures of osteogenic cells derived from bone marrow stromal cells. *J. Appl. Biomater.* **1995**, *6*, 109–116. [CrossRef] [PubMed]

metals

MDPI

Article

Ultrasound Assisted Casting of an AM60 Based Metal Matrix Nanocomposite, Its Properties, and Recyclability

Hajo Dieringa [1,*], Lydia Katsarou [1], Ricardo Buzolin [1], Gábor Szakács [1], Manfred Horstmann [1], Martin Wolff [1], Chamini Mendis [1,2], Sergey Vorozhtsov [3] and David StJohn [4]

[1] Institute of Materials Research, Helmholtz-Zentrum Geesthacht, Max-Planck-Str. 1, 21502 Geesthacht, Germany; lydiapvnrt@gmail.com (L.K.); ricardo.buzolin@tugraz.at (R.B.); gabor.szakacs@hzg.de (G.S.); manfred.horstmann@hzg.de (M.H.); martin.wolff@hzg.de (M.W.); chamini.mendis@brunel.ac.uk (C.M.)

[2] BCAST-Brunel Centre for Advanced Solidification Technology, Brunel University, Uxbridge, Middlesex UB8 3PH, UK

[3] Faculty of Physics and Engineering, National Research Tomsk State University, Tomsk 634050, Russia; vorn1985@gmail.com

[4] Centre for Advanced Materials Processing and Manufacturing, School of Mechanical and Mining Engineering, The University of Queensland, Brisbane, QLD 4072, Australia; d.stjohn@uq.edu.au

* Correspondence: hajo.dieringa@hzg.de; Tel.: +49-4152-871955

Received: 16 August 2017; Accepted: 19 September 2017; Published: 22 September 2017

Abstract: An AM60 magnesium alloy nanocomposite reinforced with 1 wt % of AlN nanoparticles was prepared using an ultrasound (US) assisted permanent-mould indirect-chill casting process. Ultrasonically generated cavitation and acoustic streaming promoted de-agglomeration of particle clusters and distributed the particles throughout the melt. Significant grain refinement due to nucleation on the AlN nanoparticles was accompanied by an exceptional improvement in properties: yield strength increased by 103%, ultimate tensile strength by 115%, and ductility by 140%. Although good grain refinement was observed, the large nucleation undercooling of 14 K limits further refinement because nucleation is prevented by the formation of a nucleation-free zone around each grain. To assess the industrial applicability and recyclability of the nanocomposite material in various casting processes, tests were performed to determine the effect of remelting on the microstructure. With each remelting, a small percentage of effective AlN nanoparticles was lost, and some grain growth was observed. However, even after the third remelting, excellent strength and ductility was retained. According to strengthening models, enhanced yield strength is mainly attributed to Hall-Petch strengthening caused by the refined grain size. A small additional contribution to strengthening is attributed to Orowan strengthening.

Keywords: nanoparticles; metal matrix nanocomposite (MMNC); AlN; magnesium alloy AM60; strengthening mechanisms

1. Introduction

Magnesium alloys have been in use for over 90 years in weight critical applications. During the last few decades, disadvantages such as vulnerability to corrosion, limited high temperature strength, and creep resistance have been overcome by the development of advanced magnesium alloys containing rare earth elements [1–3], calcium [4,5], strontium [6,7], tin [8,9], or barium [10]. However, improvements achieved through novel alloy development are limited. This limit can only be overcome through reinforcing magnesium alloys with particles or fibres, similar to metal matrix composites (MMCs). Use of microscale reinforcements significantly improves the strength [11,12], wear [13], creep resistance [14–16], and fatigue strength [17,18]. However, these improvements require the addition of 10–40% of reinforcement. This high amount usually increases the density of the material significantly and lowers the ductility making it unsuitable for light weight applications.

The addition of 0.5 wt % AlN particles of less than 5 µm size to a Mg-3Al alloy reduced the grain size to 120 µm from 450 µm without AlN addition [19]. Defining a grain size reduction (GSR) by [20] GSR $= \left(\frac{1}{\sqrt{D_{MMNC}}} - \frac{1}{\sqrt{D_0}} \right)$, a value of $4.41 \times 10^{-2} \, \mu m^{-1/2}$ has been achieved. The premise for comparing GSR values is that the casting parameters are similar and there are no other influences that may affect grain size.

For several years now, nano sized particles have been used as reinforcement for metal matrix nanocomposites (MMNCs), and the number of publications on MMNC research is increasing, see Figure 1. The main reason for this increase is the dramatic reduction in the price of nanoparticles. Particles with a diameter below 100 nm have an ideal size for Orowan strengthening, even if only small amounts are added when uniformly distributed in the matrix. Paramsothy and Gupta published a study on the addition of 1.5 vol % AlN particles of 10 nm–20 nm in size to an AZ91/ZK60 hybrid alloy [21,22]. This material was processed by Disintegrated Melt Deposition (DMD) followed by hot extrusion. Although they did not observe any effect on grain size, the ductility slightly decreased, and tensile yield strength slightly increased. An overview of nanoparticle reinforced magnesium alloys is given in [23].

It is difficult to uniformly distribute nanoparticles in metallic melts because of their high surface area and the poor wettability of nanoparticles by the metallic melt. A uniform distribution can be easily produced with powder metallurgical processes but these processes are costly and cannot easily be used for mass production. Other melt metallurgical processes such as DMD [24,25] or an evaporation of magnesium after casting [26] have a scientific focus, but are not commercially viable. Conventional casting processes which require only slight modifications for distributing nanoparticles, are, therefore, a field of interest for research as well as for industry. For this reason, the European Project ExoMet was established to explore novel grain refining and nanoparticle additions in conjunction with melt treatment by means of external fields (electromagnetic, ultrasonic, and mechanical). These external fields provide an effective and efficient method to disperse the nanoparticles into the melt with uniform distribution in the as-cast material [27]. Of these fields, ultrasonic treatment is ideal as the ultrasound waves and cavitation under the ultrasound probe promote de-agglomeration of particle clusters and particle wetting. Also, acoustic streaming facilitates vigorous convection transporting the released particles throughout the melt. Magnesium alloy Elektron21 has been successfully reinforced with 1 wt % AlN nanoparticles and it was shown that AlN-reinforced Elektron21 has significantly improved creep resistance at 240 °C [28]. At low stresses, the minimum creep rate is nearly one order of magnitude lower compared to that of the unreinforced Elektron21, although Elektron21 is already one of the most creep resistant commercially available magnesium alloys. SEM and TEM investigations showed that the AlN nanoparticles are located in the eutectic region and in primary magnesium grains close to the eutectic region. The nanoparticles seem to strengthen the eutectic region, which ultimately results in creep strengthening of the nanocomposite. The reason for this may be that the particles tend to prevent the material from grain boundary sliding by strengthening the eutectic and grain boundary regions [28].

In this paper, we investigate microstructural features, strength, and recyclability of conventional magnesium High Pressure Die Casting (HPDC) alloy AM60 reinforced with 1 wt % of AlN nanoparticles and compare it to those of unreinforced AM60 processed using ultrasound assisted casting in both cases.

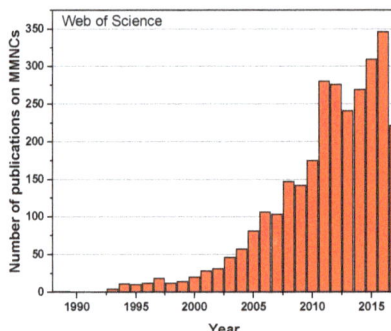

Figure 1. Number of publications on metal matrix nanocomposites (MMNCs) in Web of Science.

2. Materials and Methods

A commercial magnesium alloy AM60 was selected as it is widely used in the manufacture of high pressure die castings for the automotive industry. The nominal composition according to the supplier, MAGONTEC, of AM60 alloy is Mg-6Al-0.4Mn (wt %). The AlN nanoparticles were processed at Tomsk State University, in Russia, using electric explosion of aluminium metal wire in a nitrogen-containing atmosphere [29]. The medium particle size is 80 nm. Figure 2a shows typical particles and Figure 2b shows the particle size distribution.

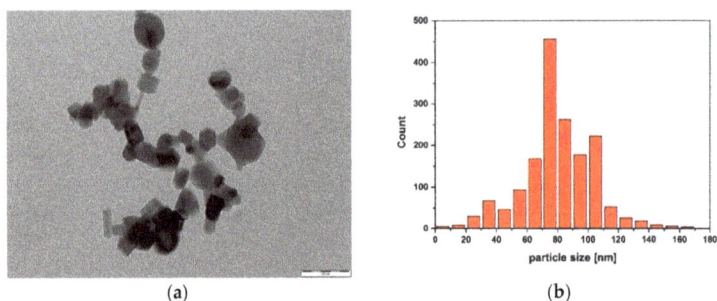

Figure 2. (**a**) Typical AlN nanoparticles (scale bar: 200 nm) and (**b**) their particle size distribution.

Approximately 3 kg of molten AM60 alloy at 720 °C was poured into a preheated cylindrical mould (450 °C), which was then placed within a three-zone resistance ring furnace (ThermConcept, Bremen, Germany) for maintaining the temperature of the melt at 670 °C. After creating a vortex by mechanical stirring (200 rpm), AlN nanoparticles wrapped in aluminium foil were introduced to the melt. As soon as the particles were no longer on the top of the melt, ultrasonic (0.3 kW, 20 kHz) stirring was applied for 5 min to disperse the particle clusters. After mixing, the stirrer and the ultrasound probe were removed from the melt and the mould was lowered mechanically into a water bath directly underneath the furnace opening. A steel mould (St52 or 1.0831) with 3 mm wall thickness was used. As the mould was lowered into the water bath at a speed of 3 mm/s, solidification initiated at the bottom and preceded upwards allowing shrinkage to be fed by the remaining melt above the solid-liquid interface producing very dense castings. This process was invented in the 1930s by the I.G. Farbenindustrie AG in Germany for production of slabs for rolling or extrusion. The cover gas, 1 vol % SF_6 with Ar, was used during the whole casting process at a constant flux.

For comparison, AM60 without nanoparticles was cast in the same way including mechanical and US assisted stirring. Cylinders of 100 mm diameter and a length of 250 mm were cast, and slices of 10 mm thickness were cut from the middle of the casting at a height of 125 mm. After extracting this slice for metallography and tensile test specimens, the remaining parts were remelted at a temperature of 670 °C in the mould. The melt was only slightly mechanically stirred and after holding for 5 min, the melt solidified as described above to obtain the first recycled cylinder from which a slice was cut out from the same area as was done in the first experiment. This process was repeated two more times in order to obtain three recycled and remelted cylinders from which the role of AlN nanoparticles during recycling could be elucidated.

Spiral casting experiments were performed with a spiral mould preheated to 375 °C and both melts of AM60 and AM60 containing AlN nanoparticles had a melt temperature of 675 °C. The length of the cast spiral was taken as a measure of relative viscosity of the melt. DSC (Differential Scanning Calorimetry) measurement was done with a DSC 2 from Mettler Toledo (Mettler-Toledo, Greifensee, Switzerland). Three heating and cooling cycles between 400 °C and 700 °C, from which the last two cycles were used for determination of the undercooling at the start of solidification. Density was determined following the Archimedean Principle by measuring the weight five times in air and ethanol with a Sartorius balance LA230S (Sartorius, Göttingen, Germany).

To study the microstructure, samples were cut from the slice at approximately 10 mm from the surface and were cold mounted using epoxy resin. Samples were ground (800, 1200, and 2500 grit) and hardness values were measured with an EMCO M1C 010 testing machine (EMCO-TEST, Kuchl, Austria) using a 5 kg load. The samples were ground with 2500 grit after the hardness tests and polished with a 1 µm diamond solution in OPS (oxide polishing suspension). The surface of the sample was etched with a 9 vol % picric acid solution and optical micrographs were recorded using polarised light microscopy with Nomarski contrast to see the grain structures.

Five micro-tensile specimens (gauge length: 9 mm, cross section 2 mm × 2 mm) were electro-discharge machined from the slices of the castings. The room temperature tensile tests were conducted using a 5 kN universal tensile testing machine (Zwick-Roell, Ulm, Germany) and the displacement was measured with a laser extensometer (Fiedler Optoelektronik GmbH, Lützen, Germany).

Scanning electron microscopy was performed using a Tescan Vega3 SEM (TESCAN ORSAY, Brno, Czech Republic) equipped with Tescan Energy dispersive X-ray (EDX) spectrometer. EDX spectra maps were measured in order to identify possible sites with a higher concentration of nitrogen. For calculations of the mismatch between AlN and Mg the crystal structures of Mg were generated with CaRIne crystallographic software [TM] (CaRIne 3.1, Software CaRIne Crystallography, Senlis, France, 2015) using data available in Pearsons crystallographic databases [30].

3. Results

3.1. Microstructure

The castings have a very uniform microstructure with less than 1% porosity and appear free of segregation. The optical micrographs in Figure 3 show that the addition of AlN refined the grain size significantly and the morphology of the grains changed from dendritic to a more equiaxed structure, Figure 3a,b. The measured grain size shows a significant reduction from 1277.0 ± 301.3 µm to 84.9 ± 6.2 µm due to the addition of nanoparticles to the AM60 alloy produced in a similar manner. Figure 3c–e and Table 1 show the microstructures and properties following recycling through remelting, with each remelting step increasing the grain size to 196.4 ± 16.0 µm after the third remelting cycle. Grain size distribution graphs of all AlN containing materials are shown in Figure 3f–i. The measured density and amount of porosity in the castings are given in Table 1. The amount of porosity was calculated assuming a theoretical density of 1.799552 g/cm^3 for AM60 + 1 wt % AlN and the results do not differ significantly.

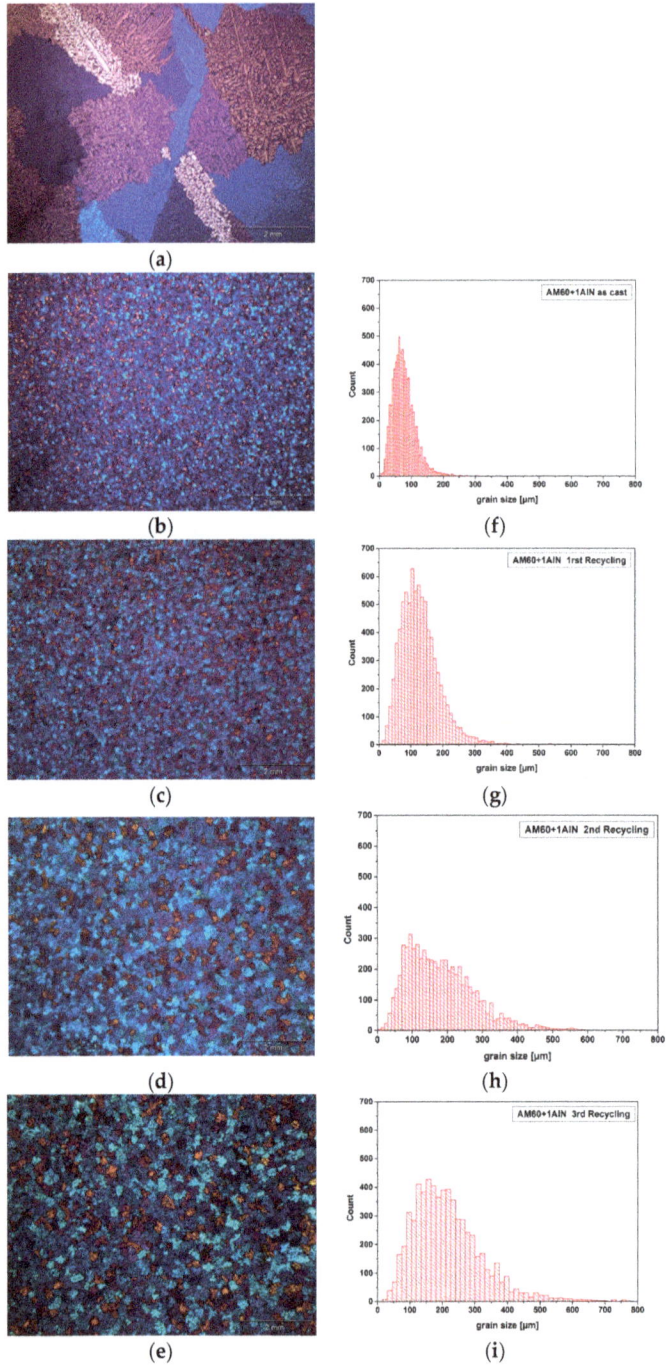

Figure 3. Microstructures of (**a**) AM60 and (**b**) AM60 + AlN. Microstructures of the first, second, and third recycled castings are shown in (**c**–**e**), respectively. Grain size distributions of (**b**–**e**) are presented in (**f**–**i**), respectively.

Table 1. Grain size, hardness, density, percent porosity, and the mechanical properties of the investigated materials. UTS = ultimate tensile strength.

Property	AM60	AM60 + AlN	1st Recycling	2nd Recycling	3rd Recycling
Grain size (μm)	1277.0 ± 301.3	84.9 ± 6.2	113.8 ± 22.1	176.8 ± 12.9	196.4 ± 16.0
Hardness (HV5)	48.0 ± 4.0	46.4 ± 6.0	50.9 ± 1.1	47.8 ± 1.0	48.5 ± 2.6
Density (g/cm³)	1.7848 ± 0.0004	1.783 ± 0	1.7842 ± 0.00075	1.7852 ± 0.0004	1.785 ± 0
Porosity (%)	-	0.919	0.853	0.797	0.808
Yield strength (MPa)	44.9 ± 6.9	91.2 ± 3.8 Δ = 46.3	73.7 ± 8.1 Δ = 28.8	70.9 ± 4.3 Δ = 26.0	69.9 ± 5.0 Δ = 25.0
UTS (MPa)	109.3 ± 19.2	235.1 ± 6.4	220.3 ± 16.8	210.0 ± 11.0	194.3 ± 12.3
Elongation (%)	6.4 ± 3.4	15.4 ± 4.2	15.4 ± 4.5	11.7 ± 2.7	10.1 ± 1.8

3.2. Mechanical Properties

The room temperature tensile tests show a remarkable increase in the yield strength (YS) and the ultimate tensile strength (UTS) after ultrasound (US) assisted addition of 1 wt % AlN nanoparticles, with both values being over twice that of the alloy without AlN nanoparticles, Table 1 and Figure 4a. There is an increase of 103% in YS and 115% in UTS. An increase of 140% was observed in the elongation to failure with the addition of AlN nanoparticles, Figure 4b. As mentioned, the reinforcement of micro-particles or fibres usually lowers the ductility significantly, but in this case, the AlN nanoparticle addition more than doubles the elongation to failure. Grain refinement often has a positive effect on ductility [31–33], although there is not a direct relationship between the two as other factors such as casting defects and the presence of intermetallic phases can also affect ductility. In this work, the absence of macroscopic cracking at twin boundaries is assumed to be the reason for improved ductility [33].

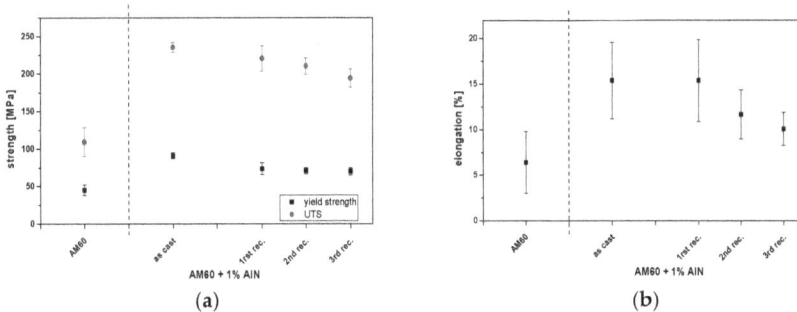

Figure 4. Mechanical properties: (**a**) yield and ultimate tensile strength and (**b**) elongation to fracture.

4. Discussion

AlN has a hexagonal crystal structure with the lattice parameters $a = 0.311$ nm and $c = 0.498$ nm, and Mg has the same lattice structure with lattice parameters of $a = 0.321$ nm and $c = 0.512$ nm [30]. The mismatch between Mg and AlN was approximately 3.1% between the $\{10\bar{1}0\}$ planes and 2.7% along the (0001) planes. The close matching between Mg and AlN is illustrated in Figure 5. Similar mismatches were observed on other major planes of AlN and Mg. The similarities in crystal structure and lattice parameter means that growth of Mg on an AlN particle is relatively easy and does not require accommodation of large strain, and the differences in lattice parameter can be overcome locally in the first layers of Mg that grow on the AlN particles.

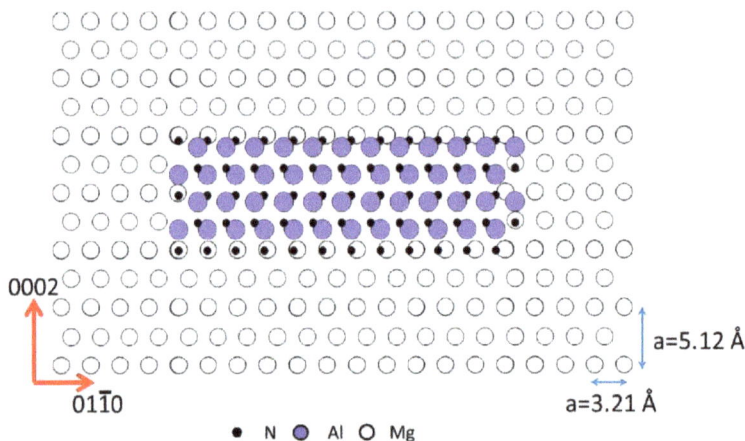

Figure 5. Crystallography of an AlN particle within a magnesium matrix.

4.1. Microstructure

The grain size can be assumed to be the average distance between successful nucleation events [34]. The number of particles added per gram of AlN powder of 80 nm average size is estimated to be 1.15×10^{15}. The largest particles were measured to be 162 nm and represent 0.003% of the total number of particles. The number of particles per gram was converted to the number of particles per volume. Assuming complete mixing of particles occurs during US, the average spacing between the largest particles would be 2.1 μm. If we also assume that the largest particles have the highest nucleation potencies [34,35] and that all of these nucleate a grain, then the grain size would be 2.1 μm compared with a measured average grain size of 84.9 μm. This indicates that only a very small fraction (approximately 0.002%) of the largest 0.003% of particles successfully nucleate a grain.

Several factors could reduce the number of nucleation events. One is particle agglomeration. Another is fading due to density differences where the higher density AlN particles sink to the bottom of the casting. These two factors are probably playing a role in the increased grain size obtained after each reheating cycle. Another cause may be particle pushing in front of a growing grain, although one might expect this to decrease the grain size. However, no localised increase in nitrogen content was observed at grain boundaries, suggesting particle pushing does not occur to a significant degree during solidification.

Considering the particle size distribution, it may be that only the very largest particles within the 162 nm band are effective nucleants. However, the alloy chemistry may have a more significant effect on grain size because of the formation of a large nucleation-free zone (NFZ) which prevents nucleation in a region around each newly formed grain. In other research [36] it was found that NFZ was the dominant factor setting the grain size of Mg-Al alloys. In fact, the final as-cast grain size was approximately equal to that in the NFZ. It was proposed that a very high density of nucleant particles naturally exist in the melt, such that a particle of suitably high potency is present at the end of NFZ where $\Delta T_{CS} = \Delta T_n$ triggers nucleation. The size of NFZ is calculated by the equation:

$$\text{NFZ} = \frac{D \cdot z \Delta T_n}{v \cdot Q} + \frac{4.6D}{v} \left(\frac{C_1^* - C_o}{C_1^* \cdot (1 - k)} \right) \qquad (1)$$

where D is the diffusion coefficient of solute in the liquid, v the velocity of the growing solid-liquid interface, Q the growth restriction factor, C_1^* the composition of the liquid at the solid-liquid interface, C_o the composition of the alloy, ΔT_n the nucleation undercooling, k the partition coefficient, and z the fraction of ΔT_n required to trigger the next nucleation event. The first term is the amount of growth required to create enough constitutional supercooling, ΔT_{CS}, to equal the nucleation undercooling of the nucleant particle, ΔT_n. The second term is the length of the constitutionally supercooled region in front of the growing grain's interface to the end of the diffusion field where $\Delta T_{CS} = \Delta T_n$. Given the number of particles of 162 nm (3.45×10^{12}), it can be assumed that there are sufficient particles to nucleate a grain as soon as $\Delta T_{CS} = \Delta T_n$ is reached (i.e., grain size equals NFZ).

ΔT_n was determined from DSC to be 14 K (16 K for AM60 without AlN present). Despite the good orientation relationship between AlN and α-Mg (above), the measured undercooling is in keeping with the Free Growth Model [35], which predicts a large nucleation undercooling due to the nanoscale size of the AlN particles. Using Equation (1) with Q of 26 K for AM60, D of 5×10^{-9} m^2/s and v of 5×10^{-6} µm/s, NFZ is of the order of 600 µm. D and v are estimates based on literature data [36]. The growth velocity v will initially be faster at an undercooling of 14 K (i.e., higher driving force) than for more potent particles. However, to reduce NFZ from 600 to about 85 µm, v would need to be about seven times faster. No data exist to verify a change of this magnitude, but given the casting rate of 3 mm/s, v may be considerably faster. On the other hand, because the grain size is relatively small, solute accumulation between the grains early in solidification will reduce the amount of constitutional supercooling which can quickly reduce v [34]. Reducing D to 7×10^{-10} m^2/s results in an NFZ (i.e., assumed to equal the grain size) of 89 µm. A possible reason for a lower diffusion coefficient is the impediment to diffusion caused by the very high density of particles (increased viscosity suggests a slower diffusion rate [37]). This effect was observed in the spiral casting tests where the length of the spiral was 96.2 cm for AM60 and 83.5 cm for AM60 + AlN. Therefore, it is possible that hypothesized changes to both v and D contribute to a reduction in the size of NFZ. Thus, the formation of NFZ during solidification prevents nucleation on many of the suitably potent particles present in the melt, reducing the potential for achieving a very fine grain size, for example, of less than 20 µm.

4.2. Mechanical Properties

A Hall-Petch diagram including the remelted nanocomposites is shown in Figure 6. It presents the relationship between the yield strength and the reciprocal of the square root of grain size (D) for AM60 and the AM60 based nanocomposites. According to the Hall-Petch relation (Equation (A1) in Appendix A) σ_0 was found to be 30.2 MPa and k_y to be 530.2 MPa µm$^{1/2}$.

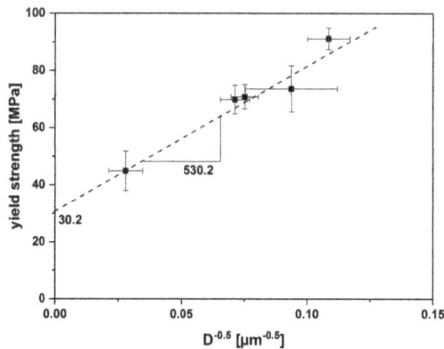

Figure 6. Hall-Petch plot of yield strength over the reciprocal of the square root of grain size D following Equation (A1) in Appendix A. (The standard deviation of the square root D is given in percentage of standard deviation of D and a minimum square root fit was applied without considering the standard deviations).

We have shown that the mechanical properties are comparable to HPDC, but without HPDC's inherent porosity (which can be up to 5%). Thus, the nanocomposites would be heat treatable and weldable, which is not possible with HPDC castings. The successful application of US in producing a uniform microstructure throughout the castings, suggests that a master alloy with, for e.g., 10 wt % nanoparticles, manufactured by this method, may feasibly be added to the base alloy when conventionally cast. Also, the results of the recycled castings make it feasible to use the initial nanocomposite materials as feedstock for casting processes where the additional cost of adding an ultrasonic system is cost prohibitive.

4.3. Comparison of Yield Strength Prediction by Models

Various models describe the increase in yield strength in nanocomposites, whereby the mechanisms for increasing the strength are based on the following microstructural effects: Orowan strengthening, grain refinement, dislocation generation due to differences in the coefficient of thermal expansion (CTE) or modulus, and the load-bearing mechanism. Kim et al. described the prediction of strengthening effects in MMNCs based on magnesium [20]. Arithmetic summation of the strengthening contributions can result in an overestimation in some cases [38,39]. This method is one of two methods used in this study because the strengthening mechanisms are considered to be independent of each other. The other method used is quadratic summation that is based on larger, micron-sized particles and assumes interaction between the individual strengthening mechanisms, which is more likely to be the case when larger particles act as reinforcement.

Under the precondition that the addition of nanoparticles reduces the grain size compared to the unreinforced alloy cast under exactly the same conditions, the improvement in yield strength can be described by Equation (A2) in Appendix A.

As mentioned above, the term in brackets is called grain size reduction, where D_{MMNC} and D_0, the grain size in the nanocomposite and the unreinforced alloy, respectively, are processed in the same way. Values for GSR are 8.05×10^{-2} $\mu m^{-1/2}$ for the as-cast nanocomposite and 6.58×10^{-2} $\mu m^{-1/2}$, 4.72×10^{-2} $\mu m^{-1/2}$, and 4.34×10^{-2} $\mu m^{-1/2}$ for castings of the first, second, and third remelting, respectively.

Taking the grain sizes from Table 1 and k_y to be 530.2 MPa $\mu m^{1/2}$, the strengthening contribution from grain size reduction according to Equation (A2) is 42.7 MPa for the as-cast nanocomposite, and 34.9 MPa, 25.0 MPa, and 23.0 MPa for the first, second, and third remelted nanocomposites. The strength increase due to grain refinement after Equation (A2) for the recycled materials is close to the measured strength increase. There is a gap of 3.6 MPa for the as-cast AM60 + AlN that may be generated by one of the other strengthening mechanisms mentioned above.

It may be assumed that Orowan strengthening contributes to the total increase in strength as well. Zhang and Chen proposed a description for the Orowan contribution $\Delta\sigma_{OR}$ to strengthening [40] given by Equation (A3) in Appendix A.

V_p is the volume fraction of 1 wt % AlN nanoparticles ($\approx 5.5214 \times 10^{-3}$), G_m is the shear modulus of the matrix alloy (16.6 GPa), d_p is the average diameter of the AlN nanoparticles (80 nm), and b the Burgers vector of the matrix (0.32 nm). Applying these structural parameters, the Orowan contribution to strengthening is calculated to be 11.9 MPa, which is larger than the gap between the measured yield strength and the contribution of Hall-Petch strengthening, when applying arithmetic summation.

Geometrically necessary dislocations (GND) contribute to strengthening too. These dislocations are created during cooling due to mismatch in the Coefficient of Thermal Expansion (CTE) between the matrix and that of the reinforcement. The strengthening contribution of CTE mismatch during cooling down from the casting temperature to room temperature can be calculated according to Equation (A5) in Appendix A [39,41], where β is considered to be 1.25 [41], $\Delta\alpha$ is the difference between the CTE of the matrix (28.5×10^{-6} K^{-1}), and AlN-particles (4.5×10^{-6} K^{-1}), and ΔT is the difference between the processing temperature and tensile test temperature (room temperature). It needs to be mentioned that particles smaller than a critical diameter d^* are not expected to contribute to CTE-strengthening. In an Al/Al_2O_3 system, Redsten et al. [42,43] proposed Equation (A6) in Appendix A for calculating the critical size. Although solidification is finished at 543 °C, all generated mismatch created at high temperatures down to a homologous temperature of 0.59 (~210 °C) or only 0.55 (176 °C) [43] is expected to relax by diffusion, so that ΔT is only 190 °C or 155 °C. Taking the mean value 172.5 °C for ΔT, the critical diameter d^* is calculated to be 87.4 nm. Having a medium particle size of 80 nm, the influence of CTE mismatch on strengthening of the material is assumed to be negligible.

Other strengthening mechanisms like modulus mismatch strengthening, which creates GND due to differences in elastic moduli of reinforcement and matrix alloy when subjected to compressive stresses, can be neglected in this case. It needs to be taken into account only when materials are post-processed, for e.g., in wrought processes, like extrusion, forging, or rolling, where compressive stresses are applied.

Load bearing strengthening can be neglected in MMNCs with low volume fraction of nanoparticles, as well. Following Equation (A7) in Appendix A [20], the strengthening contribution of the load bearing effect $\Delta\sigma_{Load}$ in the investigated system is 0.12 MPa, assuming a well bonded spherical particle in the matrix.

The above calculations show that the main strengthening contribution of 42.7 MPa comes from grain size reduction as a consequence of nanoparticle addition. Orowan strengthening contributed 11.9 MPa to yield strength improvement. Other effects can be ignored due to only small contributions to strength. Neither GND dislocations due to CTE mismatch or modulus mismatch nor the load bearing effect significantly improves yield strength. As mentioned above, either arithmetic (Equation (A8) in Appendix A) or quadratic (Equation (A9) in Appendix A) summation can be applied.

Following Equation (A8) a total strength increase of 54.8 MPa can be assumed. Experimentally we found an increase of 46.3 MPa. The arithmetic summation method, therefore, results in an overestimation of yield strength increase, but is acceptable. The quadratic summation method of Equation (A9) results in an increase of 44.3 MPa, which is slightly below the experimental yield strength increase. Thus, both methods result in acceptable estimations.

5. Conclusions

An AM60 based nanocomposite containing 1 wt % of AlN particles with an average size of 80 nm was successfully produced using an ultrasound assisted indirect chill casting process. Castings were remelted three times in order to evaluate the microstructure and mechanical properties of each cast nanocomposite. The as-cast nanocomposite is significantly grain refined compared to the AlN-free AM60 processed in the same way including stirring and ultrasonic treatment. A nucleation-free zone (NFZ) formed around each grain prevents further nucleation and, thus, limits the grain size to approximately that of the size of NFZ. The addition of AlN nanoparticles under ultrasound treatment produced remarkable improvements in mechanical properties: yield strength increased by 103%, ultimate tensile strength by 115%, and ductility by 140%. By modelling the possible strengthening mechanisms, the main contribution to yield strength is provided by Hall-Petch strengthening and a small amount from Orowan strengthening. There is a negligible contribution from the CTE mismatch mechanism. Whereas the arithmetic summation of strengthening effects overestimates the yield strength increase, the quadratic summation method slightly underestimates it. After each remelting of the as-cast nanocomposite, the yield strength, ultimate tensile strength, and ductility slightly decreased while the grain size increased. These effects are interrelated according to Hall-Petch strengthening. The small change in mechanical properties after remelting suggests the ultrasonically processed nanocomposite is suitable for use and re-use in a range of casting processes.

Acknowledgments: The authors wish to acknowledge financial support from the European Commission (ExoMet Project, 7th Framework Programme, contract FP7-NMP3-LA-2012-280421). David StJohn acknowledges the support of the Australian Research Council Discovery Grant DP140100702. Sergey Vorozhtsov acknowledges the support of Russian Science Foundation Grant (project No. 17-13-01252). The authors thank Hamdi Tek for performing the tensile tests and Günter Meister for casting.

Author Contributions: Hajo Dieringa contributed the casting of materials, strengthening calculations, idea of the paper, and coordination of authors; Lydia Katsarou, Ricardo Buzolin, and Gábor Szakács contributed metallography, grain size measurement, and hardness testing; Manfred Horstmann contributed the evaluation of tensile tests; Martin Wolff contributed the DSC tests; Chamini Mendis contributed mismatch calculations based on the crystal structure; Sergey Vorozhtsov contributed the AlN nanoparticles and their properties; and David StJohn contributed the grain refinement calculation.

Conflicts of Interest: The authors declare no conflict of interest.

Appendix A

Hall-Petch relation: (Equation (A1)), where σ_y is the yield stress, σ_0 is the friction stress that allows dislocations to move on slip planes in a single crystal in the absence of any strengthening mechanisms, k_y is the stress concentration factor, and D is the average grain size [44,45].

$$\sigma_y = \sigma_0 + k_y D^{-1/2} \tag{A1}$$

The improvement in yield strength can also be described with Equation (A2) if grain refinement by addition of nanoparticles cast under exactly the same conditions is assumed [20], where D_{MMNC} and D_0 are the grain size in the nanocomposite and the unreinforced alloy, respectively:

$$\Delta\sigma_{GR} = k_y \left(\frac{1}{\sqrt{D_{MMNC}}} - \frac{1}{\sqrt{D_0}} \right) \tag{A2}$$

Orowan strengthening: Zhang and Chen proposed a description for the Orowan contribution $\Delta\sigma_{OR}$ to strengthening [40] given by the following equation, Equation (A3):

$$\Delta\sigma_{OR} = \frac{0.13 b G_m}{\lambda} \ln \frac{d_p}{2b} \tag{A3}$$

$$\text{where } \lambda = d_p \left[\left(\frac{1}{2V_P} \right)^{1/3} - 1 \right] \tag{A4}$$

CTE mismatch: The strengthening contribution of CTE mismatch during cooling down from the casting temperature to room temperature can be calculated according to Equation (A5) [39,41].

$$\Delta\sigma_{CTE} = \sqrt{3}\beta G_m b \sqrt{\frac{12 V_p \Delta\alpha\Delta T}{b d_p}} \tag{A5}$$

Particles smaller than a critical diameter d^* are not expected to contribute to CTE-strengthening. In an Al/Al_2O_3 system, Redsten et al. [42,43] proposed Equation (A6) for calculating the critical size:

$$d^* = \frac{b}{\Delta\alpha\Delta T} \tag{A6}$$

Load bearing strengthening:

$$\Delta\sigma_{Load} = \frac{1}{2} V_p \sigma_m \tag{A7}$$

Arithmetic (Equation (A8)) or quadratic (Equation (A9)) summation of contributions to yield strength increase:

$$\Delta\sigma_{Total} = \Delta\sigma_{GR} + \Delta\sigma_{OR} + \Delta\sigma_{CTE} + \Delta\sigma_{Mod} + \Delta\sigma_{Load} \tag{A8}$$

$$\Delta\sigma_{Total} = \sqrt{\Delta\sigma_{GR}^2 + \Delta\sigma_{OR}^2 + \Delta\sigma_{CTE}^2 + \Delta\sigma_{Mod}^2 + \Delta\sigma_{Load}^2} \tag{A9}$$

References and Note

1. Gavras, S.; Zhu, S.M.; Nie, J.F.; Gibson, M.A.; Easton, M.A. On the microstructural factors affecting creep resistance of die-cast Mg-La-rare earth (Nd, Y or Gd) alloys. *Mater. Sci. Eng.* **2016**, *675*, 65–75. [CrossRef]

2. Zhu, S.M.; Gibson, M.A.; Easton, M.A.; Nie, J.F. The relationship between microstructure and creep resistance in die-cast magnesium-rare earth alloys. *Scr. Mater.* **2010**, *63*, 698–703. [CrossRef]

3. Moreno, I.P.; Nandy, T.K.; Jones, J.W.; Allison, J.E.; Pollock, T.M. Microstructural stability and creep of rare-earth containing magnesium alloys. *Scr. Mater.* **2003**, *48*, 1029–1034. [CrossRef]

4. Deming, H.; Yungui, C.; Yongbai, T.; Hongmei, L.; Gao, N. Indentation creep behavior of AE42 and Ca-containing AE41 alloys. *Mater. Lett.* **2007**, *61*, 1015–1019. [CrossRef]

5. Wan, X.; Sun, Y.; Xue, F. Microstructure and mechanical properties of ZA62 based magnesium alloys with calcium addition. *Trans. Nonferr. Met. Soc. China* **2010**, *20*, 757–762. [CrossRef]

6. Dargusch, M.S.; Zhu, S.M.; Nie, J.F.; Dunlop, G.L. Microstructural analysis of the improved creep resistance of a die-cast magnesium-aluminium-rare earth alloy by strontium additions. *Scr. Mater.* **2009**, *60*, 116–119. [CrossRef]

7. Zhao, P.; Wang, Q.; Zhai, C.; Zhu, Y. Effects of strontium and titanium on the microstructure, tensile properties and creep behavior of AM50 alloys. *Mater. Sci. Eng.* **2007**, *444*, 318–326. [CrossRef]

8. Mahmudi, R.; Moeendarbari, S. Effects of Sn additions on the microstructure and impression creep behaviour of AZ91 magnesium alloy. *Mater. Sci. Eng.* **2013**, *566*, 30–39. [CrossRef]

9. Huang, Y.; Dieringa, H.; Kainer, K.U.; Hort, N. Understanding effects of microstructural inhomogeneity on creep response—New approaches to improve the creep resistance in magnesium alloys. *J. Magnes. Alloys* **2014**, *2*, 124–132. [CrossRef]

10. Dieringa, H.; Huang, Y.; Wittke, P.; Klein, M.; Walther, F.; Dikovits, M.; Poletti, C. Compression creep response of magnesium alloy DieMag422 containing barium compared with the commercial creep-resistant alloys AE42 and MRI230D. *Mater. Sci. Eng.* **2013**, *585*, 430–438. [CrossRef]

11. Trojanova, Z.; Szaraz, Z.; Labar, J.; Lukac, P. Deformation behaviour of an AS21 alloy reinforced by short Saffil fibres and SiC particles. *J. Mater. Proc. Technol.* **2005**, *162–163*, 131–138. [CrossRef]

12. Zhang, X.; Fang, L.; Xiong, B.; Hu, H. Microstructure and Tensile Properties of Mg (AM60)/Al$_2$O$_3$ Metal Matrix Composites with Varying Volume Fractions of Fiber Reinforcement. *J. Mater. Eng. Perform.* **2015**, *24*, 4601–4611. [CrossRef]

13. Mondal, A.K.; Kumar, S. Dry sliding wear behaviour of magnesium alloy based hybrid composites in the longitudinal direction. *Wear* **2009**, *267*, 458–466. [CrossRef]

14. Sklenicka, V.; Pahutova, M.; Kucharova, K.; Svoboda, M.; Langdon, T.G. Creep Processes in Magnesium Alloys and their Composites. *Metall. Mater. Trans.* **2002**, *33*, 883–889. [CrossRef]

15. Sklenicka, V.; Svoboda, M.; Pahutova, M.; Kucharova, K.; Langdon, T.G. Microstructural processes in creep of an AZ 91 magnesium-based composite and its matrix alloy. *Mater. Sci. Eng.* **2001**, *319–321*, 741–745. [CrossRef]

16. Viswanath, A.; Dieringa, H.; Ajith Kumar, K.K.; Pillai, U.T.S.; Pai, B.C. Investigation on mechanical properties and creep behavior of stir cast AZ91-SiCp composites. *J. Magnes. Alloys* **2015**, *3*, 16–22. [CrossRef]

17. Ochi, Y.; Masaki, K.; Matsumura, T.; Wadasako, M. Effects of volume fraction of alumina short fibers on high cycle fatigue properties of Al and Mg alloy composites. *Mater. Sci. Eng.* **2007**, *468–470*, 230–236. [CrossRef]

18. Huang, Y.D.; Hort, N.; Dieringa, H.; Maier, P.; Kainer, K.U. Investigations on thermal fatigue of aluminum- and magnesium-alloy based composites. *Int. J. Fatigue* **2006**, *28*, 1399–1405. [CrossRef]

19. Fu, H.M.; Zhang, M.-X.; Qiu, D.; Kelly, P.M.; Taylor, J.A. Grain refinement by AlN particles in Mg-Al based alloys. *J. Alloys Compd.* **2009**, *478*, 809–812. [CrossRef]

20. Kim, C.-S.; Sohn, I.; Nezafati, M.; Ferguson, J.B.; Schultz, B.F.; Bajestani-Gohari, Z.; Rohatgi, P.K.; Cho, K. Prediction models for the yield strength of particle-reinforced unimodal pure magnesium (Mg) metal matrix nanocomposites (MMNCs). *J. Mater. Sci.* **2013**, *48*, 4191–4204. [CrossRef]

21. Paramsothy, M.; Gupta, M. The opposing nanoscale and macroscale effects of selected nanoparticle addition to AZ91/ZK60A hybrid magnesium alloy. *J. Nanopart. Res.* **2013**, *15*, 1938–1950. [CrossRef]

22. Paramsothy, M.; Chan, J.; Kwok, R.; Gupta, M. Nitride nanoparticle addition to beneficially reinforce hybrid magnesium alloys. *Metall. Mater. Trans.* **2013**, *44*, 1123–1138. [CrossRef]

23. Dieringa, H. Properties of magnesium alloys reinforced with nanoparticles and carbon nanotubes: A review. *J. Mater. Sci.* **2011**, *46*, 289–306. [CrossRef]

24. Tan, X.; How, W.C.K.; Weng, J.C.K.; Onn, R.K.W.; Gupta, M. Development of high-performance quaternary LPSO Mg-Y-Zn-Al alloys by Disintegrated Melt Deposition technique. *Mater. Des.* **2015**, *83*, 443–450.

25. Hassan, S.; Gupta, M. Development of a novel magnesium-copper based composite with improved mechanical properties. *Mater. Res. Bull.* **2002**, *37*, 377–389. [CrossRef]

26. Chen, L.-Y.; Xu, J.; Choi, H.; Pozuelo, M.; Ma, X.; Bhowmick, S.; Yang, J.; Mathaudhu, S.; Li, X. Processing and properties of magnesium containing a dense uniform dispersion of nanoparticles. *Nature* **2015**, *528*, 539–543. [CrossRef] [PubMed]

27. Sillekens, W.H.; Jarvis, D.J.; Vorozhtsov, A.; Bojarevics, V.; Badini, C.F.; Pavese, M.; Terzi, S.; Salvo, L.; Katsarou, L.; Dieringa, H. The ExoMet Project: EU/ESA Research on High-Performance Light-Metal Alloys and Nanocomposites. *Metall. Mater. Trans.* **2014**, *45*, 3349–3361. [CrossRef]

28. Katsarou, L.; Mounib, M.; Lefebvre, W.; Vorozhtsov, S.; Pavese, M.; Badini, C.; Molina-Aldareguia, J.M.; Cepeda Jimenez, C.; Pérez Prado, M.T.; Dieringa, H. Microstructure, mechanical properties and creep of magnesium alloy Elektron21 reinforced with AlN nanoparticles by ultrasound-assisted stirring. *Mater. Sci. Eng.* **2016**, *659*, 84–92. [CrossRef]

29. Lerner, M.; Vorozhtsov, A.; Guseinov, S.; Storozhenko, P. Metal Nanopowders Production. In *Metal Nanopowders: Production, Characterization, and Energetic Applications*; Gromov, A.A., Teipel, U., Eds.; Wiley-VCH: Hoboken, NJ, USA, 2014; pp. 79–106, ISBN 9783527680726.

30. Pearson's—Crystal Structure Database for Inorganic Compounds (on CD-ROM); Villars, P., Cenzual, K., Eds.; Release 2015/16.

31. Somekawa, H.; Mukai, T. Effect of grain refinement on fracture toughness in extruded pure magnesium. *Scr. Mater.* **2005**, *53*, 1059–1064. [CrossRef]

32. Kang, S.H.; Lee, Y.S.; Lee, J.H. Effect of grain refinement of magnesium alloy AZ31 by severe plastic deformation on material characteristics. *J. Mater. Proc. Technol.* **2008**, *201*, 436–440. [CrossRef]

33. Mukai, T.; Yamanoi, M.; Watanabe, H.; Ishikawa, K.; Higashi, K. Effect of grain refinement on tensile ductility in ZK60 magnesium allyo under dynamic loading. *Mater. Trans.* **2001**, *42*, 1177–1181. [CrossRef]

34. StJohn, D.H.; Prasad, A.; Easton, M.A.; Qian, M. The contribution of constitutional supercooling to nucleation and grain formation. *Metall. Mater. Trans.* **2015**, *46*, 4868–4885. [CrossRef]

35. Greer, A.L.; Bunn, A.M.; Tronche, A.; Evans, P.V.; Bristow, D.J. Modelling of inoculation of metallic melts: Application to grain refinement of aluminium by AlTiB. *Acta Mater.* **2000**, *48*, 2823–2835. [CrossRef]

36. StJohn, D.H.; Qian, M.; Easton, M.A.; Cao, P. The Interdependence Theory: The relationship between grain formation and grain selection. *Acta Mater.* **2011**, *59*, 4907–4921. [CrossRef]

37. Poirier, D.R. Density, viscosity, and diffusion coefficients in hypoeutectic Al-Si liquid alloys: An assessment of available data. *Metall. Mater. Trans.* **2014**, *45*, 1345–1354. [CrossRef]

38. Goh, C.S.; Wei, J.; Lee, L.C.; Gupta, M. Properties and deformation behaviour of $Mg-Y_2O_3$ nanocomposites. *Acta Mater.* **2007**, *55*, 5115–5121. [CrossRef]

39. Dai, L.H.; Ling, Z.; Bai, Y.L. Size-dependent inelastic behavior of particle-reinforced metal-matrix composites. *Compos. Sci. Technol.* **2001**, *61*, 1057–1063. [CrossRef]

40. Zhang, Z.; Chen, D.L. Consideration of Orowan strengthening effect in particulate-reinforced metal matrix nanocomposites: A model for predicting their yield strength. *Scr. Mater.* **2006**, *54*, 1321–1326. [CrossRef]

41. Vogt, R.; Zhang, Z.; Li, Y.; Bonds, M.; Browning, N.D.; Lavernia, E.J.; Schoenung, J.M. The absence of thermal expansion mismatch strengthening in nanostructured metal-matrix composites. *Scr. Mater.* **2009**, *61*, 1052–1055. [CrossRef]

42. Redsten, A.M.; Klier, E.M.; Brown, A.M.; Dunand, D.C. Mechanical properties and microstructure of cast oxide-dispersion-strengthened aluminium. *Mater. Sci. Eng.* **1995**, *201*, 88–102. [CrossRef]

43. Dunand, D.C.; Mortensen, A. On plastic relaxation of thermal stresses in reinforced metals. *Acta Metall. Mater.* **1991**, *39*, 127–139. [CrossRef]

44. Hall, E.O. The deformation and Ageing of mild steel: III Discussion of Results. *Phys. Soc.* **1951**, *64*, 747–753. [CrossRef]

45. Petch, N.J. The Cleavage Strength of Polycrystals. *J. Iron Steel Inst.* **1953**, *174*, 25–28.

Article

Study of the Influence of TiB Content and Temperature in the Properties of In Situ Titanium Matrix Composites

Cristina Arévalo [1], Isabel Montealegre-Melendez [1,*], Eva M. Pérez-Soriano [1], Enrique Ariza [2], Michael Kitzmantel [2] and Erich Neubauer [2]

[1] Department of Engineering and Materials Science and Transportation, School of Engineering, Universidad de Sevilla, Camino de los Descubrimientos s/n, 41092 Seville, Spain; carevalo@us.es (C.A.); evamps@us.es (E.M.P.-S.)

[2] RHP-Technology GmbH, Forschungs- und Technologiezentrum, 2444 Seibersdorf, Austria; enrarigal1@hotmail.com (E.A.); michael.kitzmantel@rhp-technology.com (M.K.); erich.neubauer@rhp-technology.com (E.N.)

*   Correspondence: imontealegre@us.es; Tel.: +34-954-482-278

Received: 27 September 2017; Accepted: 20 October 2017; Published: 27 October 2017

Abstract: This work focuses on the study of the microstructure, hardening, and stiffening effect caused by the secondary phases formed in titanium matrices. These secondary phases originated from reactions between the matrix and boron particles added in the starting mixtures of the composites. Not only was the composite composition studied as an influencing factor in the behaviour of the composites, but also different operational temperatures. Three volume percentages of boron content were tested (0.9 vol %, 2.5 vol %, and 5 vol % of amorphous boron). The manufacturing process used to produce the composites was inductive hot pressing, which operational temperatures were between 1000 and 1300 °C. Specimens showed optimal densification. Moreover, microstructural studies revealed the formation of TiB in various shapes and proportions. Mechanical testing confirmed that the secondary phases had a positive influence on properties of the composites. In general, adding boron particles increased the hardness and stiffness of the composites; however rising temperatures resulted in greater increases in stiffness than in hardness.

Keywords: in situ titanium composites; microstructure analysis; TiB precipitates

1. Introduction

Over the last few decades, titanium matrix composites (TMCs) have been considered as valuable materials for diverse applications in aerospace industries. This sector demands materials that can achieve high specific stiffness in addition to possessing good thermal stability at high operational temperatures, such as TMCs [1–6].

Many studies have focused on the development of TMCs by different techniques. Comparing conventional methods as ingot metallurgy with powder metallurgy routes, due to the high chemical reactivity of Ti, the conventional ingot metallurgy process has not been suitable to manufacture TMCs with ex situ additive. During the ingot metallurgy process, there could be no control in the formation of undesirable products. Furthermore, powder metallurgy (PM) technologies overcome certain problems of conventional processes: wettability between the matrix and the ceramic reinforcements, and long and complex processing steps [7]; for that reason, powder metallurgy (PM) processes have been widely used for the fabrication of TMCs. In this context, two types of PM routes have been established to produce these specific materials: ex situ and in situ methods [2]. In ex situ processes, stable ceramics such as TiC, TiB, SiC, and ZrC have been generally employed. There is little reactivity between these types of ceramics and the titanium matrix. Therefore, no compounds are synthesized during the consolidation and sintering stages. This means that the size of the ceramic particles as well as their morphology could not vary. Moreover, in such composites, the interface between the reinforcing particles and matrix is often a source of weakness since there are differences in the thermal expansion coefficients between matrix and reinforcement. Another factor of weakness could be the thin oxide layers formed on the surfaces of the reinforcements when they have been incorporated into the matrix. The bonding between particles and matrix could be affected by these oxide layers [8].

Regarding the synthesizing in situ method, thanks to the high reactivity of titanium matrix with the additive elements from compounds (Si_3N_4, TiB_2, and B_4C), stable secondary phases can be formed [9]. That is why in situ PM processes are currently considered ones of the best techniques due to the excellent properties of the produced materials. The main advantage of these kinds of composites lies in the stable interface formed between the matrix and the reinforcing phase [10–17]. Among diverse materials that could act as reactive compounds with titanium, boron (B) has been considered as a suitable candidate to start in situ secondary reinforcing phases. Many recent works have presented this non-metallic element as an ideal reactive to promote the formation of TiB_W (TiB, TiB_2) reinforcements via solid-state reactions. The significance of these borides as reinforcements is based on the fact that they are chemically compatible with the matrix, in addition to having similar densities and thermal expansion coefficients [18–21]. Moreover, it is well known that the properties of pure titanium matrix can be improved by the appearance of these TiB_W precipitates (Young's modulus of 110 GPa for pure titanium and 467 GPa for TiB) [22,23]. A previous author describes that the morphology of TiB_W reinforcements is more effective for strengthening effect when they are arranged in one direction and the growth of TiB_W takes place anisotropically within a short time at high operational temperatures due to the high reaction speed of B particles and Ti [23]. From a point of view of the size and morphology of the reinforcements, the present work also studies these characteristics and their relationship with the processing conditions and composition of the TMCs.

The employ of fast powder metallurgy methods as hot press technique, saves processing time, which could affect the size and the morphology of the TiB_W precipitates. In particular, inductive hot pressing (iHP) technology is valued for in situ TMCs manufacturing due to its short operational time (high heating rate $\leq 100\ ^{\circ}C/min$) [24–26]. The use of this technique has facilitated the investigation of TMCs' properties and the secondary phases formed at different processing temperatures [27,28]. Despite the advantages of this process, the restrictions of the specimens' size (diameters of 20 mm) limit the measurement of tensile and bending properties of the final specimens. For that reason, in this work, in addition to the iHP process, Direct Hot Pressing (dHP) technology has been employed. Through a pressure assisted sintering with direct heating of a pressing die, the consolidated composites can be formed directly from powders in a short period of time (<15 min). Direct hot pressing is also characterized by a high heating/cooling rate ($\leq 100\ ^{\circ}C/min$) [29,30].

The scope of this research is the study and evaluation of the relationship between the compositions of determinate TMCs, their processing conditions and their final properties.

2. Materials and Experimental Procedures

The starting materials were commercial Ti powders grade 1 and amorphous B particles, manufactured by TLS GmbH (Bitterfeld, Germany) and ABCR GmbH & Co. KG (Karlsruhe, Germany) respectively. The characterisation of both powders was performed to verify the information about their size and morphology supplied by the manufacturers. The particle size distribution of the starting powders was determined by laser diffraction analysis (Mastersizer 2000, Malvern Instruments, Malvern, UK). The average particle size of the titanium and amorphous boron powders are listed in Table 1.

Table 1. Particle size distributions of the starting powders.

Particle Size Distribution	Ti (µm)	Amorphous B (µm)
D10	11.88	0.74
D50	28.13	2.44
D90	51.42	14.51

Before the hot consolidation of the composites, the blends of the powders were prepared. The tested compositions and their operational parameters are shown in Table 2. The titanium powder and each different volume percentage (vol %) of the amorphous B particles were mixed by tubular machine (Sintris mixer) for 16 h with ceramic balls (ZrO_2) of 3 mm diameter. The weight ratio of ceramic balls to powder was 10:1. Moreover, the use of hexane helped towards the distribution of the fine particles of amorphous B in the metallic matrix. The powder mixture was dried and subsequently blended a second time for several minutes without the ceramic balls, to avoid possible agglomerations. This was the same blending procedure used for producing composites as in previous authors' works [28,30]. Then, the target composition of three different powder mixtures was made from titanium and 0.9, 2.5 and 5 vol % of B particles. With these compositions the predesigned values of TiB are 2.65, 7.42, and 15.02 vol % respectively. These values were calculated based on the theoretical densities: (i) 4.51 g/cm^3 for titanium, (ii) 4.56 g/cm^3 for TiB, and (iii) 2.46 g/cm^3 for boron [1].

Subsequently, the hot and rapid consolidation of the specimens was carried out. Two machines were employed to manufacture the specimens. The first was a self-made hot pressing machine, inductive Hot Pressing (iHP) equipment made by RHP-Technology GmbH & Co. KG (Seiberdorf, Austria). Its main advantage is its high heating rate due to its special inductive heating set-up.

The die used for all the ihp cycles was made from graphite (punch Ø 20 mm). It was lined with thin paper with a protective coating of boron nitride (bn) for each ihp cycle. Then, it was introduced into the hot pressing machine, with fixed processing parameters, heating rate and vacuum conditions [16].

Six specimens were consolidated by this iHP method (see Table 2). The second machine was used to fabricate specimens with suitable dimensions in order to measure their mechanical properties. Assuming the composite with low properties and TiB precipitates formation (5 vol % B at 1000 °C), a second rapid hot pressing machine (direct hot pressing dHP with larger die (Ø 80 mm)) was also used in order to measure mechanical properties.

Table 2. Composition and processing parameters for the manufacturing of titanium matrix composites (TMCs).

Amorphous B (vol %)	Temperature (°C)	Pressure (MPa)	Dwell Time (min)	Processing Method	Diameter (mm)
0.9	1100	50	15	iHP *	20
2.5	1100	50	15	iHP	20
5	1000	50	15	iHP	20
5	1000	35	15	dHP *	80
5	1100	50	15	iHP	20
5	1200	50	15	iHP	20
5	1300	50	15	iHP	20

* inductive Hot Pressing (iHP) and direct Hot Pressing (dHP).

Regarding the operational parameters, Figure 1 shows the evolution of the cycle's parameters for each of the composites manufacturing runs in terms of temperature, pressure, and displacement of the punches (uniaxial press). Figure 1a relates to the representative cycle in each of the hot pressing machines. For dHP, the starting pressure and the heating rate are lower than in iHP due to requirements of this technique.

The graphs shown in Figure 1b are drawn in order to compare all the run cycles across both iHP and dHP equipment. As it is appreciated, the temperature versus time is represented in addition to the punch displacement versus time. In all the cycles, the holding time (15 min) and the vacuum conditions (5 × 10^{-4} bar) were fixed. In particular for the iHP runs, the consolidation temperature was varied: 1000, 1100, 1200, and 1300 °C (see Figure 1b). These values of temperature were employed to investigate the effect of 100 °C increments in the microstructure and properties of specimens made from same starting powder composition. In case of the specimen fabricated by dHP, the operational conditions were similar than the iHP ones; however, only an operational temperature of 1000 °C was set. This value was fixed according to a previous authors' work in which an interesting microstructure phenomenon took place at this temperature in TMCs at similar conditions, but made from different raw materials [28].

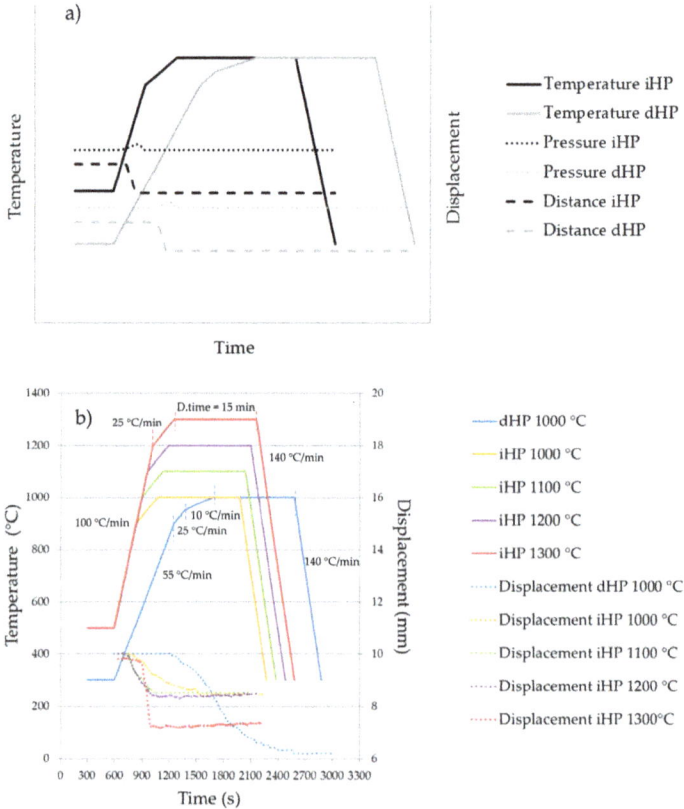

Figure 1. (**a**) Graphical representation of temperature vs. time in iHP and dHP cycles; (**b**) temperature variations vs. time and shrinkage displacements vs. time for TMCs processing cycles from the same starting powders.

Once the iHP and dHP cycles were finished, the samples with 20 and 80 mm of diameter were taken out from the respective dies and cleaned by a sand blasting machine to remove the graphite paper remains from the surfaces. Then, the characterization of all the specimens was performed. Firstly, metallographic preparation of all the specimens was carried out carefully to study the newly-formed phases and the microstructure of the TMCs. The specimens were cut in two pieces. The cross-section was polished. In this prepared cross section the X-ray (XRD) diffraction analysis was performed. Then, XRD equipment (Brunker D8 Advance A25, Billerica, MA, USA) was employed to identify the diverse crystalline phases in the composites. The microstructure characterisation was studied by optical microscope (OM), Nikon Model Epiphot 200 equipment (Tokyo, Japan), and by scanning electron microscope (SEM) JEOL 6460LV (Tokyo, Japan), integrated with electron backscatter diffraction (EBSD) detector and Energy Dispersive Spectroscopy (EDS). The measurements of the precipitates' sizes were performed using the software Image-Pro Plus 6.2 (Media Cybernetics, Rockville, MD, USA).

The density of the specimens was measured by Archimedes' method (ASTM C373-14). The results were compared to the theoretical density calculated by rule of mixtures. Hardness measurements were carried out on the polished cross-sections of the specimens. Eight indentations were done by a tester model, Struers-Duramin A300 (Ballerup, Germany), to ascertain the Vickers hardness (HV10). The estimation of the specimens' Young's Modulus was made by ultrasonic method (Olympus 38 DL, Tokyo, Japan). It was used with a pulse generator/receiver, recording the transit time (outward/return) through the thickness. This technique allowed the determination of both the longitudinal (VL) and transverse (VT) propagation velocities of acoustic waves. To correctly measure the propagation velocities of these waves, the surface of samples must be properly grinded and polished (to create samples with smooth and parallel surfaces) and the delay times of transducers minimised by following an iterative measurement protocol. The Young's Modulus was calculated from the density (g/cm^3), VL and VT [31]. Tensile tests were performed on a universal testing machine Instron 5505 (Norwood, MA, USA) with a strain rate of 1 mm/min. Additionally, the same machine was employed to carry out the flexural tests at 5 mm/min. Both properties were evaluated according to the standards UNE EN 10002-1:2002 and UNE EN ISO3325 respectively.

3. Results and Discussion

The obtained results are presented and discussed considering the two main issues of this work: (i) the influence of starting powder compositions (vol % of amorphous B) at identical processing conditions; (ii) the effect of rising temperature (1000, 1100, 1200, and 1300 °C) for the same starting powder mixture (5 vol % of amorphous B).

3.1. Microstructural Study and XRD Analysis

Firstly, taking into account the volume percentages (vol %) of the amorphous B particles added in the blend, the microstructures of the specimens are compared. Figure 2 shows the SEM images of three specimens fabricated at 1100 °C for 15 min and made from the mixtures of Ti and amorphous B with 0.9, 2.5, and 5 vol %, respectively. As it might be expected, precipitates are observed in the microstructure of the specimens, since at this temperature (1100 °C) there have been reactions between the matrix and the boron particles [10,32]. As many authors have previously described, these precipitates are supposed to be in situ formed TiB [28,33] because of this reaction (see Figure 2). It is important to highlight a clear evolution of the size and the volume of the precipitates related to the amorphous B content (vol %). As predicted, the increment of the B content in the composites causes the appearance of more boride precipitates. According to the phase diagram, alpha titanium plus TiB phases are expected in these systems [34].

TiB precipitates can be appreciated presenting the typical morphology of whiskers [18,32,35–37]. In this study, two different morphologies of TiB precipitates have been considered through an evaluation of the length/width ratio: whiskers and rounded hexagonal shapes (Figure 2). The processing time is short (15 min) in hot press techniques. Therefore, the studied microstructures revealed, besides high length/width ratio whiskers, also precipitates in which the aspect ratio was close to 1.

The reinforcements with high aspect ratio (whiskers) can form anisotropically during higher sintering temperature due to the high reaction speed of B particles and Ti [23]. All of the TiB$_W$ precipitates are randomly distributed in the matrix.

To verify the composition of both types of precipitates, EDS analysis has been performed in the marked spots in Figure 2b. Using 0.9 vol % of B particles, the size of the round precipitates is the smallest one compared to the rest of the precipitates' size formed in TMCs, with B contents of 2.5 and 5 vol %. It is observed that increasing the content of B, the size of the precipitates also increased.

The whiskers' lengths remain generally constant although the widths increase slightly by increasing the B content. Moreover, the round hexagonal precipitates become larger. In Figure 2c, there are some darker grey areas where the B particles are believed to remain in the titanium matrix without reacting.

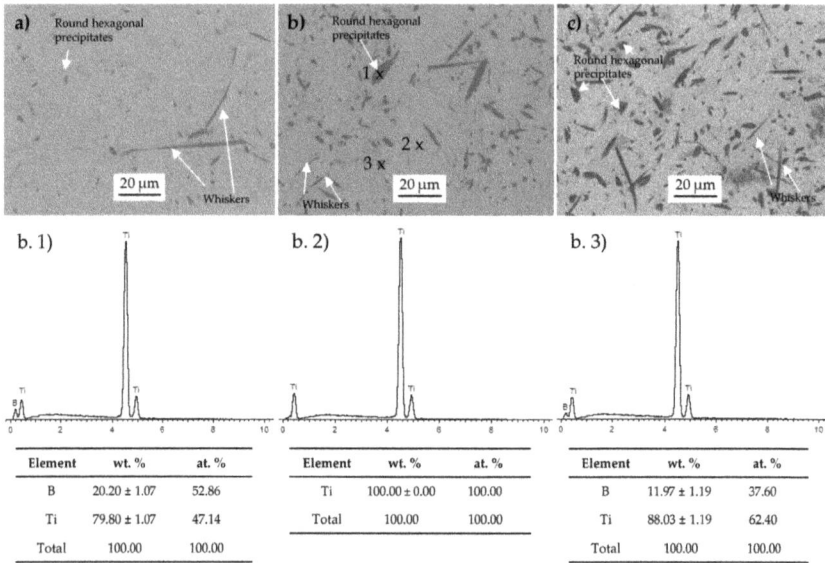

Element	wt. %	at. %
B	20.20 ± 1.07	52.86
Ti	79.80 ± 1.07	47.14
Total	100.00	100.00

Element	wt. %	at. %
Ti	100.00 ± 0.00	100.00
Total	100.00	100.00

Element	wt. %	at. %
B	11.97 ± 1.19	37.60
Ti	88.03 ± 1.19	62.40
Total	100.00	100.00

Figure 2. Scanning electron microscope (SEM) images of TMCs manufactured at 1100 °C with different content of amorphous B in their starting powders: (**a**) 0.9 vol %; (**b**) 2.5 vol %; and (**c**) 5 vol %; (**b1**) EDS spectra spot (1); (**b2**) EDS spectra spot (2); and (**b3**) EDS spectra spot (3).

The second target parameter of the study is the processing temperature for specimens fabricated from identical starting powder (5 vol % of amorphous B). There are relevant changes in the microstructures of the composites caused by increasing the consolidation temperature from 1000 to 1300 °C. This phenomenon can be observed in Figure 3. The lower the temperature, the fewer the number of formed precipitates. Titanium grains can be clearly recognised (see spot 2 in Figure 3a). Additionally, there are possible agglomerations of the reinforcing phases located in these grain boundaries. Regarding the reaction between Ti and B at 1000 °C, the time (15 min) and operational temperature are insufficient to promote an atomic diffusion phenomenon of boron into the matrix grains. EDS analysis reveals grey areas corresponding to such B particles agglomeration (see in Figure 3). However, only one increment of 100 °C (from 1000 to 1100 °C) causes the origin of TiB precipitates in the matrix with the two different morphologies previously mentioned. When the operational temperature rises from 1100 to 1200 °C, the size of the round hexagonal shapes becomes bigger (see Figure 3b,c). In relation to the whisker morphology, the changes caused by the increment of the temperature are more easily appreciated in their thickness than in their length.

The tendency to thicken the size of the precipitates remains constant up to 1200 °C. Comparing the microstructure of Figure 3c,d, variations in size of the precipitates are not visible. In both SEM images (TMCs fabricated at 1200 and 1300 °C) the thickness of the whiskers, 1.06 and 1.26 μm, respectively, are larger than the ones formed at 1100 °C, 0.92 μm. The size of the round hexagonal precipitates, 1.51 μm at 1200 °C and 1.67 μm at 1300 °C, is a little bigger than in TMCs processed at 1100 °C, 1.24 μm.

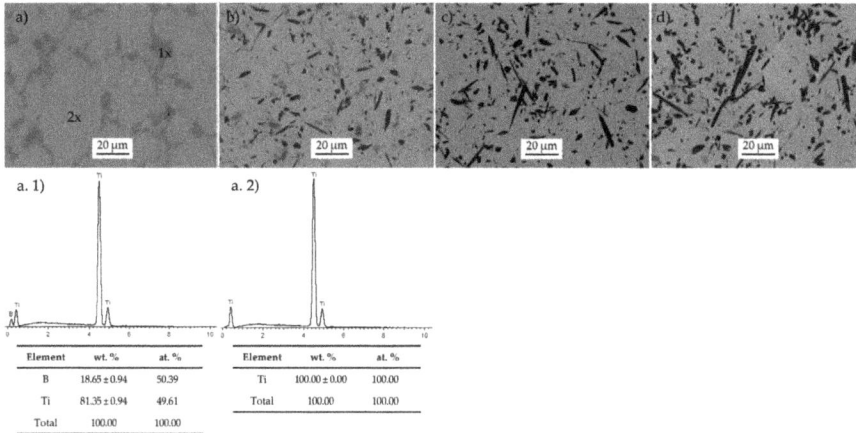

Figure 3. SEM images of TMCs made from starting powder with 5 vol % of amorphous B and manufactured at: (**a**) 1000 °C; (**b**) 1100 °C; (**c**) 1200 °C; and (**d**) 1300 °C. (**a1**) EDS spectra spot (1), (**a2**) EDS spectra spot (2).

To go into detail about precipitates, a semi-quantitative study of both types of precipitates (whiskers and round shapes), considering their size and frequency/image, were developed by image analysis (using ten SEM images at the same magnifications for each specimen). It is important to note that, in the specimens where their precipitates are located at the grain boundaries, this image analysis could not be carried out. The main parameters evaluated after the image analysis are: (i) mean length and mean width of the whiskers; (ii) maximum length and maximum width of the whiskers; and (iii) mean and maximum diameter for the round shapes precipitates. The results of the measurements are represented in Figure 4.

In general, there are more whisker precipitates than round shapes, independent of the composite compositions and processing temperature.

Concerning the dimensions of the whisker precipitates, the increase of the temperature and the boron content in the starting blend involves an increase of both mean values (length and width). Figure 4c shows a gradual growth in whisker size at higher temperatures and at higher volumes of boron. Each increment of 100 °C drives the growth of whisker mean length size by approximately 15% at the same composite composition. However, the effect of the processing temperature and the boron addition is not the same in the maximum length and width of the whiskers. Despite increasing both temperature and composition, a dimensional limit around 26 μm exists in their maximum length. This means that there are not whiskers measured with length higher than 26 μm independently of these two factors. Regarding the maximum width, affected only by temperature, maintains a value of around 0.12 μm across different compositions.

With respect to the round-shaped precipitates (Figure 4d), the higher the temperature and the volume of boron content are, the higher mean and maximum diameters of the precipitates. An increment of 100 °C, from 1100 to 1200 °C, produces a 21% increase to the mean diameters of these precipitates.

Related to the manufacturing processes, iHP and dHP, similar microstructures of composites were observed in specimens from the same starting powder in spite of their different fabrication methods. This means that at the same processing conditions (1000 °C for 15 min) but in different hot pressing machines, the microstructural properties are alike (see Figure 5). It could therefore be argued that the results obtained could be reproduced using either machine. At 1000 °C, there is insufficient diffusion time and temperature to end the boron source to form TiB.

The phenomenon of the agglomeration of B particles at the grain boundaries and inhomogeneous microstructure was observed in both composites.

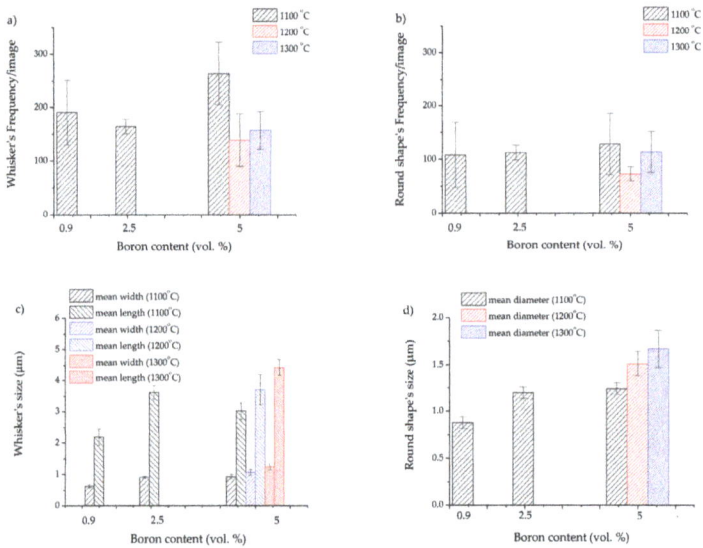

Figure 4. (**a**) Whisker frequency; (**b**) Round shape precipitate frequency vs. Boron content; (**c**) Whiskers size; and (**d**) Round shape precipitate size vs. Boron content.

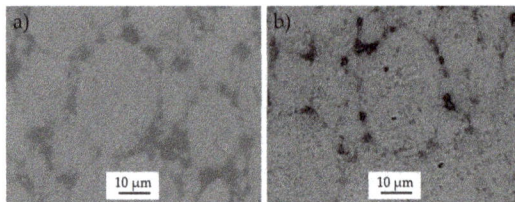

Figure 5. SEM images of TMCs made from starting powder with 5 vol % of amorphous B at 1000 °C for 15 min via (**a**) iHP; (**b**) dHP.

The results of XRD analyses confirm the TiB formation as a product of the reaction between the matrix and the B particles. Figures 6 and 7 show the XRD patterns of the composites made from several starting powders (vol % of B) and processing at different temperatures, respectively. In general, no recordable peaks of TiB_2 are observed in all of the XRD patterns. Comparing the effect of the starting powder compositions, the lower the B content is, the lower the observed peak of TiB is (see Figure 6a). However, at the same operational conditions, increasing the vol % of B to 5%, there are sharper peaks for TiB phase (see in Figure 6c). The influence of temperature on the formation of the TiB phase is quite clear as shown by the XRD patterns in Figure 7. It is well known that, by increasing temperature, the reaction between Ti and B tends to be more complete. This can be seen clearly in Figure 7. XRD peaks of TiB in composite produced at 1000 °C are slightly weaker than the ones in the pattern of composite produced at 1100 °C. Moreover, there are two new weak peaks corresponding to the TiB phase when the temperature increases from 1100 to 1200 °C.

However, there is very little variation in the pattern of composites processed at 1300 °C compared to specimens produced at 1200 °C. The reason is related to the diffusion phenomena in both cases being high enough (at 1200 and 1300 °C). At these high temperatures, the content of TiB could be higher if the time were to be increased.

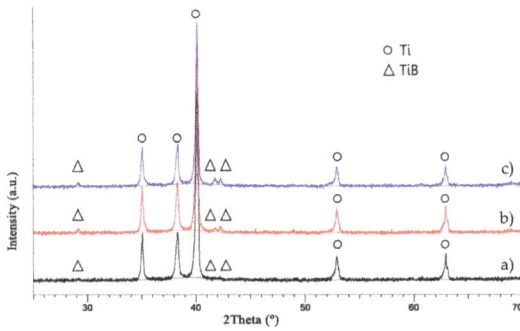

Figure 6. XRD patterns of composites manufacture at 1100 °C for 15 min via iHP with different % volume of B: (**a**) 0.9 vol %; (**b**) 2.5 vol %; and (**c**) 5 vol %.

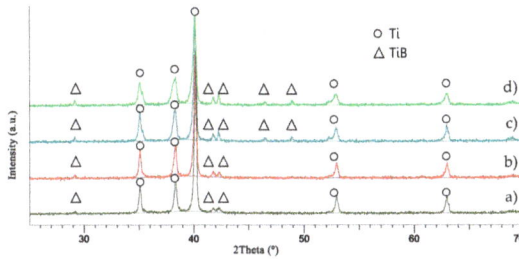

Figure 7. XRD patterns of composites manufacture with 5 vol % of B for 15 min via iHP at different temperatures: (**a**) 1000 °C; (**b**) 1100 °C; (**c**) 1200 °C; and (**d**) 1300 °C.

The semi-quantitative analyses, made by the Reference Intensity Ratio (RIR) method, allowed for the determination of TiB fractions (see Figure 8). The calculated values of TiB (vol %) are lower than the theoretical values of in situ formed TiB (considering full reaction between the matrix and the B particles). The incomplete reaction between the Ti and B could be the responsible of such differences.

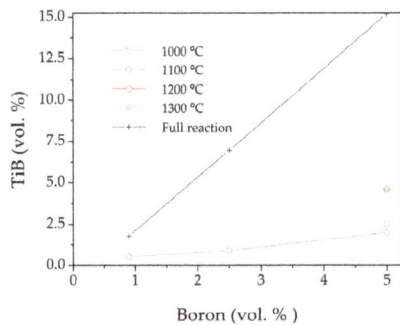

Figure 8. Volume percentage of TiB formed by full reaction Ti-B vs. volume percentage of Boron added as starting material.

The higher the B content, the greater the differences are between the in situ formed TiB at 1100 °C and the TiB content calculated theoretically. Increasing the temperature from 1000 to 1100 °C leads to a slight increase of the amount of TiB formed. However, when the temperatures reach 1200 and 1300 °C there are fewer differences between the in situ formed TiB and the theoretical one.

3.2. Density, Hardness, Young's Modulus, and Mechanical Properties

In general, the relative density of the TMCs reach values of 98% (Table 3). There is an improving effect in the densification resulting from the increase of the processing temperature. Furthermore, the density was affected by the TMCs composition; the higher the content of reinforcements the higher the densification.

Table 3. Densification values of processed TMCs.

Specimen	Temperature (°C)	Densification (%)
Ti + 0.9 vol %	1100	96.61
Ti + 2.5 vol %	1100	98.10
Ti + 5 vol %	1000	98.42
Ti + 5 vol % (dHP)	1000	98.73
Ti + 5 vol %	1100	98.46
Ti + 5 vol %	1200	98.91
Ti + 5 vol %	1300	99.35

The hardening and stiffening effects induced by the TiB precipitates, in addition to the grain refinement due to the reinforcement content, are two phenomena described by previous authors [7,32]. Hardening induced by TiB precipitates and a slight grain refinement are observed in the in situ TMCs fabricated from the three powder mixtures (0.9%, 2.5%, and 5% volume of boron) and at different temperatures (1000, 1100, 1200, and 1300 °C). On one hand, as shown in Figure 9a, the higher the B content the higher the hardness and the Young's Modulus. The hardness increases by 5% from 0.9 to 2.5 vol % B content and by 32% from 0.9 to 5 vol %. This is closely related to the in situ formed TiB as shown in Figure 8. Clearly, the content of the reinforcement particles and secondary phases contribute to the hardening effect. The tendency of the Young's Modulus to increase is less pronounced than the hardness' trend with values of 5% and 18% respectively (see in Figure 9a).

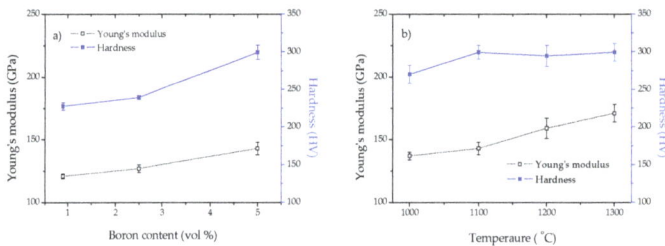

Figure 9. (**a**) Hardness and Young's Modulus vs. volume percentage of Boron (vol %), (**b**) Hardness and Young's Modulus vs. operational temperatures.

On the other hand, increasing the temperature also promotes the variations in the hardness and Young' Modulus due to the number of precipitates. In this case, the increase of the temperature more greatly affects the enhancement of the Young's Modulus than the hardness of the specimens. In general, the values of hardness remain around 300 HV while the Young's modulus values present increases related to the specimens processed at 1000 °C; 4%, 10%, and 25% for 1100, 1200, and 1300 °C, respectively (see Figure 9b). Even though the microstructure seems similar in specimens produced at

1200 and 1300 °C, the increase of stiffness could be the result of ongoing reactions in the matrix and a higher densification value for the one produced at 1300 °C.

To study the specimen's microstructure and its mechanical behaviour, the composite produced from the powder mixture with 5% of volume percentage of boron and hot pressed at 1000 °C was also fabricated via dHP to carry out tensile and bending tests. Table 4 shows a summary of the tensile properties tested at room temperature and at 250 °C; as ultimate tensile stress (σ_{UTS}), as well as the deformation (ε). Additionally, the flexural behaviour of this kind of composite is also presented (ultimate bending strength, σ_{UBS}, and deformation, ε).

Table 4. Mechanical properties of in situ TMCs produced at 1000 °C with 5 vol % of Boron.

Material	Tensile Properties				Bending Properties	
	Room temperature		250 °C		Room temperature	
Ti + 5 vol % of B	σ_{UTS} (MPa)	ε (%)	σ_{UTS} (MPa)	ε (%)	σ_{UBS} (MPa)	ε (%)
	780	1.94	533	6.18	1454.38	3.44

With respect to the tensile properties measured at room temperature, there is an increase in the σ_{UTS} (MPa) due to the boron addition (values are compared to the σ_{UTS} of pure Ti grade 1, from 240 to 345 MPa [38,39]. However, the ductility behaviour is significantly lower compared to the reference values of pure Ti grade 1 (20%). The distribution of the reinforcement in the matrix and the in situ formed TiB increases the strength of the material. The location of the reinforcement of some particles around the matrix grains blocks the dislocation motion promoting the embrittlement of the matrix and improving the strength of the material. When the tensile test is carried out at 250 °C, there is an increase in the percentage of the maximum deformation of the material. However, the σ_{UTS} measured at this temperature shows a lower value than σ_{UTS} measured at room temperature. As expected, the motion of the dislocation was encouraged by the increase of the temperature during the tensile test. There is a considerable enhancement to the σ_{UBS}.

From the point of view of the microstructural behaviour, specimens in which the distribution of precipitates is homogenous inside the matrix, better mechanical behaviour can be expected with respect to density, hardness, and Young's Modulus.

4. Conclusions

The following conclusions can be drawn:

- High densification composites are produced. The influence of the in situ formed TiB and processing conditions on the material behaviour is verified.
- The microstructural study reveals changes in the composites depending on the operational temperatures. In the range of 1000 to 1100 °C, the location of the precipitates and the boron particles evolves from the grain boundaries into the matrix. Up to 1100 °C, two different morphologies of TiB precipitates have been considered: whiskers and round hexagonal shapes. Increasing the temperature promotes a gradual growth of the TiB phases. At the same composite composition, the TiB precipitates remained relatively constant even if the temperature rose from 1200 to 1300 °C.
- Relating to the boron addition, variations of the sizes of these secondary phases were also observed. Although the addition of more boron involved greater formation of precipitates, the proportions between the boron content and the TiB formed were lower at the highest boron content in the starting mixture. The formed TiB and the boron particles significantly contributed to the hardening and stiffness effects. Increasing the temperature helped to increase the stiffness of the composites more than its hardness.

Acknowledgments: We thank the Microscopy and the X-Ray Laboratory Services of CITIUS (University of Seville). Furthermore, the authors wish to thank Daniel García Luque, and Rayner Simpson for his assistance with the English translation.

Author Contributions: All the authors have been collaborating with each other to obtain high quality research work. Isabel Montealegre-Meléndez performed the materials selection, analysed the data, and designed the structure of the paper. Cristina Arévalo has been responsible of microstructure characterization for specimens: optical and electron microscopy, and the relation between processing parameters and materials properties. Enrique Ariza has done the mechanical properties and references selection. Eva M. Perez-Soriano has performed the metallographic preparation and the relation between processing parameters and materials properties. Michael Kitzmantel has controlled the fabrication process. Erich Neubauer has optimized the equipment and applications.

Conflicts of Interest: The authors declare no conflict of interest.

References

1. Ravi Chandran, K.S.; Panda, K.B.; Sahay, S.S. TiBw-reinforced Ti composites: Processing, properties, application prospects, and research needs. *JOM* **2004**, *56*, 42–48. [CrossRef]

2. Kondoh, K. 16–Titanium metal matrix composites by powder metallurgy (PM) routes. In *Titanium Powder Metallurgy*; Elsevier: Amsterdam, The Netherlands, 2015; pp. 277–297.

3. Zadra, M.; Girardini, L. High-performance, low-cost titanium metal matrix composites. *Mater. Sci. Eng. A* **2014**, *608*, 155–163. [CrossRef]

4. Campbell, F.C. Chapter 9: Metal matrix composites. In *Manufacturing Technology for Aerospace Structural Materials*; Elsevier Science: Oxford, UK, 2006; pp. 419–457, ISBN 978-1-85-617495-4.

5. Kainer, K.U. *Metal Matrix Composites*; Wiley-VCH: Weinheim, Germany, 2006; ISBN 978-3-527-31360-0.

6. Peters, M.; Kumpfert, J.; Ward, C.H.; Leyens, C. Titanium Alloys for Aerospace Applications. *Adv. Eng. Mater.* **2003**, *5*, 419–427. [CrossRef]

7. Sabahi Namini, A.; Azadbeh, M.; Shahedi Asl, M. Effect of TiB_2 content on the characteristics of spark plasma sintered Ti–TiBw composites. *Adv. Powder Technol.* **2017**, *28*, 1564–1572. [CrossRef]

8. AlMangour, B.; Grzesiak, D.; Yang, J.M. In-situ formation of novel TiC-particle-reinforced 316L stainless steel bulk-form composites by selective laser melting. *J. Alloys Compd.* **2017**, *706*, 409–418. [CrossRef]

9. Attar, H.; Bönisch, M.; Calin, M.; Zhang, L.-C.; Scudino, S.; Eckert, J. Selective laser melting of in situ titanium-titanium boride composites: Processing, microstructure and mechanical properties. *Acta Mater.* **2014**, *76*, 13–22. [CrossRef]

10. Zhang, J.; Ke, W.; Ji, W.; Fan, Z.; Wang, W.; Wang, H. Microstructure and properties of in situ titanium boride (TiB)/titanium (TI) composites. *Mater. Sci. Eng. A* **2015**, *648*, 158–163. [CrossRef]

11. Munir, K.S.; Zheng, Y.; Zhang, D.; Lin, J.; Li, Y.; Wen, C. Microstructure and mechanical properties of carbon nanotubes reinforced titanium matrix composites fabricated via spark plasma sintering. *Mater. Sci. Eng. A* **2017**, *688*, 505–523. [CrossRef]

12. Ozerov, M.N.; Stepanov, N.; Kolesnikov, A.; Sokolovsky, V.; Zherebtsov, S. Brittle-to-ductile transition in a Ti–TiB metal-matrix composite. *Mater. Lett.* **2017**, *187*, 28–31. [CrossRef]

13. Jia, L.; Wang, X.; Chen, B.; Imai, H.; Li, S.; Lu, Z.; Kondoh, K. Microstructural evolution and competitive reaction behavior of Ti-B_4C system under solid-state sintering. *J. Alloys Compd.* **2016**, *687*, 1004–1011. [CrossRef]

14. Popov, V.A.; Shelekhov, E.V.; Prosviryakov, A.S.; Presniakov, M.Y.; Senatulin, B.R.; Kotov, A.D.; Khomutov, M.G. Particulate metal matrix composites development on the basis of in situ synthesis of TiC reinforcing nanoparticles during mechanical alloying. *J. Alloys Compd.* **2017**, *707*, 365–370. [CrossRef]

15. Jiao, Y.; Huang, L.J.; Wang, S.; Li, X.T.; An, Q.; Cui, X.P.; Geng, L. Effects of first-scale TiBw on secondary-scale Ti_5Si_3 characteristics and mechanical properties of in-situ (Ti_5Si_3 + TiBw)/Ti6Al4V composites. *J. Alloys Compd.* **2017**, *704*, 269–281. [CrossRef]

16. Ariza, E.; Montealegre-Meléndez, I.; Arévalo, C.; Kitzmantel, M.; Neubauer, E. Ti/B_4C Composites Prepared by In Situ Reaction Using Inductive Hot Pressing. *Key Eng. Mater.* **2017**, *742*, 121–128. [CrossRef]

17. Lu, L.; Fuh, J.Y.H.; Chen, Z.D.; Leong, C.C.; Wong, Y.S. In situ formation of TiC composite using selective laser melting. *Mater. Res. Bull.* **2000**, *35*, 1555–1561. [CrossRef]

18. Hu, Y.; Zhao, B.; Ning, F.; Wang, H.; Cong, W. In-situ ultrafine three-dimensional quasi-continuous network microstructural TiB reinforced titanium matrix composites fabrication using laser engineered net shaping. *Mater. Lett.* **2017**, *195*, 116–119. [CrossRef]

19. Oghenevweta, J.E.; Wexler, D.; Calka, A. Sequence of phase evolution during mechanically induced self-propagating reaction synthesis of TiB and TiB$_2$ via magnetically controlled ball milling of titanium and boron powders. *J. Alloys Compd.* **2017**, *701*, 380–391. [CrossRef]

20. Schmidt, J.; Boehling, M.; Burkhardt, U.; Grin, Y. Preparation of titanium diboride TiB$_2$ by spark plasma sintering at slow heating rate. *Sci. Technol. Adv. Mater.* **2007**, *8*, 376–382. [CrossRef]

21. Lu, H.; Zhang, D.; Gabbitas, B.; Yang, F.; Matthews, S. Synthesis of a TiBw/Ti6Al4V composite by powder compact extrusion using a blended powder mixture. *J. Alloys Compd.* **2014**, *606*, 262–268. [CrossRef]

22. Gorsse, S.; Petitcorps, Y.L.; Matar, S.; Rebillat, F. Investigation of the Young's modulus of TiB needles in situ produced in titanium matrix composite. *Mater. Sci. Eng. A* **2003**, *340*, 80–87. [CrossRef]

23. Lieberman, S.I.; Gokhale, A.M.; Tamirisakandala, S.; Bhat, R.B. Three-dimensional microstructural characterization of discontinuously reinforced Ti64–TiB composites produced via blended elemental powder metallurgy. *Mater. Charact.* **2009**, *60*, 957–963. [CrossRef]

24. Bose, A.; Eisen, W.B. *Hot Consolidation of Powders and Particulates*; MPIF: Princeton, NJ, USA, 2003; ISBN 1-878954-495-4.

25. Radhakrishna Bhat, B.V.; Subramanyam, J.; Bhanu Prasad, V.V. Preparation of Ti-TiB-TiC & Ti-TiB composites by in-situ reaction hot pressing. *Mater. Sci. Eng. A* **2002**, *325*, 126–130. [CrossRef]

26. Montealegre-Meléndez, I. *Development of Titanium Metal Matrix Composites via Powder Metallurgy*; Technische Universitat Wien: Wien, Viena, 2009.

27. Liu, B.X.; Huang, L.J.; Geng, L.; Wang, B.; Cui, X.P. Fracture behaviors and microstructural failure mechanisms of laminated Ti–TiBw/Ti composites. *Mater. Sci. Eng. A* **2014**, *611*, 290–297. [CrossRef]

28. Arevalo, C.; Montealegre-Meléndez, I.; Ariza, E.; Kitzmantel, M.; Rubio-Escudero, C.; Neubauer, E. Influence of Sintering Temperature on the Microstructure and Mechanical Properties of In Situ Reinforced Titanium Composites by Inductive Hot Pressing. *Materials* **2016**, *9*, 919. [CrossRef] [PubMed]

29. Montealegre-Meléndez, I.; Neubauer, E.; Arévalo, C.; Rovira, A.; Kitzmantel, M. Study of Titanium Metal Matrix Composites Reinforced by Boron Carbides and Amorphous Boron Particles Produced by Direct Hot Pressing. *Key Eng. Mater.* **2016**, *704*, 85–93. [CrossRef]

30. Montealegre-Meléndez, I.; Neubauer, E.; Danninger, H. Consolidation of titanium matrix composites to maximum density by different hot pressing techniques. *Mater. Sci. Eng. A* **2010**, *527*, 4466–4473. [CrossRef]

31. Davis, J.R. Nondestructive Evaluation and Quality Control. In *ASM Handbook*; ASM-International: Novelty, OH, USA, 1989.

32. Attar, H.; Ehtemam-Haghighi, S.; Kent, D.; Okulov, I.V.; Wendrock, H.; Bönisch, M.; Volegov, A.S.; Calin, M.; Eckert, J.; Dargusch, M.S. Nanoindentation and wear properties of Ti and Ti-TiB composite materials produced by selective laser melting. *Mater. Sci. Eng. A* **2017**, *688*, 20–26. [CrossRef]

33. Wang, B.; Huang, L.J.; Geng, L.; Yu, Z.S. Modification of microstructure and tensile property of TiBw/near-α Ti composites by tailoring TiBw distribution and heat treatment. *J. Alloys Compd.* **2017**, *690*, 424–430. [CrossRef]

34. Murray, J.L.; Liao, P.K.; Spear, K.E. The B-Ti (Boron-Titanium) System. *Bull. Alloy Phase Diagr.* **1986**, *7*, 550–551. [CrossRef]

35. Yan, Z.; Chen, F.; Cai, Y.; Zheng, Y. Microstructure and mechanical properties of in-situ synthesized TiB whiskers reinforced titanium matrix composites by high-velocity compaction. *Powder Technol.* **2014**, *267*, 309–314. [CrossRef]

36. Choi, B.J.; Kim, I.Y.; Lee, Y.Z.; Kim, Y.J. Microstructure and friction/wear behavior of (TiB + TiC) particulate-reinforced titanium matrix composites. *Wear* **2014**, *318*, 68–77. [CrossRef]

37. Zhang, W.; Feng, Y.; Chen, W.; Yang, J. Effects of heat treatment on the microstructure and mechanical properties of in situ inhomogeneous TiBw/Ti6Al4V composite fabricated by pre-sintering and canned powder extrusion. *J. Alloys Compd.* **2017**, *693*, 1116–1123. [CrossRef]

38. Lütjering, G.; Williams, J.C. *Titanium*, 2nd ed.; Springer: Berlin, Germany, 2007.

39. Leyends, C.; Peters, M. *Titanium and Titanium Alloys: Fundamentals and Applications*, 1st ed.; Wiley-VCH Verlag GmbH & Co. KGaA: Weinheim, Germany, 2003.

metals

MDPI

Article

Reduced Graphene Oxide Reinforced 7075 Al Matrix Composites: Powder Synthesis and Mechanical Properties

Youhong Sun [1,2], Chi Zhang [1,2], Baochang Liu [1,2,*], Qingnan Meng [1,2,*], Shaoming Ma [1,2] and Wenhao Dai [1]

[1] School of Construction Engineering, Jilin University, Changchun 130026, China; syh@jlu.edu.cn (Y.S.); zhangchi15@mails.jlu.edu.cn (C.Z.); masm14@mails.jlu.edu.cn (S.M.); daiwh2414@mails.jlu.edu.cn (W.D.)

[2] Key Laboratory of Drilling and Exploitation Technology in Complex Conditions, Ministry of Land and Resources, No. 938 Ximinzhu Street, Changchun 130026, China

* Correspondence: liubc@jlu.edu.cn (B.L.); qingnanmeng@jlu.edu.cn (Q.M.); Tel./Fax: +86-431-8850-2357 (B.L. & Q.M.)

Received: 25 September 2017; Accepted: 7 November 2017; Published: 13 November 2017

Abstract: Reduced graphene oxide (rGO) reinforced 7075 Al matrix composites were fabricated by electrostatic self-assembly and powder metallurgy. 7075 Al powders were surface modified by introducing a cetyl trimethyl ammonium bromide (CTAB) membrane on the surface, which was able to form a strong bonding with graphene oxide (GO) through electrostatic interaction. During the vacuum sintering process, CTAB was effectively removed and GO was thermally reduced into rGO. Morphologies of GO nanosheets, GO/7075 Al powders, microstructures, and tensile fractographs of the composites were observed. The effect of rGO content on mechanical properties of rGO/7075 Al composites was investigated. The results show that a good bonding between rGO and matrix is achieved. With the rGO content increasing, the hardness increases gradually, while the ultimate tensile strength and yield strength initially increase and later decrease. The improvement in strength of rGO/7075 Al composites was attributed to stress transfer and dislocation strengthening. With rGO content reaching 0.50 wt %, the excessive addition of rGO gave rise to a weakening in the enhancement of the tensile properties due to the increasing amounts of brittle Al_4C_3 and cracks.

Keywords: metal matrix composites; 7075 Al alloy; reduced graphene oxide; mechanical properties; strengthening mechanism

1. Introduction

Today, aluminum matrix composites (AMCs) reinforced with nanoparticles, fibers, or whiskers are in high demand due to their attractive characteristics, such as low density combined with high strength, large matrix selectable range, various production processes, and considerable improvement in mechanical properties after heat treatment [1,2]. Due to these desirable properties, AMCs are widely applied as structural materials in many industries, such as automobiles, aerospace, as well as drill pipe material used in extra-deep oil drilling [3–6]. Among aluminum alloys, 7000-series aluminum alloys possess the highest strength-to-weight ratio, which makes them an attractive candidate for reinforcement by a second phase to further improve their properties [7–9].

As the perfect layer structure of two-dimensional (2D) sp2-hybridized carbon atoms, graphene has been extensively investigated in recent years due to its outstanding properties, such as high Young's modulus, high fracture strength, and excellent thermal conductivity [10–14]. Due to these excellent mechanical properties and high specific surface area, graphene is expected to be an ideal reinforcement phase interacting with aluminum matrix, even in a small concentration. However, the research on AMCs reinforced with graphene is still in its infancy because the dispersion method, preparation technology, and harmful interface reaction have become the main problems that restrict its development [15].

Powder metallurgy is widely used for the preparation of AMCs due to its low processing temperature, which is beneficial for mitigating a harmful interface reaction. The uniform dispersion of graphene has been regarded as the most important factor in order to achieve high strength. Ball milling (BM) is normally carried out for the dispersion of graphene in aluminum. Wang et al. [16] have fabricated aluminum composites reinforced with graphene nanosheets (GNS/Al composite) through a feasible methodology based on flake powder metallurgy, and found that 0.30 wt % GNSs give rise to 62% enhancement over pure Al. However, Bartolucci et al. [17] have investigated the mechanical properties of 0.1 wt % GNSs/Al composite through BM followed by hot isostatic pressing and hot extrusion, and obtained a remarkable decrease in tensile strength and elongation for 0.1 wt % GNS/Al composite compared with pure Al. In addition, BM produces a great deal of heat that can easily cause an explosion.

In this work, graphene oxide (GO) nanosheets, rather than graphene, are used as a raw material, because many hydroxyl and epoxy groups existing on the surface of GO make it much easier to realize the uniform adsorption of GO on 7075 Al powders free of BM via an electrostatic self-assembly [18]. The effect of electrostatic self-assembly process on achieving the uniform adsorption of GO on 7075 Al powders is investigated in detail. In this process, CTAB is used to introduce a cationic membrane on the surface of 7075 Al powders, which ensures a strong bonding between CTAB-modified 7075 Al powder and GO with a negative charge. The in situ removal of CTAB and the reduction of GO are achieved during the sintering process. Furthermore, the influences of reduced graphene (rGO) on hardness and tensile properties of the composites are investigated. The strengthening mechanisms of rGO reinforcing 7075 Al are elaboratedd.

2. Experimental Procedures

2.1. Research Materials

Graphite oxide having particle size of 5 μm with 99% purity was provided by Hengqiu Graphene Technology Co. Ltd., Suzhou, China. CTAB obtained from Shandong West Asia Chemical Industry Co. Ltd., Jinan, China was in powder form with an ignited residue content less than 0.1 wt %. As a starting material, spherical 7075 Al powders (99% purity, ~9 μm) were supplied by Chaowei Nanotechnology Co. Ltd., Shanghai, China. The nominal composition of the as-received 7075 Al powder in this investigation is listed in Table 1.

Table 1. The chemical composition of the as-received 7075 Al powder.

Elements	Zn	Mg	Cu	Fe	Si	Mn	Others	Al
(wt %)	5.720	2.310	1.560	0.092	0.087	0.080	0.050	bal.

2.2. Composites Preparation

Figure 1 elaborates the fabrication procedure of the rGO/7075 Al composites used in this investigation. Four principle steps are involved in the fabrication process:

(1) Preparation of GO aqueous dispersion: To strip the as-received graphite oxides into GO nanaosheets with several-layers structure, the graphite oxide was added into deionized water and then ultrasonicated for 2 h to obtain a brown dispersion with no residual sediment. A 1 mg/mL GO aqueous dispersion was finally prepared. Figure 2 displays an AFM image of GO nanosheets obtained from the GO aqueous dispersion. As can be seen, the thickness of the GO nanosheets was ~5 nm. Considering the thickness of monolayer GO nanosheets was ~1 nm due to the

attachment of oxygen functionalities [19], the GO nanosheets used in this investigation were no more than five layers.

(2) Modifying 7075 Al powders with CTAB: 50 g 7075 Al powders and 300 mL CTAB aqueous solution (0.8 wt %) were magnetically stirred for 2 h, filtered, and then rinsed with deionized water to obtain the CTAB-modified 7075 Al powders.

(3) Adsorption of GO onto the 7075 Al powders: A powder slurry was prepared through adding CTAB-modified 7075 Al powders (~50 g) into deionized water. The GO aqueous dispersion was added drop by drop. The mixed slurry was magnetically stirred until the color changed to transparent, and was then filtered and rinsed to obtain the composite powders. The composite powders were finally vacuum dried at 70 °C for 8 h.

(4) CTAB removal and rGO/7075 Al composites fabrication. No particular heating treatment was used to remove the CTAB and reduce the GO, because the sintering temperature (560 °C) was high enough to achieve the purpose. The green billets (30 mm in diameter and 40 mm in height) were prepared through compacting the composite powders under 140 MPa at room temperature. Subsequently, the green billets were heated in a vacuum furnace at 560 °C for 2 h, followed by hot pressing under 70 MPa for 10 min to ensure the density. After that, slabs with cross-sections 12 mm in width and 4 mm in thickness were obtained by hot extrusion at 450 °C. The extrusion ratio and ram speed used in present study were 14.7 and 1 mm/s, respectively. Subsequently, the slabs were solution-treated in a resistance furnace at 470 °C for 2 h, followed by water quenched, and then aged at 120 °C for 24 h. The contents of rGO in the composites were 0.15, 0.30, and 0.50 wt %, respectively. For comparison, a 7075 Al sample was also prepared using the same method without adding GO.

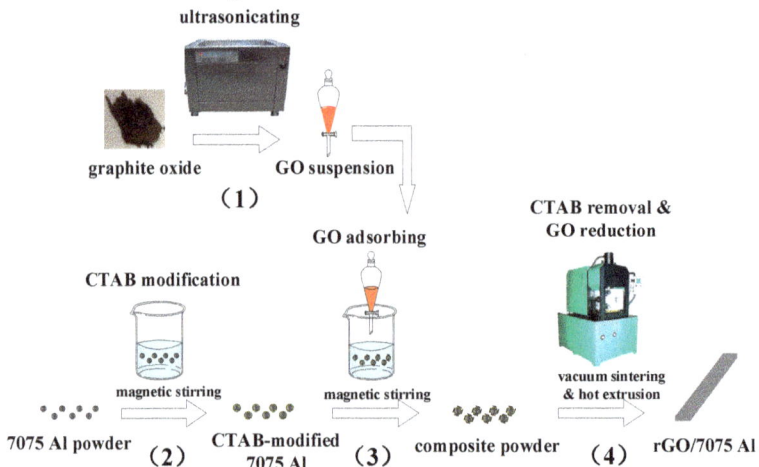

Figure 1. Fabrication procedures for rGO/7075 Al composites.

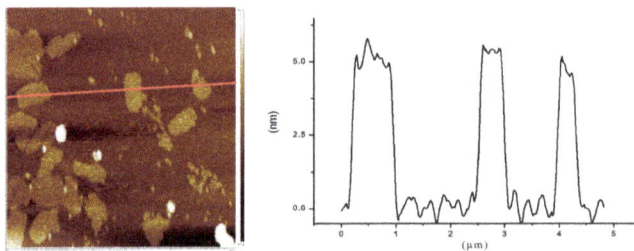

Figure 2. AFM image of the GO nanosheets and depth profile of the line on the GO nanosheets.

2.3. Mechanical Properties and Density Measurements

The specimens were prepared in accordance with ASTM Standard E-8/E8M-09 [20] parallel to the extrusion direction with a 10 mm gauge length. The tensile test was performed after polishing the samples in air at room temperature using an electronic universal test machine (DDL 100, CIMACH, Changchun, China) operated at a constant crosshead speed with an initial strain rate of 5×10^{-4} s^{-1}. At least three samples for each composite were measured to ensure the accuracy. The hardness of the composites and 7075 Al samples were tested by a microhardness tester (1600-5122VD Microment 5104, Buehler Ltd., Chicago, IL, USA) under an applied load of 100 g for 15 s. At least seven measurements were performed for each condition to ensure the accuracy of the results. The relative density of the as extruded composites and 7075 Al samples was measured by Archimedes' principle.

2.4. Microstructure Characterizations

The microstructures were characterized by a scanning electron microscopy (SEM; S-4800, Hitachi Ltd., Tokyo, Japan). Fourier transform infrared spectroscopy (FTIR; Thermal Scientific Nicolet iS10, Nicolet Ltd., Madison, WI, USA) was used to identify functional groups in GO, CTAB-modified 7075 Al powders, and composite powders. Thermal analysis was carried out using a simultaneous thermal analyzer (STA 499C, Netzsch Ltd., Bavaria, Germany) to analyze the thermal reduction of GO and CTAB removal during sintering process. The microstructures of rGO/7075 Al composites were characterized by optical microscopy (OM; Carl Zeiss–Axio Imager A2m, Gottingen, Germany) and transmission electron microscopy (TEM; JEOL-2000EX, Tokyo, Japan). The phase constituents of rGO/7075 Al composites were identified by X-ray diffraction (XRD; D/Max 2500PC, Rigaku Ltd., Tokyo, Japan) using Cu Kα radiation in step mode from 20° to 80° with a scanning speed of 5°/min.

3. Results

3.1. Improvement in Adsorption Uniformity of GOs by CTAB Modification

The significant effect of CTAB on improving the dispersion uniformity of GO can be proven by comparing the mixture of GO and 7075 Al with and without CTAB modification. Figure 3 shows the mixture of the GO suspension with the unmodified 7075 Al and the CTAB-modified 7075 Al slurry after magnetic stirring. For the mixture of GO with unmodified 7075 Al, the upper layer of the mixture remains brown, indicating that most of the GO was not absorbed onto the 7075 Al surface and was still in the suspension. Some brown flocs lie on the surface of unmodified 7075 Al slurry, indicating the agglomeration of GO. By contrast, the upper layer of the GO and CTAB-modified 7075 Al mixture is nearly transparent, suggesting that most of the GO in the mixture was absorbed onto the 7075 Al surface.

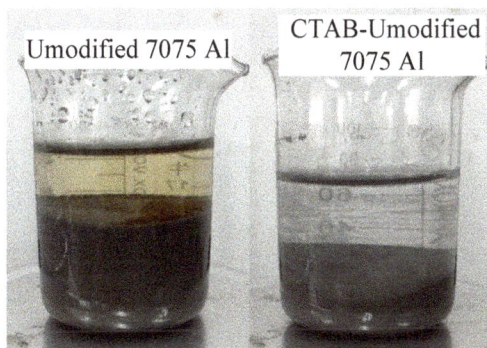

Figure 3. The image of mixture after 7075 Al powders absorbing GO.

The FTIR spectrum displays the difference of CTAB-modified 7075 Al powder and unmodified 7075 Al powder in terms of surface structure. As shown in Figure 4, the FTIR spectrum of the CTAB-modified 7075 Al powder shows many additional bands compared to that of unmodified 7075 Al powder. The band at 1000 cm^{-1} is caused by the C-N stretch vibration, a characteristic of CTAB. The band at 1471 cm^{-1} conveys the message that CTA$^+$ cations are in a liquid-like molecular environment where the alkyl chain of CTA$^+$ rotates freely around its long axis [21]. The strong absorption bands at 2918 cm^{-1} and 2849 cm^{-1} correspond to the C-H symmetrical and asymmetrical stretch vibrations of methyl and methylene, respectively [22]. The FTIR analysis confirms that the 7075 Al surface was coated with the CTAB membrane.

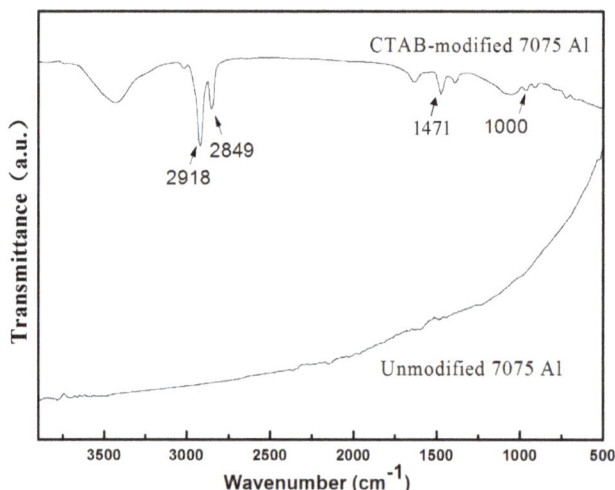

Figure 4. FTIR spectrum of unmodified and CTAB-modified 7075 Al powders.

Typical SEM images of the composite powders are given in Figure 5. For the unmodified 7075 Al powders (Figure 5a), a large area is observed with no adsorption of GO. On the other hand, the severe GO clustering is found on the edge of the powder, indicating an unacceptable dispersion uniformity of GO without CTAB modification. Figure 5b shows the uniform adsorption achieved by CTAB-modified 7075 Al powders. It can be seen that many wrinkles are distributed on the surface of 7075 Al powder and many wrappings of GO are found dispersed at the edge, and no obvious agglomeration of the GO is observed between the particles, indicating a uniform dispersion of GO on the 7075 Al surface.

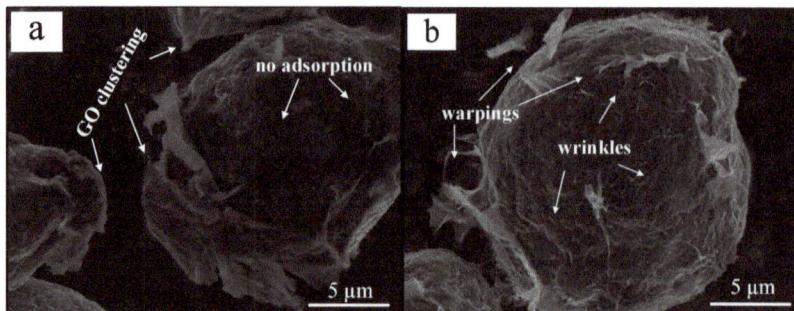

Figure 5. The SEM images of the composite powders: (**a**) GO clustering in unmodified 7075 Al powders; and (**b**) uniform adsorption of GO achieved by CTAB modification.

3.2. Removal of CTAB and Thermal Reduction of GO during Sintering Process

In order to study the CTAB removal and the GO reduction during sintering process, TGA was used to analyze the thermal stability of CTAB and GO up to 560 °C, with a heating rate of 10 °C/min. As shown in Figure 6a, CTAB shows little mass loss before 200 °C, revealing a favorable thermal stability of CTAB at low temperature. The curve displays a significant mass loss at ~90% from 200 °C to 350 °C, attributed to the CTAB pyrolysis. In the last stage, the curve displays a stable mass remaining at ~10%, indicating the end of CTAB pyrolysis. The remaining mass could be the residual carbon. The mass remaining curve of GO presented in Figure 6b also displays three discrete stages. Firstly, a slight mass loss (~10%) occurs up to 150 °C. In the second stage, an additional significant mass loss (~32%) occurs up to 260 °C. Finally, an 8% mass loss can be observed up to 560 °C.

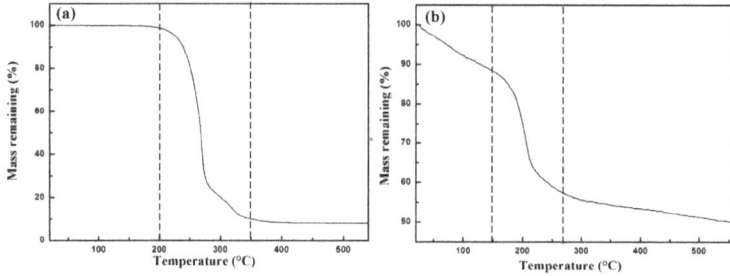

Figure 6. Thermogravimetric analysis (TGA) of (**a**) CTAB and (**b**) GO from 20 °C to 560 °C at the heating speed of 10 °C/min.

The XRD patterns (Figure 7a) show the difference for composite powders before and after vacuum heating. For the composite powders before heating, apart from the four peaks corresponding to Al (PDF#65-2869), three peaks corresponding to CTAB in the composite powder are observed at 21.48°, 22.51°, and 24.51° (PDF#48-2454). Meanwhile, no obvious peak corresponding to GO is observed, which might result from the low content of GO. For the composite powders after heating, only four Al peaks are observed from the XRD pattern, while CTAB peaks have all vanished, revealing the removal of CTAB. Figure 7b also shows a typical FTIR spectra of GO before and after heating. For the FTIR spectra of GO before heating, there is a broad and intense band of O–H stretching vibration at 3418 cm^{-1}, as well as the bands of C=O, C–O–C stretching vibration at 1722 cm^{-1} and 1052 cm^{-1}. There is also a band of O–H deformation vibration at 1395 cm^{-1} [16]. After vacuum heating, the O–H band of the stretching vibration and the C–O–C band shift to 3441 cm^{-1} and 1096 cm^{-1}, respectively, and the breadth and intensity fall dramatically. Moreover, the C=O stretching vibration band also disappears.

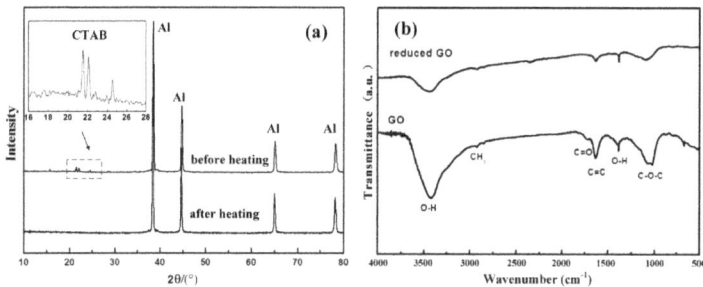

Figure 7. (**a**) XRD patterns for the composite powders before and after heating; and (**b**) FTIR spectra of GO and rGO after thermal reduction.

3.3. Microstructure

The representative OM images for the ND-TD surface of rGO/7075 Al composites are shown in Figure 8. The microstructure shows a typical fibrous structure attributed to the hot extrusion. Due to this microstructure, it is very difficult to make a quantitative measurement on grain size. In order to determine the phase composition of the composites, the X-ray diffraction was performed. The XRD patterns for 7075 Al and rGO/7075 Al composites are given in Figure 9. According to the XRD results, major aluminum peaks are observed at 38.47° (1 1 1), 44.71° (2 0 0), 65.09° (2 2 0), and 78.22° (3 1 1) (PDF#65-2869). A $MgZn_2$ (2 0 0) peak at 40.17° (PDF#65-3578) and an AlCu (−7 1 2) peak at 58.11° (PDF#26-0016) are also observed. These second phases in the aluminum matrix were mainly precipitated in the process of aging treatment and benefitted to improve the mechanical properties of 7075 Al alloys through precipitation strengthening mechanism [23]. Based on the XRD patterns and Scherrer's equation [24], the average crystal size of 7075 Al, 0.15, 0.30, and 0.50 wt % rGO/7075 Al composites are 91.3 nm, 64.7 nm, 51.3 nm and 46.0 nm respectively. No aluminum carbide (Al_4C_3) peak or graphene peak is found in the XRD patterns, different from the detection results obtained by Rashad et al. [15] and Li et al. [25]. However, it cannot be concluded that no Al_4C_3 phase was formed during the sintering process, since the rGO content (up to 0.5 wt %) is probably beyond the detection limit of XRD. As the interfacial phase (Al_4C_3) and the bonding condition between rGO and matrix are very significant to the mechanical properties of composites, it is necessary to analyze the rGO/7075 Al composites by TEM.

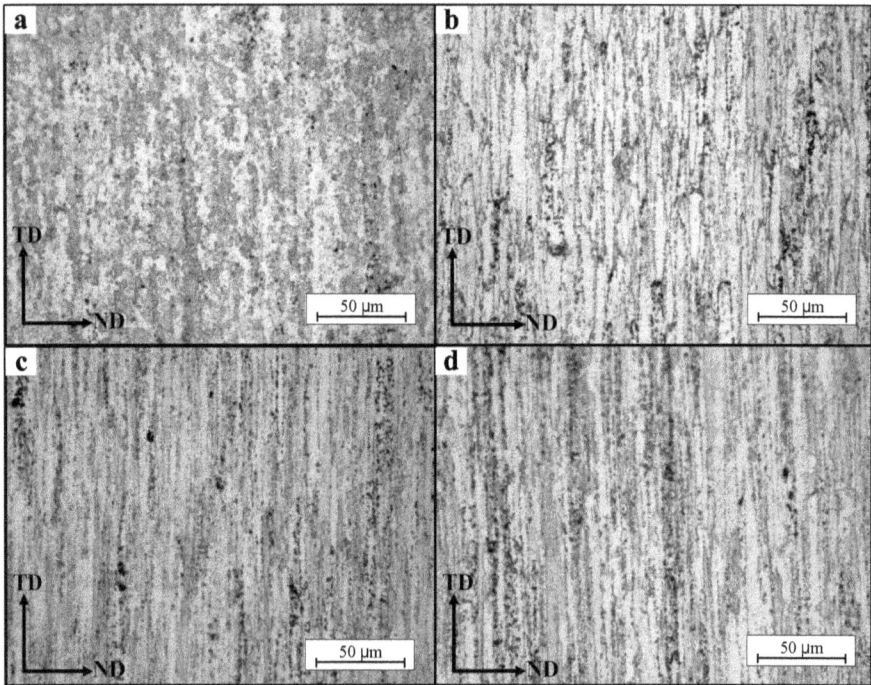

Figure 8. OM images of ND-TD surface for (**a**) 7075 Al; (**b**) 0.15 wt % rGO/7075 Al; (**c**) 0.30 wt % rGO/7075; and (**d**) 0.50 wt % rGO/7075.

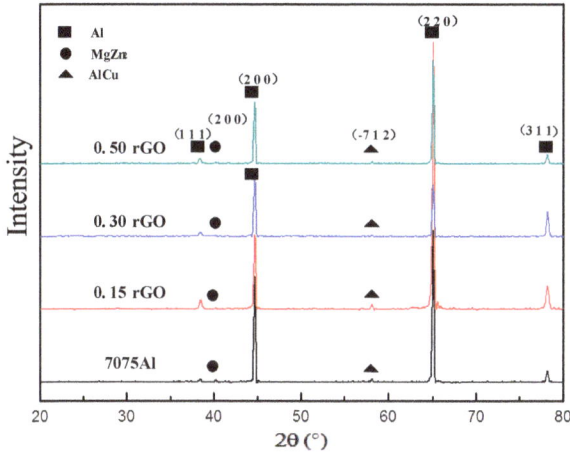

Figure 9. XRD patterns for 7075 Al alloys without and with different contents of rGO addition.

TEM morphology of the 0.30 rGO/7075 Al composite is shown in Figure 10. As shown by the white arrows, several wrinkled edges are observed as a typical feature of rGO, indicating that the rGO was not destroyed by thermal reduction. It can also be concluded that a good exfoliation of GO occurred through sonication. For the bonding condition, rGO is closely embedded in the Al grain boundary free from defects, porosity, or impurities. Although no Al_4C_3 phase is detected via X-ray diffraction (Figure 9), some black interfacial products (as shown by white arrows) are found at the interface. These products are further identified as Al_4C_3 phase via selected area electron diffraction (SAED) analysis, which is in agreement with the result obtained by Li et al. [25]. The TEM result reveals that the disappearance of Al_4C_3 phase in XRD patterns is due to the detection limit of the X-ray diffraction equipment.

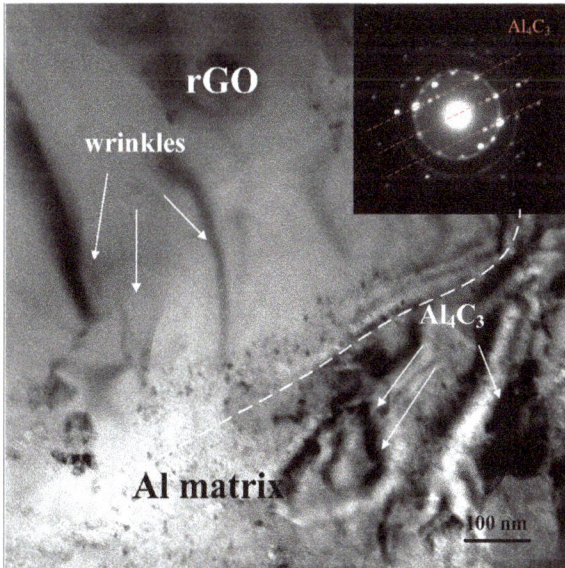

Figure 10. TEM image of 0.30 rGO/7075 Al composite.

3.4. Mechanical Properties

The Vickers hardness results of 7075 Al and composites are presented in Figure 11a. As shown in Figure 11a, the Vickers hardness value increases with the increasing rGO content. This improvement is attributed to the uniformly distributed rGO provided to transfer load from the 7075 Al matrix and restrains the dislocation movement during indentations. On the other hand, the density of composites also plays a vital role in affecting their hardness. Table 2 shows the density and the relative density of 7075 Al and composites. All composites display a high relative density at ~99.40%, though with rGO addition the relative density of composites slightly decreases. This phenomenon indicates that an outstanding densification is achieved through sintering and extrusion. Compared to the density, the reinforcement rGO plays a more important role in improving the hardness.

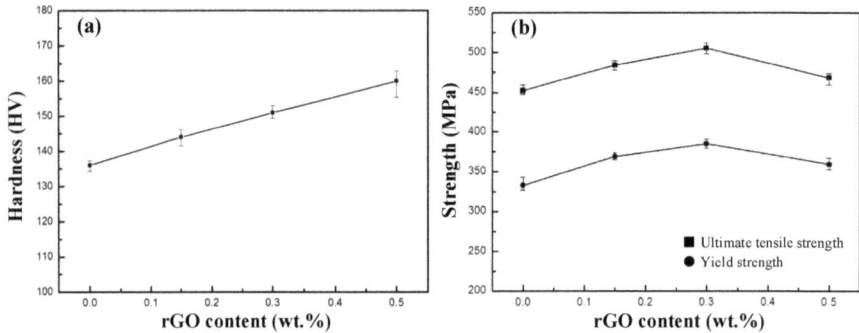

Figure 11. The variation tendencies of (**a**) hardness and (**b**) yield strength, and the ultimate tensile strength with rGO content.

Table 2. Densities of 7075 Al and rGO/7075 Al.

Materials	Density (g·cm^{-3})	Theoretical Density (g·cm^{-3})	Relative Density (%)
7075 Al	2.7863	2.8000	99.51
0.15 rGO/7075 Al	2.7830	2.7989	99.43
0.30 rGO/7075 Al	2.7823	2.7979	99.44
0.50 rGO/7075 Al	2.7795	2.7966	99.39

The tensile properties of composites with different addition contents of rGO are presented in Table 3. As shown in Table 3, rGO exhibits a significant effect on the tensile properties of 7075 Al alloys. Compared with 7075 Al having 452 MPa tensile strength, 333 MPa yield strength, the composites all show higher tensile properties. Figure 11b displays the variation tendencies of yield strength and ultimate tensile strength with increasing rGO content. The yield strength and the ultimate tensile strength both display an increase followed by a decrease, and reach the peak value at 385 MPa and 505 MPa, respectively, with 0.30 wt % rGO addition. Compared with 7075 Al, the yield strength and ultimate tensile strength are increased by 15.6% and 11.7%. For the elongation, the composites experience a decrease in elongation with rGO content increasing.

Table 3. Tensile properties of 7075 Al and rGO/7075 Al composites.

Materials	Yield Strength (MPa)	UTS (MPa)	Elongation
7075 Al	333^{+5}_{-3}	452^{+7}_{-5}	$13.5^{+0.5}_{-0.7}$
0.15 rGO/7075 Al	369^{+1}_{-1}	484^{+2}_{-2}	$11.8^{+0.2}_{-0.2}$
0.30 rGO/7075 Al	385^{+2}_{-2}	505^{+3}_{-3}	$10.7^{+0.5}_{-0.7}$
0.50 rGO/7075 Al	359^{+4}_{-3}	468^{+6}_{-9}	$8.0^{+0.4}_{-0.8}$

4. Discussion

The significant difference in the dispersion uniformity of GO between the unmodified 7075 Al powder and the CTAB-modified 7075 Al powder reveals the critical effect of CTAB modification on improving the absorption uniformity of GO. The FTIR spectrum for GO (Figure 7b) reveals that GO is fabricated with large amounts of hydrophilic oxygen-containing groups, such as –OH, –COOH, and –C=O, endowing the GO with hydrophilic surface property, are very similar to the –COOH functionalized carbon nanotubes [26]. However, due to the naturally-formed alumina film, 7075 Al powder is identified as hydrophobic [26]. The contradictory water wettability determines that the GO can hardly form a homogeneous distribution with the unmodified 7075 Al powders through stirring.

CTAB is a water soluble cationic surfactant that shows good coordination with anionic and amphoteric surfactants [27]. The FTIR spectrum for CTAB-modified 7075 Al powder (Figure 4) reveals that a considerable number of positive charges (CTA$^+$) were introduced on the surface of 7075 Al powder. The CTAB modification leads to the formation of a thin cationic membrane due to its hydrophobic long carbon chains and electropositive polar groups. The compatible water wettability of alumina film plays a key role in absorbing the long carbon chains through van der Waals force to form a stable CTAB membrane [18]. The CTAB membrane also introduces a large number of positive charges onto the surface of the 7075 Al powder. Due to the negative charged nature, GO tends to absorb onto the cationic CTAB membrane through electrostatic attraction and, thus, a uniform distribution of GO in the CTAB-modified 7075 Al powders can be effectively achieved.

To achieve the removal of impurity (CTAB in this research) and the reduction of GO, Wang et al. [16] and Jiang et al. [26] employed the extra thermal treatment. However, the TGA result for CTAB (Figure 6a) shows a remarkable mass loss (~90%) up to 560 °C, indicating a complete decomposition of CTAB during the sintering process, which can also be proven by the XRD patterns for the composite powders before and after heating (Figure 7a). Meanwhile, the TGA result for GO up to 560 °C (Figure 6b) displays the mass loss in three discrete stages. In accordance with the analysis obtained by Tegou et al. [28], the mass loss (~10%) in the first stage is mainly caused by the elimination of physisorbed and interlamellar water molecules. The significant mass loss (~32%) in the second stage is primarily attributed to the thermal decomposition of covalently bonded oxygen. Finally, in the last stage the mass loss (~8%) is caused by the removal of more stable oxygen-containing functional groups. However, compared to the TGA results obtained by Tegou et al. [28], the 150 °C decomposition temperature here is approximately 30 °C higher, while the ending temperature of the second stage is 260 °C, 60 °C higher. The difference is probably due to the ambient atmosphere adopted in literature [28], while the TGA in this investigation was carried out in vacuum atmosphere. The FTIR spectrum (Figure 7b) also demonstrate the disappearance of the oxygen functional groups on GO, suggesting the thermal reduction of GO into rGO during the sintering period.

The strengthening mechanisms of well dispersed rGO reinforcing rGO/7075 Al composites are generally explained by grain refinement, stress transfer, and dislocation strengthening: as a nano-reinforcement, rGO is expected to impede the grain coarsening during thermal processing, resulting in higher mechanical properties [16]. Although the results of XRD spectra shows that the average crystal size decreases with rGO content increasing, the OM images of rGO/7075 Al composites and 7075 Al show little difference in grain size. According to the investigation of Li et al. [25], the effect of grain refinement is negligible due to the hot-pressed sintering, which employs a low temperature avoiding 7075 Al changing into a liquid state. The enhancement of strength for composites is strongly depended on the interfacial bonding between rGO and matrix. TEM results show a good interfacial bonding condition obtained in this study. A good interfacial bonding condition allows the tensile stress to transfer from matrix to reinforcements and create an interfacial shear stress [29]. The contribution of yield strength increase for composites can be calculated from shear lag model [15]:

$$\sigma_{SL} = \sigma_m + f_v \sigma_m \left(\frac{S}{2}\right) \tag{1}$$

where σ_{SL} is yield strength of composite calculated by shear stress model, f_v is volume fraction of rGO, σ_m is yield strength of Al matrix, and S is aspect ratio of rGO. The wrinkles on the rGO surface were also reported to be beneficial to the formation of the mechanical bond between rGO and the pure Al matrix [30].

Orowan looping is an important model related to dislocation strengthening. The addition of rGO contributes to the strengthening through restricting dislocation movements. A uniform dispersion of rGO could offer a large number of nano-particles to take part in this strengthening mechanism [25]. Plastic deformation leads to form residual dislocation loops around rGO and these loops produce back stress restricting dislocation movement and, thus, increasing the strength [15]. On the other hand, the significant mismatch in coefficient of thermal expansion (CTE) between rGO and the 7075 Al matrix ($CTE_{rGO} = 0.9 \times 10^{-6} K^{-1}$, $CTE_{7075\ Al} = 23.6 \times 10^{-6} K^{-1}$) can form a large number of dislocations at the interface. Dislocation density is determined by the surface area of reinforcement. Smaller reinforcement particles lead to a higher dislocation density which results in the strengthening of the composites [31]. Due to its unique two-dimensional structure, rGO is known for its very large surface area (theoretically as much as 2630 m^2/g) [32]. Therefore, rGO is able to contribute to a rather high dislocation density. The yield strength of composites σ_c can be calculated by the following equation [33]:

$$\sigma_c = (1 + 0.5f_v)(\sigma_m + \Delta\sigma_{CTE} + \Delta\sigma_O + \frac{\Delta\sigma_O \Delta\sigma_{CTE}}{\sigma_m}) \tag{2}$$

where:

$$\Delta\sigma_O = \frac{0.13Gb}{d_p[(\frac{1}{2f_v})^{\frac{1}{3}} - 1]} \ln(\frac{d_p}{2b}) \tag{3}$$

$$\Delta\sigma_{CTE} = 1.25Gb\sqrt{\frac{12\Delta T \Delta C f_v}{bd_p}} \tag{4}$$

where G is the shear modulus of the 7075 Al matrix, b is the Burgers vector of matrix (0.286 nm for 7075 Al), ΔT is the difference between the processing temperature and the testing temperature, ΔC is the difference in CTE between the matrix and rGO, f_v is the volume fraction of rGO, and d_p is the mean particle size of rGO.

Figure 12 displays the comparison of yield strength obtained from different models and the present study. The experimental data matches closely with the theoretical prediction by the Orowan strengthening mechanism, indicating that the Orowan strengthening mechanism is predominant in the rGO/7075 Al composites. Uniform dispersion of rGO in the matrix restricts the dislocation motion in the matrix and leads to high dislocation density at the rGO/7075 Al interface [31]. Higher surface area and smaller particle size of rGO will lead to higher dislocation density.

However, it is worth noting that the excessive addition of rGO (0.50 wt %) gives rise to a weakening in the enhancement of the tensile properties compared with 0.30 rGO/Al composite, which is different from the prediction by Orowan strengthening mechanism. The decrease in tensile properties might be caused by the formation of Al_4C_3 in the composites, an interfacial product between the rGO and Al matrix formed at high sintering temperature via the following reaction:

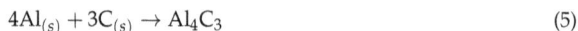

$$4Al_{(s)} + 3C_{(s)} \rightarrow Al_4C_3 \tag{5}$$

Moreover, Koratkar et al. [17] reported that wrinkles and folds on the surface of rGO provide massive defect sites, which could become reaction sites for the Al_4C_3 formation. Al_4C_3 is known as a brittle phase having harmful effects on tensile properties of composites [25]. Thus, the increasing amount of Al_4C_3 is likely to be the primary factor leading to the weakening in the enhancement of tensile properties for 0.50 rGO/7075 Al.

The fracture morphology of 7075 Al and composites is presented in Figure 13. A large number of dimples and transgranular fracture surfaces are observed in 7075 Al and all composites, indicating the ductile failure. For the composites with different contents of rGO addition, the pull-out rGO is found at the edge of dimples. Meanwhile the amount of dimples decreases with increasing rGO content, leading to the decrease in elongation. Apart from the pull-out rGO, cracks are detected on the fracture surface of 0.30 rGO/7075 Al and 0.50 rGO/7075 Al, and the number of cracks increases with increasing rGO content from 0.30 to 0.50 wt %. At the edge of the cracks, some pull-out rGO is also observed. This phenomenon might result from the weak bonding force of the few-layered rGO used in this study (see AFM in Figure 2). The cracks are most likely to originate at the interlayer of rGO and propagate along the tensile direction [18]. Therefore, the increasing rGO content also leads to an increase in cracks, having a negative effect on the tensile properties and elongation. When the rGO content reaches a critical value, the negative effect of cracks becomes predominant and causes the tensile properties of composites to decline. It is anticipated, with further increasing rGO content (larger than 0.50 wt %), that the excess rGO will give rise to a lower tensile strength even in comparison to 7075 Al.

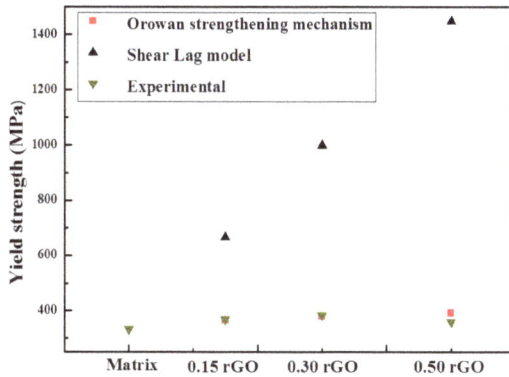

Figure 12. Comparison of yield strength calculated by different models and experimental data.

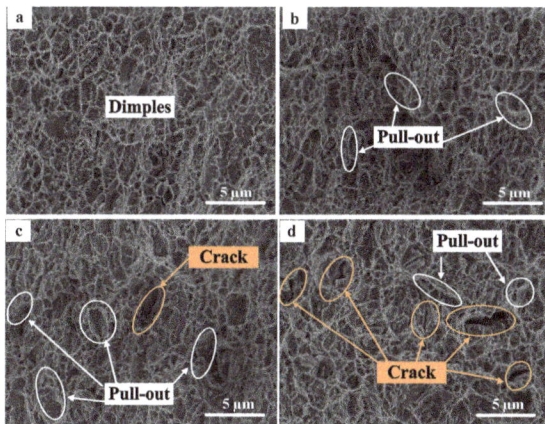

Figure 13. Tensile fracture morphologies of (**a**) 7075 Al; (**b**) 0.15 rGO/7075 Al; (**c**) 0.30 rGO/7075 Al; and (**d**) 0.50 rGO/7075 Al.

5. Conclusions

We have fabricated 7075 Al matrix nanocomposites based on powder metallurgy and electrostatic self-assembly processes. The effect of rGO content on the mechanical properties of rGO/7075 Al composites were studied. The results are summarized as follows:

(1) The significant improvement in the adsorption uniformity of GO is attributed to the formation of the cationic CTAB membrane on 7075 Al powders. Due to the negatively-charged nature, GO tends to absorb onto the cationic CTAB membrane through electrostatic attraction and, thus, a uniform distribution of GO in the CTAB-modified 7075 Al powders is effectively achieved.

(2) During the sintering process, CTAB is effectively removed before 350 °C so that it would have little negative effect on the mechanical properties of composites. On the other hand, the thermal reduction of GO is demonstrated by the results of TGA and FTIR.

(3) The Vickers hardness of rGO/7075 Al composites increases with the increase in rGO content. rGO has a significant effect on the tensile properties of 7075 Al alloys. Compared with 7075 Al, the composites all show higher tensile properties. The yield strength and the ultimate tensile strength both reach the peak values for rGO/7075 Al composite with 0.30 wt % rGO addition. The yield strength and ultimate tensile strength of 0.30 rGO/7075 Al are increased by 15.6% and 11.7% compared with 7075 Al. The improvement in strength of rGO/7075 Al composites is attributed to stress transfer and dislocation strengthening. With rGO content reaching 0.50 wt %, however, the excessive addition of rGO gives rise to a weakening in the enhancement of the tensile properties compared with 0.30 rGO/Al composite, due to the increasing amounts of brittle Al_4C_3 and cracks.

Acknowledgments: Financial supports from SinoProbe-09-05 (project No. 201011082) and International S&T Cooperation Program of China (grant No. 2013DFR70490).

Author Contributions: Youhong Sun, Chi Zhang, and Baochang Liu conceived and designed the experiments; Chi Zhang and Wenhao Dai performed the experiments; Shaoming Ma and Qingnan Meng analyzed the data; Chi Zhang, Qingnan Meng, and Wenhao Dai contributed reagents, materials, and analysis tools; and Chi Zhang wrote the paper.

Conflicts of Interest: The authors declare no conflict of interest

References

1. Miracle, D.B. Metal matrix composites-from science to technological significance. *Compos. Sci. Technol.* **2005**, *65*, 2526–2540.
2. Surappa, M.K. Aluminum matrix composites: Challenges and opportunities. *Sadhana* **2003**, *28*, 319–334.
3. Williams, J.C.; Strake, E.A., Jr. Process in structural materials for aerospace systems. *Acta Mater.* **2003**, *51*, 5775–5799.
4. Li, B.; Pan, Q.L.; Chen, C.P.; Wu, H.H.; Yin, Z.M. Effects of solution treatment on microstructural and mechanical properties of Al-Zn-Mg alloy by microalloying with Sc and Zr. *J. Alloys Compd.* **2016**, *664*, 553–564.
5. Davydov, V.G.; Rostova, T.D.; Zakharov, V.V.; Filatov, Y.A.; Yelagin, V.I. Scientific principles of making an alloying addition of scandium to aluminum alloys. *Mater. Sci. Eng. A* **2000**, *280*, 30–36.
6. Jones, M.J.; Humphreys, F.J. Interaction of recrystallization and precipitation: The effect of Al3Sc on the recrystallization behavior of deformed aluminum. *Acta Mater.* **2003**, *51*, 2149–2159.
7. Shen, Q.; Wu, C.D.; Luo, G.Q.; Fang, P.; Li, C.Z.; Wang, Y.Y.; Zhang, L.M. Microstructure and mechanical properties of Al-7075/B4C composites fabricated by plasma activated sintering. *J. Alloys Compd.* **2014**, *588*, 265–270.
8. Panigrahi, S.K.; Jayaganthan, R. Effect of ageing on microstructure and mechanical properties of bulk, cryorolled, and room temperature rolled Al 7075 alloy. *J. Alloys Compd.* **2011**, *509*, 9609–9616.
9. Taleghani, M.A.J.; Ruiz Navas, E.M.; Torralba, J.M. Microstructural and mechanical characterization of 7075 aluminium alloy consolidated from a premixed powder by cold compaction and hot extrusion. *Mater. Des.* **2014**, *55*, 674–682.
10. Seiner, H.; Ramirez, C.; Koller, M.; Sedlak, P.; Landa, M.; Miranzo, P.; Belmonte, M.; Osendi, M.I. Elastic properties of silicon nitride ceramics reinforced with graphene nanofillers. *Mater. Des.* **2015**, *87*, 675–680.
11. Wang, H.D.; Kurata, K.; Fukunaga, T.; Zhang, X.; Takamatsu, H. Width dependent intrinsic thermal conductivity of suspended monolayer graphene. *Int. J. Heat Mass Transf.* **2017**, *105*, 76–80.

12. Bolotin, K.I.; Sikes, K.J.; Jiang, Z.; Klima, M.; Fudenberg, G.; Hone, J.; Kim, P.; Stormer, H.L. Ultrahigh electron mobility in suspended graphene. *Solid State Commun.* **2008**, *146*, 351–355.

13. Geim, A.K.; Novoselov, K.S. The rise of graphene. *Nat. Mater.* **2007**, *6*, 183–191. [PubMed]

14. Begum, K.R.; Sankeshwar, N.S. Electronic thermal conduction in suspended graphene. *Phys. E Low-Dimens. Syst. Nanostruct.* **2015**, *73*, 27–34.

15. Rashad, M.; Pan, F.S.; Tang, A.T.; Asif, M. Effect of Graphene Nanoplatelets addition on mechanical properties of pure aluminum using a semi-powder method. *Prog. Nat. Sci. Mater. Int.* **2014**, *24*, 101–108.

16. Wang, J.Y.; Li, Z.Q.; Fan, G.L.; Pan, H.H.; Chen, Z.X.; Zhang, D. Reinforcement with graphene nanosheets in aluminum matrix composites. *Scr. Mater.* **2012**, *66*, 594–597.

17. Bartolucci, S.F.; Paras, J.; Rafiee, M.A.; Rafiee, J.; Lee, S.; Kapoor, D.; Koratkar, N. Graphene-aluminum nanocomposites. *Mater. Sci. Eng. A* **2011**, *528*, 7933–7937.

18. Gao, X.; Yue, H.Y.; Guo, E.J.; Zhang, H.; Lin, X.Y.; Yao, L.H.; Wang, B. Preparation and tensile properties of homogeneously dispersed graphene reinforced aluminum matrix composites. *Mater. Des.* **2016**, *94*, 54–60.

19. De Silva, K.K.H.; Huang, H.H.; Joshi, R.K.; Yoshimura, M. Chemical reduction of graphene oxide using green reductants. *Carbon* **2017**, *119*, 190–199.

20. *ASTM Standard E8/E8M, 2009. Standard Test Methods for Tension Testing of Metallic Materials*; ASTM International: Montgomery County, PA, USA, 2009. [CrossRef]

21. Zhu, J.X.; He, H.P.; Zhu, L.Z.; Wen, X.Y.; Deng, F. Characterization of organic phases in the interlayer of montmorillonite using FTIR and ^{13}C NMR. *J. Colloid Interface Sci.* **2005**, *286*, 239–244. [PubMed]

22. Yu, W.H.; Ren, Q.Q.; Tong, D.S.; Zhou, C.H.; Wang, H. Clean production of CATB-montmorillonite: Formation mechanism and swelling behavior in xylene. *Appl. Clay Sci.* **2014**, *97–98*, 222–234.

23. Han, N.M.; Zhang, X.M.; Liu, S.D.; He, D.G.; Zhang, R. Effect of solution treatment on the strength and fracture toughness of aluminum alloy 7050. *J. Alloys Compd.* **2011**, *509*, 4138–4145.

24. Meng, Q.N.; Wen, M.; Mao, F.; Nedfors, N.; Jansson, U.; Zheng, W.T. Deposition and characterization of reactive magnetron sputtered zirconium carbide films. *Surf. Technol.* **2013**, *232*, 876–883.

25. Li, G.; Xiong, B.W. Effect of graphene content on microstructures and tensile property of graphene-nanosheets/aluminum composites. *J. Alloys Compd.* **2017**, *697*, 31–36.

26. Jiang, L.; Fan, G.L.; Li, Z.Q.; Kai, X.Z.; Zhang, D.; Chen, Z.X.; Humphries, S.; Heness, G.; Yeung, W.Y. An approach to the uniform dispersion of a high volume fraction of carbon nanotubes in aluminum powder. *Carbon* **2011**, *49*, 1968–1971.

27. Yang, Y.J.; Li, W.K. CTAB functionalized graphene oxide/multiwalled carbon nanotube composite modified electrode for the simultaneous determination of ascorbic acid, dopamine, uric acid and nitrite. *Biosens. Bioelectron.* **2014**, *56*, 300–306. [PubMed]

28. Tegou, E.; Pseiropoulos, G.; Filippidou, M.K.; Chatzandroulis, S. Low-temperature thermal reduction of graphene oxide films in ambient aptmosphere: Infra-red spectroscopic studies and gas sensing applications. *Microelectron. Eng.* **2016**, *159*, 146–150.

29. Dixit, S.; Mahata, A.; Mahapatra, D.R.; Kailas, S.V.; Chattopadhyay, K. Multi-layer graphene reinforced aluminum—Manufacturing of high strength composite by friction stir alloying. *Compos. Part B Eng.* **2018**, *136*, 63–71.

30. Bisht, A.; Srivastava, M.; Kumar, R.M.; Lahiri, I.; Lahiri, D. Strengthening mechanism in graphene nanoplatelets reinforced aluminum composite fabricated through spark plasma sintering. *Mater. Sci. Eng. A* **2017**, *695*, 20–28.

31. Arsenault, R.J.; Shi, N. Dislocation generation due to differences between the coefficients of thermal expansion. *Mater. Sci. Eng.* **1986**, *81*, 175–187.

32. Yin, C.M.; Wang, J.F. Effects of activation temperature on the deoxygenation specific surface area and supercapacitor performance of graphene. *Carbon* **2016**, *109*, 558–565. [CrossRef]

33. Zhang, Z.; Chen, D.L. Consideration of Orowan strengthening effect in particulate-reinforced metal matrix nanocomposites: A model for predicting their yield strength. *Scr. Mater.* **2006**, *54*, 1321–1326. [CrossRef]

![metals logo] *metals*

MDPI

Article

New Nanocomposite Materials with Improved Mechanical Strength and Tailored Coefficient of Thermal Expansion for Electro-Packaging Applications

Abdollah Saboori [1,*], Seyed Kiomars Moheimani [2], Matteo Pavese [1], Claudio Badini [1] and Paolo Fino [1]

[1] Department of Applied Science and Technology, Politecnico Di Torino, Corso Duca Degli Abruzzi 24, 10129 Torino, Italy; matteo.pavese@polito.it (M.P.); claudio.badini@polito.it (C.B.); paolo.fino@polito.it (P.F.)
[2] Department of Materials Engineering, Najafabad Branch, Islamic Azad University, Najafabad 8514143131, Iran; kiomars.moheimani@gmail.com
* Correspondence: abdollah.saboori@polito.it; Tel.: +39-011-090-4762

Received: 31 October 2017; Accepted: 28 November 2017; Published: 1 December 2017

Abstract: In this research, copper nanocomposites reinforced by graphene nanoplatelets (GNPs) were fabricated using a wet mixing method followed by a classical powder metallurgy route. In order to find the best dispersion technique, ball milling and wet mixing were chosen. Qualitative evaluation of the structure of the graphene after mixing indicated that the wet mixing is an appropriate technique to disperse the GNPs. Thereafter, the influence of graphene content on microstructure, density, hardness, elastic modulus, and thermal expansion coefficient of composites was investigated. It was shown that by increasing the graphene content the aggregation of graphene is more obvious and, thus, these agglomerates affect the final properties adversely. In comparison with the unreinforced Cu, Cu–GNP composites were lighter, and their hardness and Young's modulus were higher as a consequence of graphene addition. According to the microstructural observation of pure copper and its composites after sintering, it was concluded that grain refinement is the main mechanism of strengthening in this research. Apart from the mechanical characteristics, the coefficient of thermal expansion of composites decreased remarkably and the combination of this feature with appropriate mechanical properties can make them a promising candidate for use in electronic packaging applications.

Keywords: metal matrix nanocomposite; copper; graphene; dispersion; powder metallurgy; thermal expansion coefficient; thermal conductivity; electrical resistance

1. Introduction

Nowadays, Metal Matrix Nanocomposites (MMNCs) are materials which are used extensively in different applications such as automotive, aerospace, and electronic packaging industries [1–4]. For instance, consider the field of electronic packaging: with the progress of technology and the corresponding increase in the performance of electronic devices, the devices release more heat, and thermal considerations in the design and material selection thus become seriously important factors [5]. Indeed, the functioning and useful life of packaged electronic assembly depend seriously on advanced electronic packaging materials. Electrical conduction, electrical insulation, mechanical support, environmental protection, and thermal conduction and dissipation are the main functions that should be considered [6,7]. Heat sinks, which are the cooling systems for electronic devices, have been developed to solve this issue. It has been indicated that materials with a thermal conductivity higher than 150 W/(m·K) and coefficient of thermal expansion (CTE) between 4×10^{-6} K^{-1} and 9×10^{-6} K^{-1} are desirable materials for use in this application [8].

To select a proper material to be employed in electronic packaging applications, different considerations should be taken into account to strike a balance between function, performance, manufacturability, reliability, and cost [9]. Among the metallic materials, copper is one of the most interesting candidates which can be used in electronic packaging applications, owing mainly to its high thermal conductivity and cost. However, its high thermal expansion coefficient ($16.9 \times 10^{-6} \, K^{-1}$) at ambient temperature and low tensile strength have limited its application. Thus, in order to address these drawbacks and broaden the application of copper and its alloys, it is necessary to improve these characteristics through either fabrication of composites or heat treatment. However, in order to develop the application of copper composites, it is necessary to fabricate new composites with high thermal and electrical conductivity, low coefficient of thermal expansion, and high mechanical properties. Thus, it is essential to design the material to have a uniform dispersion of reinforcing material as well as a strong interfacial bonding between the reinforcement and matrix. Generally speaking, copper matrix composites can be produced by the addition of stable and non-soluble particles into the copper matrix. These non-soluble and stable particles can be different based on the target application, and could be oxides (Al_2O_3, SiO_2, etc.), borides (TiB_2, ZrB_2, etc.), nitrides (TiN, ZrN, etc.), carbides (SiC, B_4C, TiC, etc.), or carbonaceous materials (CNTs, graphite, graphene, diamond) [10–12]. According to the literature, due to the poor wettability between molten copper and reinforcing particles, conventional casting techniques are not an appropriate fabrication technique for this kind of composite. Thus, other manufacturing routes such as powder metallurgy techniques have been developed to produce the copper-based composites [12]. The main areas of interest for copper and its composites are in the electronic packaging and heat sink industries, as well as for structural and frictional applications [12–14]. It should be noted that the electrical properties of copper composites can be seriously affected by impurities: some of them may precipitate during the heat treatment, and these precipitates deteriorate the electrical conductivity [13]. For instance, Caron et al. have reported that the electrical conductivity of copper can be lowered to only 86% IACS (International Annealed Copper Standard) as a consequence of 0.023% Fe addition during the fabrication process [13]. Moreover, it is reported that the presence of 0.3% Zr, 1.25% Al, or 0.1% P can lower the electrical conductivity of copper to 85%, 70%, or 50% IACS, respectively [12].

On the other hand, graphene nanoplatelets, due to their unique characteristics such as very high thermal conductivity ($\approx$5000 W/(m·K)) and very low coefficient of thermal expansion in the in-plane direction ($-1 \times 10^{-6} \, K^{-1}$), could attract considerable attention for use as reinforcing materials [15–18].

The main aim of this study is not only to investigate the effect of graphene nanoplatelets on the physical and mechanical characteristics of copper composites, but also to develop new Cu/GNP composite materials with improved mechanical strength, appropriate thermal and electrical conductivity, and tailored CTE.

2. Materials and Methods

2.1. Starting Materials

The starting materials which were used to produce the copper matrix nanocomposites with differing graphene nanoplatelet content included:

- Copper (Cu) powder, with 150–240 μm size and 99.5% purity (Figure 1a). This powder has an irregular morphology with a porous surface (spongy structure) and was supplied by Alfa Aesar, Ward Hill, MA, USA.
- Graphene nanoplatelets (GNPs) with 99% purity, ~100 nm thickness, 25 μm width, and surface area of 700 m^2/g, supplied by abcr GmbH, Karlsruhe, Germany (Figure 1b).

The densities of the copper powder and graphene nanoplatelets (GNPs) were 8.96 g/cm^3 and 2.2 g/cm^3, respectively.

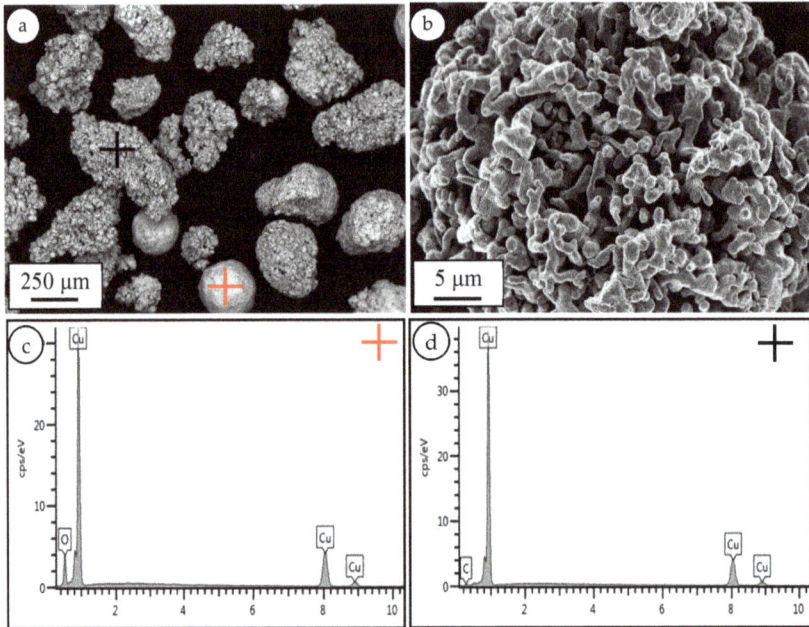

Figure 1. (**a,b**) SEM images of pure Cu powder with (**c,d**) the corresponding EDS analysis; the red cross and black cross belong to the copper oxide and copper particle, respectively.

2.2. Mixing

In principle, in order to take the advantage of the unique mechanical, thermal, and electrical properties of graphene nanoplatelets, it is necessary to achieve a uniform dispersion of GNPs within the metallic matrix. Therefore, in this work, the distribution of GNPs within the metal matrix was performed by means of mechanical milling and a novel wet mixing method proposed by Rashad et al. [19]; these techniques are described in the following sections.

2.2.1. Ball Milling

The nanocomposite powder preparation was carried out by a mechanical milling technique in a Pulverisette 5 planetary ball mill. The mixing of the starting powder was conducted in a hardened steel jar, and the milling media were 5 mm hardened stainless steel balls. The ball-to-powder ratio (BPR) was chosen to be equal to 10:1. During the milling process, methanol was used as a process control agent (PCA) to prevent the excessive cold welding of metallic powders, and an argon atmosphere was chosen in order to avoid oxidation during the milling. Mechanical milling was carried out by the agitation of a ball mill at a rotating speed of 200 rpm for 90 min, using a 30 min stop step after every 30 min of agitation in order to control the temperature during the milling. Reference samples of pure copper were mechanically milled under the same condition.

2.2.2. Wet Mixing

In this method, at first, graphene nanoplatelets and copper powders were separately dispersed in ethanol by means of ultrasonication. After ultrasonication for 45 min, the graphene nanoplatelet (GNP) suspension was added dropwise into the copper powder slurry to achieve the final content of GNPs. By adjusting the GNP content, composite powders with 4.0 vol % and 8.0 vol % graphene were produced. The mixture was ultrasonicated for 60 min to obtain a homogeneous dispersion. Afterwards, the mixture was filtered and dried at 80 °C for 6 h to achieve the final nanocomposite powder.

2.3. Consolidation and Sintering

Die compaction of the powders is an economical and highly productive technique for fabricating the metal matrix nanocomposites [20]. Thus, as-prepared nanocomposite powders were compacted in a stainless steel die with 30 mm diameter at room temperature. The green cylindrical samples of Cu/GNP nanocomposites were sintered at 950 °C for 2 h under an inert atmosphere (N2). The corresponding bulk Cu/GNP composites were denoted as Cu–4.0 vol % GNPs and Cu–8.0 vol % GNPs, respectively.

2.4. Characterization

The morphology and microstructure of the starting and nanocomposite materials were characterized by means of a field-emission scanning electron microscope (FESEM; Merlin-Zeiss, München, Germany, operating at 15 kV) equipped with an energy-dispersive X-ray spectrometer (EDS). The theoretical densities of composites were calculated according to the rule of mixtures, whereas their relative densities were measured using the Archimedes method. X-ray diffraction (XRD) analysis of the nanocomposite was performed on a Philips X'Pert diffractometer using Cu Kα radiation, a tube voltage of 40 kV, and a tube current of 40 mA. In this research, Raman spectroscopy was used to evaluate the graphene structure during the fabrication process. Hence, the Raman spectra of graphene were obtained by using a Renishaw InVia Reflex micro-Raman spectrometer (Renishaw plc, Wotton-under-Edge, UK) equipped with a cooled charge-coupled device camera. The laser beam (λ = 514.5 nm) was focused using a 50× objective lens before irradiating the samples. In order to avoid damage to the samples, the integration time was 50 s and the laser power was reduced to 5 mW. The hardness measurement was performed by a Vickers hardness tester at a load of 5 kg for 15 s on a polished surface. The hardness values are the average of five different measurements on the same surface. Thermal expansion measurements were carried out by means of a SETSYS Evolution TMA with a heating rate of 5 °C/min. These measurements were carried out in the temperature range of 25–750 °C with a heating rate of 5° C/min in an argon atmosphere. The linear CTE was measured according to the following equation:

$$\alpha = \varepsilon \frac{1}{\Delta T} = \frac{\Delta L}{L_0} \frac{1}{\Delta T} \tag{1}$$

where, α is the CTE, ΔT is the temperature interval (T-298), L_0 is the starting length of the specimen, $\Delta L = L - L_0$ and $\varepsilon = \Delta L/L_0$. Cylindrical samples with a dimension of ϕ 5 mm × 120 mm were produced directly by means of a conventional powder metallurgy technique. Before starting the measurements, the TMA apparatus was calibrated according to DIN 51045 with NBS-Pt.

The elastic modulus of the composites was determined through a nondestructive test based on the impulse excitation vibration method. The reported results for the Young's modulus are the average of three different measurements. A laser-flash method was used to measure the thermal diffusivity, using a FLASHLINETM apparatus (Anter Corporation, Pittsburgh, PA, USA). The thermal conductivity of specimens was measured at 30 °C. In order to evaluate the electrical conductivity, firstly, the electrical resistivity of samples was measured; then, the electrical conductivity of specimens was calculated by using the inverse relationship of electrical resistivity and conductivity.

3. Results and Discussion

Figure 1 illustrates a representative FESEM image of pure copper powder used in this work; it can be seen that the copper particles have an irregular morphology with a porous structure. Moreover, in the copper starting powder, some dense particles (brighter particles) were revealed. According to the EDS analysis (red cross), they turned out to be copper oxide.

Figure 2a shows the FESEM images of the as-received GNPs. As can be seen in this image, the lateral size and thickness of the GNPs were approximately 25 μm and 100 nm, respectively. Figure 2b shows the Raman spectrum of the as-received graphene nanoplatelets. It can be seen that three main bands, corresponding to the D, G (+D′), and 2D vibrational modes, are present in the spectrum.

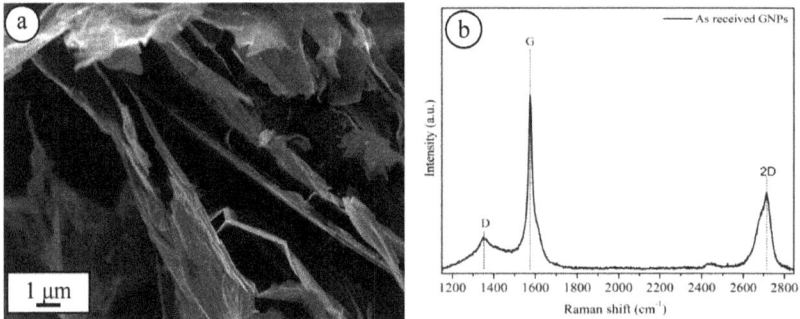

Figure 2. (**a**) SEM image and (**b**) Raman spectrum of as-received graphene nanoplatelets (GNPs).

In principle, the Raman spectrum of graphene-based material shows several bands in the range of 1200–2800 cm^{-1}. The first peak (D-band) at ~1350 cm^{-1} is due to a second-order scattering process which involves a graphene defect and a phonon. The second band at ~1580 cm^{-1} (G-band) is related to the E2g phonon at the center of the Brillouin zone [21]. The last important vibrational fingerprint deals with the 2D-band, also defined as the G′-band, which is spectrally located at ~2695 cm^{-1} for single-layer graphene [22].

According to the previous research, mixing a metal matrix with reinforcement by means of ball milling is the simplest and most common way [23–25]. Even if this technique shows great potential in the dispersion of ceramic particles within the metallic matrix, the following results provided some proof that advise against the ball milling technique as a dispersion method in the Cu/GNP nanocomposites. The Raman data, such as I_D/I_G and I_{2D}/I_G for as-received graphene and graphene in the composites after ball milling and wet mixing, are presented in Table 1.

Table 1. Raman data of as-received GNPs and Cu/GNPs after wet mixing and ball milling.

Composition	Ball Milled		Wet Mixed	
	I_D/I_G	I_{2D}/I_G	I_D/I_G	I_{2D}/I_G
As-received GNPs	0.112	0.511	0.112	0.511
Cu–8 vol % GNPs	0.893	0.584	0.127	0.525

The increase of I_D/I_G ratio signifies the introduction of defects into the carbon lattice as well as the reduction of the crystal size of the graphene platelets. Here, it is evident that these two undesirable phenomena have taken place during the ball milling and, accordingly, it was not possible to take advantage of the superior properties of graphene within the matrix. Surprisingly, the I_{2D}/I_G ratio did not change significantly; this means that the number of layers in the graphene did not change. It can be thus inferred that, after ball milling, the graphene was still multilayer. On the contrary, the I_D/I_G ratio of GNPs after wet mixing did not alter with respect to the as-received ones, implying that wet mixing could be a safe method to disperse the GNPs within the copper matrix.

Apart from the structural investigations by Raman spectroscopy, the morphology of the powder and the dispersion of graphene within the copper matrix were studied by means of a field-emission scanning electron microscope (FESEM). As can be seen in Figure 3, the distribution of the GNPs within

Cu–8 vol % GNPs nanocomposite powder was comparatively homogeneous. Moreover, it is evident that the graphene nanoplatelets were located in the open porosities on the copper particles and that most of the porosities on the surface of the copper particles were filled by GNPs.

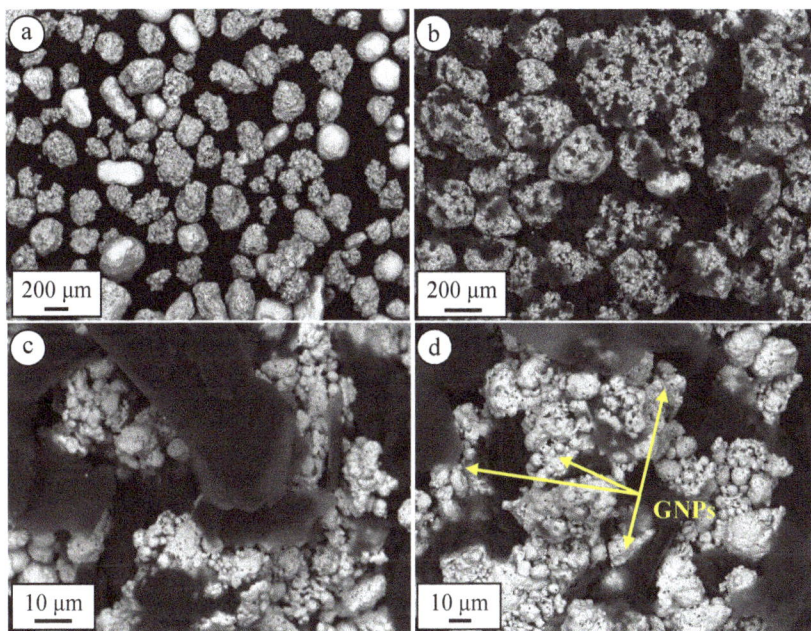

Figure 3. SEM images of (**a**) pure Cu and (**b–d**) Cu–8 Vol % GNPs composite powder mixture.

It is, therefore, likely that such a connection exists between the mixing technique and dispersion of GNPs as well as the properties of graphene after the mixing. It can be thus concluded that, through wet mixing, not only can a uniform dispersion be achieved but also the structure of the graphene can remain unchanged with respect to the starting graphene.

Figure 4a,b shows the microstructure of Cu–4.0 vol % GNPs and Cu–8.0 vol % GNPs, respectively, parallel to the compacting direction. According to these images, the dispersion of graphene seems to be homogeneous after the compaction and sintering process. However, it appears that, by increasing the graphene content up to 8 vol % GNPs (apart from the locating of graphene at the grain boundaries), the graphene platelets tend to form clusters and the number of agglomerates increases noticeably (red dashed circles). The average grain sizes of pure copper and composite were calculated based on the intercept method, and the results show that the addition of GNPs results in the reduction of grain size from 14 μm to 10 μm and 8 μm for Cu–4 vol % GNPs and Cu–8 vol % GNPs, respectively. The morphology of Cu–8.0 vol % GNPs perpendicular to the compacting direction is shown in Figure 4c. As can be seen, the graphene nanoplatelets are oriented mainly perpendicular to the compacting direction and this can result in anisotropic properties in the composite.

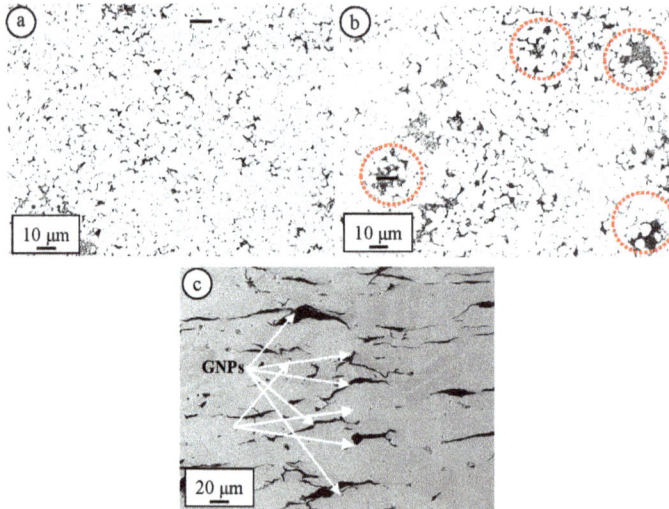

Figure 4. SEM images of sintered (**a**) Cu–4 vol % GNPs, (**b**) Cu–8 vol % GNPs parallel to the compaction direction, and (**c**) Cu–8 vol % GNPs perpendicular to the compaction direction.

The X-ray diffraction pattern of the Cu–8.0 vol % GNPs bulk sample after sintering at 950 °C for 2.5 h is presented in Figure 5. This pattern displays three clear peaks of copper ((111), (200), (220)) and an obvious peak of copper oxide, which provides further evidence of the presence of copper oxide in the starting copper powder. Furthermore, as can be seen in the XRD spectra, the peak at the 2θ value of 26.4° corresponds to the reflection of the graphitic structure (002). In the end, according to the XRD pattern, it could be deduced that no new phase was formed during the fabrication process.

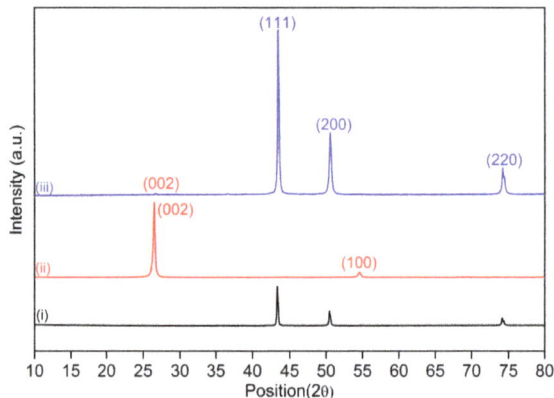

Figure 5. X-ray diffraction pattern of (i) pure copper, (ii) as-received GNPs, and (iii) Cu–8 vol % GNPs after sintering at 950 °C for 2.5 h.

Figure 6 illustrates the relative density and Vickers hardness of the copper nanocomposites as a function of graphene content. As can be seen, the relative density of copper and its composites decreased as a function of GNP content, particularly in the Cu–8 vol % GNPs sample. Thus, it is possible to conclude that the GNP content has a negative influence on the final density of composites. The decrease in the density of the composites was more pronounced than the theoretical value;

this discrepancy is probably due to the presence of extra porosity inside the composite. As shown earlier, GNPs tend to aggregate at higher content and result in the formation of steric obstacles in consolidation of the composites, restricting matrix material flow into the agglomerates [26].

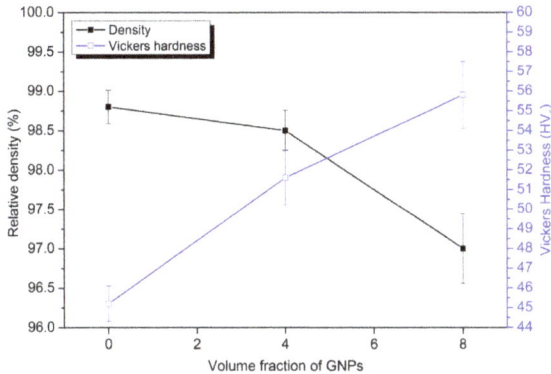

Figure 6. Vickers hardness and relative density of copper composites as a function of GNPs.

On the other hand, by increasing the GNP content, the Vickers hardness of the composites mainly increased, owing to the grain refinement effect of GNPs within the matrix.

The coefficient of thermal expansion values show the level of mismatch strains introduced into the matrix, either as a consequence of reinforcement addition with different properties or due to microstructural changes [27,28]. In fact, the grains of each phase have their physical properties, shapes, or volumes, leading to the residual stresses at the boundaries after thermal treatment [27,29]. These residual stresses at the boundaries can be transmitted without expansion of copper or debonding of the boundaries [30]. In this work, two different volume fractions of graphene were used as reinforcement in order to improve the mechanical properties and reduce the CTE of the copper composite. As shown previously, GNPs tend to orient preferentially perpendicular to the compacting direction, which leads to anisotropic properties. Due to a significant difference between the CTEs of graphene in the out-of-plane ($26 \times 10^{-6}/°C$) and in-plane ($1.0 \times 10^{-6}/°C$) directions, the anisotropic behavior in CTE of the composite was predictable. In the current research, all the CTE measurements for the Cu/GNP composite were carried out in the direction parallel to the compaction force.

As can be observed in Figure 7, the coefficient of thermal expansion of copper composites was decreased by increasing the GNP content. This phenomenon could be related to the more effective drag force on the grain boundary motion applied by GNPs. Moreover, GNPs exerted a higher compression force on the grain boundaries during the expansion of copper as a consequence of their very low CTE. This force during the expansion can limit this feature of copper grains, and resulted in the decrease in the CTE of the composites. Moreover, the measured CTEs in this work were in good agreement with those reported in the literature [31]. According to the existing literature on the CTE of composite materials, there are a number of available models to estimate the CTE of the composites [32–34]. As described earlier, the uniaxial compaction technique was used to fabricate the composites and, during the compaction, GNPs show a preferred orientation with respect to the compaction direction. Therefore, this kind of preferred orientation may affect the properties of composite materials in different axes and, thus, it should be taken into account for the modeling. This anisotropy property of the composite was considered by a simple model which was proposed by Schapery [35].

$$\alpha = \frac{\alpha_i E_i V_i + \alpha_o E_o V_o + \alpha_{Cu} E_{Cu} V_{Cu}}{E_i V_i + E_o V_o + E_{Cu} V_{Cu}} \tag{2}$$

Here, α is the coefficient of thermal expansion, E is the Young's modulus, and V is the volume fraction. The subscripts "i" and "o" represent the in-plane and out-of-plane properties of graphene, whereas the "Cu" subscript denotes the properties of copper.

As shown earlier, by increasing the GNPs content from 4.0 vol % to 8.0 vol %, the graphene started to agglomerate, and these clusters resulted in the lower interface area. Moreover, these agglomerates can affect the CTE of composites owing to either the anisotropic properties of graphene or the introduction of internal porosity. Substantially, due to the anisotropy of graphene, the whole volume of graphene can be sorted in two parts: one part with the positive contribution of graphene in overall CTE which is attributed to the Cu/GNP interface, and another one with a negative contribution of graphene in the CTE of the composite. Therefore, it can be defined that $V_j = c_j V_G$ (j = i and o) and c_a and c_c are the fractions of graphene volume which play the negative and positive roles in the final CTE. However, it should be mentioned that $c_i + c_o = 1$ and $V_i + V_o = V_G$. These are the main assumptions used in this model, and by inserting these assumptions in the previous equation the model could be as follows [8]:

$$\alpha = \frac{(\alpha_i E_i V_i + \alpha_o E_o (1 - c_i)) V_G + \alpha_{Cu} E_{Cu} V_{Cu}}{(E_i c_i + E_o (1 - c_i)) V_G + E_{Cu} V_{Cu}}. \tag{3}$$

The parameters, such as the CTE of graphene in two directions and its mechanical properties, together with the data of copper used in this model, are presented in Table 2. However, the elastic modulus of copper was considered according to our experimental measurements.

Table 2. Parameters used in this model [8].

α_i (10^{-6} K^{-1})	α_o (10^{-6} K^{-1})	E_i (GPa)	E_o (GPa)	E_{Cu} (GPa)	α_{Cu} (10^{-6} K^{-1})
−1	26	1000	20	75	17.1

In order to have insight into the contribution of GNPs with a negative CTE to the thermal expansion of the composite, different volume fractions of graphene in the range of 4.0–8.0 were considered.

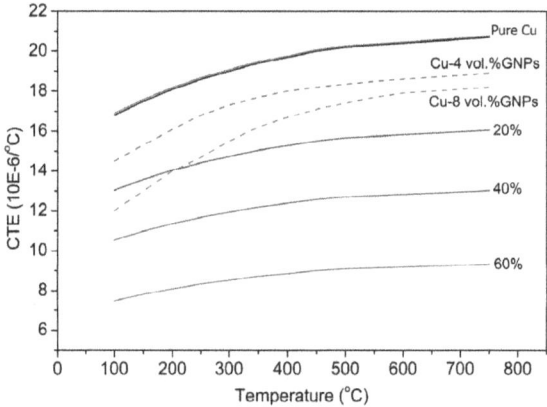

Figure 7. The theoretical predictions and experimental coefficient of thermal expansion of pure copper, Cu–4 vol % GNPs, and Cu–8 vol % GNPs as a function of temperature.

The measured CTEs of pure Cu, Cu–4.0 vol % GNPs, and Cu–8.0 vol % GNPs (dashed lines), together with the calculated CTEs at different volume fractions of graphene with negative CTE, are shown in Figure 7. The volume fraction of graphene with negative CTE can also be defined as the volume fraction of GNPs which are aligned perpendicular to the compaction. As can be seen in Figure 7, if 60% of GNPs are oriented perpendicular to the compaction force, the effect of GNPs with a negative CTE would be significant; this effect changes as a function of the volume percentage of GNPs in this direction. By comparison with the measured CTE results of Cu–4.0 vol % GNPs and Cu–8.0 vol % GNPs, it seems that a less than 20% volume percentage of GNPs were aligned perpendicular to the compaction force. Nonetheless, it could conceivably be hypothesized that, if there were strong bonding between the Cu and GNPs, the CTE of the composite would be lower. Conversely, a lower efficiency of GNPs in terms of CTE reduction was found in Cu–8.0 vol % GNPs with respect to the Cu–4.0 vol % GNPs. This low efficiency could be attributed to the aggregation of GNPs at higher graphene content, poor interfacial bonding, and porosity.

Figure 8 compares the experimental data and theoretical calculations of the Young's modulus of Cu/GNPs composites as a function of GNP content. Previous research has established several models in order to predict the effect of reinforcement addition on the elastic behavior of MMNCs. Halpin and Tsai proposed a model considering the distribution and aspect ratio of reinforcement as the main assumptions; the model can be expressed as follows [36]:

$$E_c = E_m \left[\frac{3}{8} \times \frac{1 + 2/3 \eta_L p V_G}{1 - \eta_L V_r} + \frac{5}{8} \times \frac{1 + 2 \eta_T V_G}{1 - \eta_T V_r} \right], \tag{4}$$

$$E_{||} = E_m \left[\frac{1 + 2/3 \eta_L p V_G}{1 - \eta_L V_G} \right], \tag{5}$$

$$\eta_L = \frac{E_G/E_m - 1}{E_G/E_m + 2/3p}, \tag{6}$$

$$\eta_T = \frac{E_G/E_m - 1}{E_G/E_m + 2}, \tag{7}$$

where E is the elastic modulus, V is the volume fraction, and p is the aspect ratio of graphene. The subscripts "c", "G", and "m" denote the composite, graphene, and matrix. This model considers the random distribution and preferred orientations of GNPs within the matrix. It would be possible to predict this anisotropic effect by Equations (4) and (5) for random distribution (E_c) and preferred orientation ($E_{||}$). The aspect ratio of the GNPs is calculated by dividing the length of graphene by its thickness. In the calculations, the elastic modulus of Cu was considered to be 75 GPa (according to our measurement), and the elastic modulus of graphene was 1.0 TPa (reported from the previous works [37]). The lower Young's modulus of copper with respect to the theoretical value can be related to the purity of the starting powder and the porosity content of sintered specimens. As can be seen in Figure 8, the elastic properties of the composite can be improved through graphene addition, but it is clear that the practical improvement is lower than the predictions. This discrepancy suggests that the elastic properties of the composite could be improved more through the improvement of the process parameters and quality of graphene. This difference between the models and practical data could be attributed to the following reasons: (i) since as-received graphene used in this research is not a defect-free graphene, its real elastic modulus might be lower than the theoretical one, (ii) poor interfacial bonding between Cu and GNPs, and (iii) random distribution of graphene with various orientations could disturb the unidirectional load-transfer mechanism and reduce the composite properties. These findings clearly imply that the dispersion of graphene at higher content is a key challenge. Furthermore, this comparison shows that further improvement in the mechanical properties can be achieved through either the optimization of the process parameters or by using some post-processing steps.

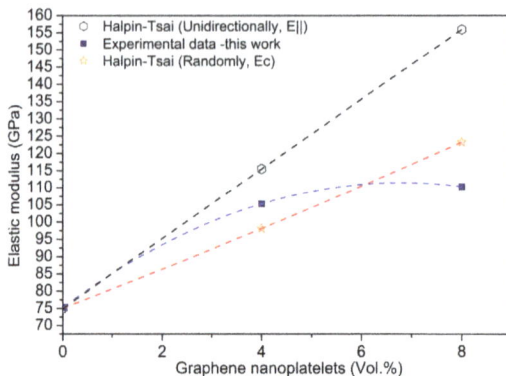

Figure 8. Comparison between the measured CTE and theoretical calculations of elastic modulus (Ec and E||) of Cu/GNPs composites as a function of GNP content.

Table 3 shows the characteristics of some metals and alloys which are used in electronic packaging, together with the experimental results which were achieved in this research. As can be seen in this table, in spite of the proper thermal expansion of Cu–Mo and Cu–W at ambient temperatures, the medium thermal conductivity and high density of these two materials limits their application as electronic packaging materials. In the case of Cu–Ni–Sn, although the electrical resistivity of this alloy is increased with respect to pure copper, its thermal conductivity is significantly decreased, and the thermal expansion of this alloy is still high. These characteristics either limit its application or reduce its heat dissipation efficiency. As can be seen in Table 3, the thermal and electrical conductivity of copper composites were decreased by increasing the GNP content. These reductions in thermal and electrical conductivities, which are affected significantly by activities of free electrons, can be related to several reasons. In this work, poor interfacial bonding (interfacial porosity) and grain refinement can be considered as two main sources of these reductions. However, Chen et al. have reported that the high O content produced in the PM (Powder Metallurgy) process can be another reason [38,39]. It can be thus concluded that grain refinement, high oxygen content interface, and porosity within the matrix increase electron scattering and consequently decrease electron and thermal conductivities. In comparison with these commonly used materials in heat sink applications, the high thermal conductivity and low thermal expansion of Cu/GNP composites, which were developed within this work, make them interesting materials for use in the electronic packaging industry.

Table 3. Typical properties of some metallic materials used in electronic packaging.

Metals and Alloys	Thermal Conductivity at 0–100 °C (W/m·K)	Electrical Resistivity at 20 °C (Ω m $\times 10^8$)	Thermal Expansion (1/k $\times 10^6$)	Density at 20 °C (g/cm^3)	Reference
Aluminium (Al)	238	2.8	23.5	2.7	[9]
Cu–Ni–Sn	50	14	17.1	8.88	[9]
Copper (Cu)	393	1.7	17	8.9	[40]
Red Brass (15% Zn)	151	4.66	18.7	8.8	[40]
CuW	180–220	–	7–9	>9	[40]
CuMo	160–185	–	7–9	9	[41]
Copper (Cu)	350	1.45	18	8.9	Current work
Cu–4 vol % GNPs	304	1.49	14	8.7	Current work
Cu–8 vol % GNPs	262	1.55	12	8.3	Current work

4. Conclusions

The Cu/GNP nanocomposites were successfully fabricated by wet mixing followed by a classical powder metallurgy route. Raman analysis of the GNPs and microstructural observations confirm that ball milling is not an appropriate technique to disperse the GNPs within the copper matrix, whereas the wet mixing method has great potential to be used as a dispersion technique. Through wet mixing, the graphene nanoplatelets were randomly distributed within the matrix and most of the graphene was located at the copper grain boundaries, thus leading to grain refinement. During the compaction of the composite powder, the graphene nanoplatelets oriented preferentially perpendicular to the compaction direction, which implies that the produced composite might show anisotropic properties. By increasing the graphene content up to 8 vol %, apart from the locating of graphene at the grain boundaries, the nanoplatelets tend to form clusters and the number of agglomerates increases noticeably. By increasing the graphene content, the Vickers hardness of the composites increased as well as their elastic modulus, mainly owing to the grain refinement effect. The variation in the coefficient of thermal expansion of copper composites at different temperatures from 100 °C to 750 °C showed that the coefficient of thermal expansion of copper composites decreased by increasing the GNP content. On the other hand, it is found that the combination of grain refinement and porosity content were responsible for decreased electron and thermal conductivities. In addition, it was found that, through the addition of GNPs, it would be possible to produce a novel composite with almost high thermal and electrical conductivity together with low thermal expansion and density. Since the achieved results are desirable characteristics for the electronic packaging industry, this fabricated composite is one of the most interesting candidates for use in heat sink applications. All in all, it can be concluded that the fabrication of composites reinforced by GNPs faces several challenges, such as the dispersion of graphene at higher content levels, interfacial bonding between the copper and GNPs, and the preferred orientation of graphene during the fabrication process.

Author Contributions: In this work, Abdollah Saboori and Seyed Kiomars Moheimani conceived and designed the experiments and thereafter performed the experiments. Abdollah Saboori and Matteo Pavese analyzed the data. Claudio Badini and Paolo Fino contributed reagents, materials, and analysis tools. Abdollah Saboori wrote the paper and the other authors revised the paper.

Conflicts of Interest: The authors declare no conflict of interest.

References

1. Macke, A.; Schultz, B.F. Metal matrix composites offer the automotive industry and opportunity to reduce vehicle weight, improve performance. *Adv. Mater. Process.* **2012**, *170*, 19–23.
2. Deng, C.F.; Ma, Y.X.; Zhang, P.; Zhang, X.X.; Wang, D.Z. Thermal expansion behaviors of aluminum composite reinforced with carbon nanotubes. *Mater. Lett.* **2008**, *62*, 2301–2303. [CrossRef]
3. Chen, Y.; Zhang, X.; Liu, E.; He, C.; Han, Y.; Li, Q.; Nash, P.; Zhao, N. Fabrication of three-dimensional graphene/Cu composite by in-situ CVD and its strengthening mechanism. *J. Alloys Compd.* **2016**, *688*, 69–76. [CrossRef]
4. Saboori, A.; Pavese, M.; Badini, C.; Fino, P. A Novel approach to enhance the mechanical strength and electrical and thermal conductivity of Cu-GNP nanocomposites. *Metall. Mater. Trans. A* **2017**, 1–13. [CrossRef]
5. Du, H.; Lu, D.; Qi, J.; Shen, Y.; Yin, L.; Wang, Y.; Zheng, Z.; Xiong, T. Heat dissipation performance of porous copper with elongated cylindrical pores. *J. Mater. Sci. Technol.* **2014**, *30*, 934–938. [CrossRef]
6. Memis, I. Electronic Packaging Materials: Properties and Selection. In *Reference Module in Materials Science and Materials Engineering*; Elsevier B.V.: Amesterdam, The Netherlands, 2016, ISBN 978-0-12-803581-8.
7. Huang, D.-S.; Tu, W.-B.; Zhang, X.-M.; Tsai, L.-T.; Wu, T.-Y.; Lin, M.-T. Using Taguchi method to obtain the optimal design of heat dissipation mechanism for electronic component packaging. *Microelectron. Reliab.* **2016**, *65*, 131–141. [CrossRef]
8. Ková čik, J.; Emmer, Š. Thermal expansion of Cu/graphite composites: Effect of copper coating. *Kov. Mater.* **2011**, *49*, 411–416.

9. Kutz, M. (Ed.) Materials in Electronic Packaging. In *Hand Book of Materials Selection*; John Wiley & Sons Inc.: New York, NY, USA, 2002.

10. Chen, B.; Bi, Q.; Yang, J.; Xia, Y.; Hao, J. Tribology international tribological properties of solid lubricants (graphite, h-BN) for Cu-based P/M friction composites. *Tribol. Int.* **2008**, *41*, 1145–1152. [CrossRef]

11. Hanada, K.; Matsuzaki, K.; Sano, T. Thermal properties of diamond particle-dispersed Cu composites. *J. Mater. Process. Technol.* **2004**, *153–154*, 514–518. [CrossRef]

12. Davis, J.R. *Powder Metallurgy: Copper and Copper Alloys*. ASM Specialty Handbook; ASM International: Geauga County, OH, USA, 2001, ISBN 0-87170-726-8.

13. Caron, R.N. *Copper Alloys: Properties and Applications*, 2nd ed.; Elsevier B.V.: Amesterdam, The Netherlands, 2001.

14. Kundig, K.J.A.; Cowie, J.G. *Mechanical Engineers' Handbook: Materials and Mechanical Design*, 3rd ed.; Kutz, M., Ed.; John Wiley & Sons, Inc.: New York, NY, USA, 2006, ISBN 9780471449904.

15. Rashad, M.; Pan, F.; Asif, M.; Tang, A. Powder metallurgy of Mg–1%Al–1%Sn alloy reinforced with low content of graphene nanoplatelets (GNPs). *J. Ind. Eng. Chem.* **2014**, *20*, 4250–4255. [CrossRef]

16. Liu, L.; Qing, M.; Wang, Y.; Chen, S. Defects in graphene: Generation, healing, and their effects on the properties of graphene: A review. *J. Mater. Sci. Technol.* **2015**, *31*, 599–606. [CrossRef]

17. Van Chuc, N.; Thanh, C.T.; Van Tu, N.; Phuong, V.T.Q.; Thang, P.V.; Thi, N.; Tam, T. A simple approach to the fabrication of graphene-carbon nanotube hybrid films on copper substrate by chemical vapor deposition. *J. Mater. Sci. Technol.* **2015**, *31*, 479–483. [CrossRef]

18. Saboori, A.; Pavese, M.; Badini, C.; Fino, P. Microstructure and thermal conductivity of Al-Graphene composites fabricated by powder metallurgy and hot rolling techniques. *Acta Metall. Sin. (Engl. Lett.)* **2017**, *30*, 675–687. [CrossRef]

19. Rashad, M.; Pan, F.; Tang, A.; Asif, M.; Hussain, S.; Gou, J.; Mao, J. Improved strength and ductility of magnesium with addition of aluminum and graphene nanoplatelets (Al + GNPs) using semi powder metallurgy method. *J. Ind. Eng. Chem.* **2015**, *23*, 243–250. [CrossRef]

20. Saboori, A.; Novara, C.; Pavese, M.; Badini, C.; Giorgis, F.; Fino, P. An investigation on the sinterability and the compaction behavior of aluminum/graphene nanoplatelets (GNPs) prepared by powder metallurgy. *J. Mater. Eng. Perform.* **2017**, *26*, 993–999. [CrossRef]

21. Ferrari, A.C.; Robertson, J. Resonant Raman spectroscopy of disordered, amorphous, and diamondlike carbon. *Phys. Rev. B* **2001**, *64*, 075414. [CrossRef]

22. Mohiuddin, T.M.G.; Lombardo, A.; Nair, R.R.; Bonetti, A.; Savini, G.; Jalil, R.; Bonini, N.; Basko, D.M.; Galiotis, C.; Marzari, N.; et al. Uniaxial strain in graphene by Raman spectroscopy: G peak splitting, Grüneisen parameters, and sample orientation. *Phys. Rev. B* **2009**, *79*, 205433. [CrossRef]

23. Perez-Bustamante, R.; Bolanos-Morales, D.; Bonilla-Maetinez, J.; Estrada-Guel, I. Microstructural and hardness behavior of graphene-nanoplatelets/aluminum composites synthesized by mechanical alloying. *J. Alloys Compd.* **2014**, *615*, S578–S582. [CrossRef]

24. Pérez-Bustamante, R.; Gómez-Esparza, C.D.; Estrada-Guel, I.; Miki-Yoshida, M.; Licea-Jiménez, L.; Pérez-García, S.A.; Martínez-Sánchez, R. Microstructural and mechanical characterization of Al–MWCNT composites produced by mechanical milling. *Mater. Sci. Eng. A* **2009**, *502*, 159–163. [CrossRef]

25. Tabandeh-khorshid, M.; Ferguson, J.B.; Schultz, B.F.; Kim, C.; Cho, K.; Rohatgi, P.K. Strengthening mechanisms of graphene- and Al_2O_3-reinforced aluminum nanocomposites synthesized by room temperature milling. *Mater. Des.* **2016**, *92*, 79–87. [CrossRef]

26. Chu, K.; Jia, C.C. Enhanced strength in bulk graphene-copper composite. *Phys. Status Solidi A* **2014**, *211*, 184–190. [CrossRef]

27. Levy-Tubiana, R.; Baczmanski, A.; Lodini, A. Relaxation of thermal mismatch stress due to plastic deformation in an Al/SiCp metal matrix composite. *Mater. Sci. Eng. A* **2003**, *341*, 74–86. [CrossRef]

28. Qian, L.H.; Wang, S.C.; Zhao, Y.H.; Lu, K. Microstrain effect on thermal properties of nanocrystalline Cu. *Acta Mater.* **2002**, *50*, 3425–3434. [CrossRef]

29. Kassner, M.E.; Hayes, T.A. Creep cavitation in metals. *Int. J. Plast.* **2003**, *19*, 1715–1748. [CrossRef]

30. Weidenmann, K.; Tavangar, R.; Weber, L. Mechanical behaviour of diamond reinforced metals. *Mater. Sci. Eng. A* **2009**, *523*, 226–234. [CrossRef]

31. Wang, X.; Li, J.; Wang, Y. Improved high temperature strength of copper-graphene composite material. *Mater. Lett.* **2016**, *181*, 309–312. [CrossRef]

32. Hale, D.K. The physical properties of composite materials. *J. Mater. Sci.* **1976**, *11*, 2105–2141. [CrossRef]

33. Turner, P.S. Thermal-expansion stresses in reinforced plastics. *J. Res. Natl. Bur. Stand.* **1946**, *37*, 239–250. [CrossRef]
34. Hashin, Z.; Shtrikman, S. A variational approach to the theory of the elastic behaviour of multiphase materials. *J. Mech. Phys. Solids* **1963**, *11*, 127–150. [CrossRef]
35. Schapery, R.A. Thermal expansion coefficients of composite materials based on energy principles. *J. Compos. Mater.* **1968**, *2*, 380–404. [CrossRef]
36. Saboori, A.; Pavese, M.; Badini, C.; Fino, P. Development of Al- and Cu-based nanocomposites reinforced by graphene nanoplatelets: Fabrication and characterization. *Front. Mater. Sci.* **2017**, *11*, 171–181. [CrossRef]
37. Lee, C.; Wei, X.; Kysar, J.W.; Hone, J. Measurement of the elastic properties and intrinsic strength of monolayer graphene. *Science* **2008**, *321*, 385–388. [CrossRef] [PubMed]
38. Chen, B.; Li, S.; Imai, H.; Jia, L.; Umeda, J.; Takahashi, M.; Kondoh, K. Carbon nanotube induced microstructural characteristics in powder metallurgy Al matrix composites and their effects on mechanical and conductive properties. *J. Alloys Compd.* **2015**, *651*, 608–615. [CrossRef]
39. Chen, B.; Jia, L.; Li, S.; Imai, H.; Takahashi, M.; Kondoh, K. In situ synthesized Al_4C_3 nanorods with excellent strengthening effect in aluminum matrix composites. *Adv. Eng. Mater.* **2014**, *16*, 972–975. [CrossRef]
40. Tong, X.C. Electronic Packaging Materials and Their Functions in Thermal Managements. In *Advanced Materials for Thermal Management of Electronic Packaging*; Springer: New York, NY, USA, 2011; pp. 131–167, ISBN 978-1-4419-7759-5.
41. Wilson, J. *Thermal Conductivity of Common Alloys in Electronics Packaging*; Electronics Cooling: Montgomery County, PA, USA, 2007.

metals

MDPI

Article

Mechanical Properties of Thixoforged In Situ Mg$_2$Si$_p$/AM60B Composite at Elevated Temperatures

Suqing Zhang [1,2], Tijun Chen [1,*], Jixue Zhou [2], Dapeng Xiu [2], Tao Li [2] and Kaiming Cheng [2]

[1]   State Key Laboratory of Advanced Processing and Recycling of Nonferrous Metals,
     Lanzhou University of Technology, Lanzhou 730050, China; zhangsuqing1985@163.com
[2]   Shandong Key Laboratory for High Strength Lightweight Metallic Materials,
     Qilu University of Technology (Shandong Academy of Science), Jinan 250014, China; zhoujx@sdas.org (J.Z.);
     xiudapeng@126.com (D.X.); litao@sdas.org (T.L.); chengkm@sdas.org (K.C.)
*   Correspondence: chentj@lut.cn; Tel.: +86-931-297-6573

Received: 8 December 2017; Accepted: 29 January 2018; Published: 2 February 2018

Abstract: The mechanical behaviors of the thixoforged in situ Mg$_2$Si$_p$/AM60B composite at elevated temperatures were evaluated. The results indicated that the thixoforged composite exhibits higher UTS (ultimate tensile strength) than that of the thixoforged AM60B at the cost of elongation. As the testing temperature rises from 25 to 300 °C, the UTS of both these two materials decreases while their elongations increases. The enhanced dislocation motion ability, the softened eutectic β phase at 120 °C, the activated non-basal slipping and the dynamic recovery and recrystallization mechanisms at 150 °C are responsible for the change in tensile properties with testing temperatures. The fracture mode transforms from the ductile into the brittle as the initial strain rate increases from 0.01 to 0.2 s^{-1} at 200 °C.

Keywords: thixoforging; magnesium-based composite; fracture

1. Introduction

In the past few years, magnesium alloys have been widely used in the fields of automotion, electronic products, portable tools, sporting goods, and aerospace vehicles owing to their light weight, excellent castability, damping capacity, machinability, and so on [1]. However, the rapid loss of strength at temperatures above 120 °C limits their extended applications [2]. In order to solve this problem, several new heat resistant magnesium alloys have been recently developed [3]. The design in heat resistant magnesium alloys mainly abides by the following ideas: strengthening the α-Mg matrix or/and limiting the cross-slip of dislocations and migration of the grain boundary [4]. Most of the heat resistant magnesium alloys with high performances are achieved through the addition of rare earth elements [5–8]. Unfortunately, the formidable cost of rare earth elements limits the development of the magnesium industry. The requirements of high-performance and light-weight materials in automotive and aerospace fields have become increasingly urgent in recent years [1], leading the development of heat-resistant magnesium alloys with cost-effective technologies.

Combining the mechanical properties of magnesium alloys with ceramics, magnesium-based composites exhibit a higher service temperature with affordable cost [9–12]. Therefore, the magnesium-based composites become attractive candidates for the applications at elevated temperatures [13]. However, most of the related investigations focus on the fabrication of magnesium-based composites [14]. The mechanical behaviors of magnesium-based composites at elevated temperatures dramatically influence their performance. Unfortunately, the investigations involving this subject are lagging far behind with the rare earth-contained heat-resistant magnesium alloys [7,15,16]. Generally, there are three strengthening mechanisms for metal-matrix composites: load transfer, Orowan looping, and dislocation strengthening. Every mechanism is changing as the temperature changes. However, this has not been discussed in detail. It can be expected that due to the difference in microstructural constituents and the processing technique, the related fundamental knowledge for magnesium-based composites should also be different. Therefore, the mechanical behaviors of magnesium-based composites at elevated temperatures become an indispensable part for their applications and require sustained research effort.

Thixoforging is a combination of casting and forging, in which the semisolid ingot is injected steadily and solidified under applied pressure [17]. It has been pointed out that thixoforging is especially suitable for the forming of aluminum and magnesium alloys [17–19]. Thus, it is theoretically expected that thixoforging should be a maneuverable method to fabricate magnesium-based composites. In the authors' previous investigations [20–22], the ultimate tensile strength (UTS) of the thixoforged in situ Mg_2Si_p/AM60B composite is 35.6% higher than the thixoforged AM60B alloy at the cost of the elongation. However, the mechanical behaviors at elevated temperatures of this composite have not been studied.

In this experimental paper, the mechanical behaviors at elevated temperatures of the thixoforged in situ Mg_2Si_p/AM60B composite will be discussed in detail. The informative results are compared to the thixoforged AM60B alloy. The strengthening mechanisms of the Mg_2Si particle will also be investigated.

2. Materials and Methods

The raw materials used for this work were commercial AM60B, Al-30 wt % Si, and pure Mg, melted at 790 °C in an electric resistance furnace (Shanghai Shiyan Electric Furnace Co., Ltd., Shanghai, China). In order to avoid oxidation, the melting was covered by RJ-2 (Hongguang Co., Ltd., Shanghai, China), which was designed for magnesium alloys. Then, 0.5 wt % Sr (using Mg-30Sr master alloy) was added into the melting in order to modify the Mg_2Si phase. After the melt was held for 20 min, 0.2% SiC_p (using pressed cake of Mg_p-$25SiC_p$ mixture powders) was introduced and stirred for 3 min for purpose of refining the α-Mg phases. Subsequently, the melt was degassed using C_2Cl_6 and pouring into a steel mold with a cavity of ф 50 mm × 500 mm. Thus, the as-cast ingots were obtained.

Some small ingots were cut from the as-cast ingots, with dimensions of ф 42 mm × 30 mm. Those ingots were reheated in a resistance furnace at 600 °C for 60 min and the semisolid ingots were obtained. The semisolid ingots were transferred into the bottom die, which was preheated to 300 °C with a cavity of ф 50 mm × 40 mm. Then the die was closed, driven by the hydraulic pressing machine (Tianjin Tianduan Press Co., Ltd., Tianjin, China). The pressure of 192 MPa forced on the semisolid ingot increased to the setting value within 5 s and held for 20 s. Repeating the above experiment, thixoforged composites were obtained. The thixoforged AM60B alloys were also prepared by this method.

According to the standard of GB/T 4338-2006, some tensile specimens were machined using the wire-cut machine (Fangzheng CNC Machine Tool Co., Ltd., Taizhou, China) from the center of the thixoforged products and parallel to the pressure. The gauge dimensions of the specimens were 10 mm × 1.2 mm × 2.5 mm. The tensile testing was conducted in a universal material testing machine with a heating device, of which the temperature control precision is ±1 °C. The tensile specimen was heated at the setting temperature for 10 min, and then the test was conducted. In view of the servicing temperature of the magnesium alloys [2,3], which is no more than 300 °C, the tensile testing was carried out under the temperatures of 25 (room temperature), 100, 150, 200, 250, and 300 °C at cross head speeds of 1.0 mm·s^{-1}, and at cross head speeds of 0.1, 0.5, 1.5, and 2 mm·s^{-1} under 200 °C, respectively. The corresponding initial strain rates were 0.01, 0.05, 0.1, 0.15, and 0.2 s^{-1}. The average of at least five testing values was taken as the tensile properties of a thixoforged composite. Some typical fracture surfaces, and side views of them, were also observed on the scanning electron microscope (SEM; NEC Electronics Corporation, Tokyo, Japan) and the optical microscope (OM; Nikon Instruments Co., Ltd., Shanghai, China). Microstructural specimens were cut from the center of each of the thixoforged products. One cross-section parallel to the pressure direction was prepared under standard metallographic procedures. Microstructural characterization was also carried out with this OM. The typical images were quantitatively examined by Image-Pro Plus 5.0 software (5.0, Media Cybernetics Co., Ltd., Silver Spring, MD, USA), the fraction of the Mg_2Si particles to the whole was taken as its volume fraction.

3. Results

3.1. Microstructures of As-Cast, Semisolid, and Thixoforged Composite and Thixoforged AM60B Alloy

In order to identify the microstructures of the thixoforged composite, the initial as-cast microstructures and its solidifying process should be first clarified. As shown in Figure 1a,b, the microstructure of the as-cast composite contains primary Mg_2Si particles, primary α-Mg dendrites, eutectic Mg_2Si particles, and eutectic structures. Together with the well-known solidifying process of AM60B, the solidification of this composite begins with the precipitation of the primary Mg_2Si particles and then the primary α-Mg phases. The first-formed Mg_2Si particles are pushed by the growing primary α-Mg phases and distribute in the residual eutectic liquid. Finally, the eutectic liquid solidifies into eutectic structures. During the last stage of the solidification, the eutectic Mg_2Si phases separate out previously from the eutectic liquid, then the eutectic α-Mg phases preferentially grow up on the surface of the primary α-Mg dendrites, and the eutectic β phases ($Mg_{17}Al_{12}$) are left in the interdendritic regions.

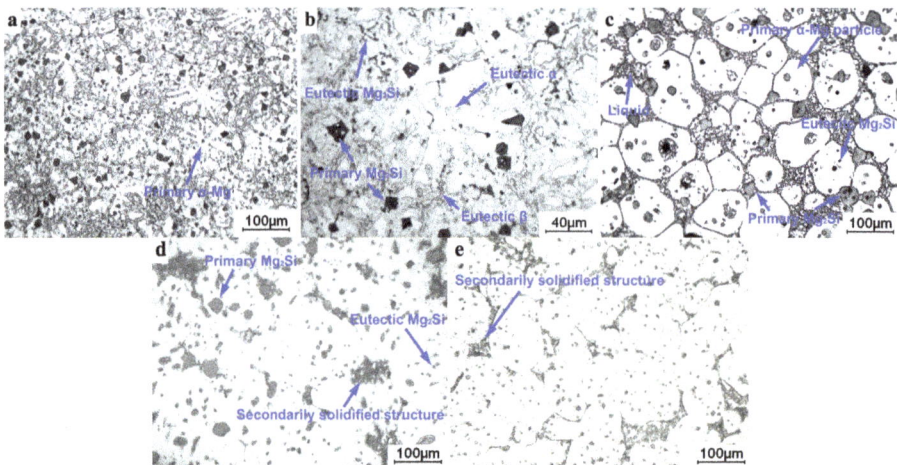

Figure 1. Microstructures of: (**a,b**) as-cast composite; (**c**) semisolid composite; (**d**) thixoforged composite; (**e**) thixoforged AM60B.

After the composite is reheated at 600 °C for 60 min, the α-Mg dendrites evolve into the α-Mg particles, and the sharp edges and corners of the Mg_2Si particles are obviously blunted. Both of them are suspended in the liquid (shown in Figure 1c). The microstructural evolution of α-Mg dendrites and Mg_2Si particles during partial remelting have been discussed in detail in the previous works [23]. Subsequently, the semisolid ingot is thixoforged. Therefore, the liquids solidify into secondary solidified structures. As shown in Figure 1d,e, the microstructures of the thixoforged composite and AM60B alloy consist of primary α-Mg particles (in bright color) and secondary solidified structures. The amount of the secondary solidified structures is significantly less than that of the liquids in the semisolid microstructures. This phenomenon results from the solidified characteristics of the semisolid ingot. The secondary α-Mg phases (to differentiate from the primary α-Mg particles, the α-Mg phases solidified from the liquids is named as secondary α-Mg phases) preferentially grow on the surfaces of the primary α-Mg particles without boundaries. In this case, it is difficult to separate these two phases from each other and results in the secondary solidified structures seemingly less than the liquids in the semisolid ingot. Additionally, the thixoforged composite contains the reinforcement of Mg_2Si particles, of which the volume fraction is about 10 vol % (Figure 1d). There are two kinds of Mg_2Si particles in Figure 1d. The larger ones are the primary Mg_2Si particles, the size of which ranges from 20 to 40 μm. The smaller ones are eutectic Mg_2Si particles, the size of which is less than 10 μm. Most of the primary Mg_2Si particles locate at the boundary of the α-Mg particles, and only a few of them locate inside. The eutectic Mg_2Si particles mainly distribute within the α-Mg particles. The amount of the primary Mg_2Si particle is obviously larger than that of the eutectic Mg_2Si particles. These are attributed to the following reasons: The first is the amount of the formed eutectic Mg_2Si particles is less than the primary Mg_2Si particles, since the eutectic point of the Mg-Si system is 1.38% [24,25]. On the other hand, the morphological evolution of the Mg_2Si particles obey the Ostwald ripening mechanism during partial remelting, which refers to the dissolution of small-sized grains and the growth of large particles [26]. Owing to the dissolution and subsequent reprecipitation, the amount of eutectic Mg_2Si particles decreases and the primary Mg_2Si particles becomes blunt. Based on the abovementioned, the distribution of the Mg_2Si particles can easily understood.

3.2. Tensile Properties of the Thixoforged Composite at Different Testing Temperatures

Figure 2 reveals the tensile properties of the thixoforged composite and AM60B at different testing temperatures. It indicates that the UTS of the thixoforged composite is always higher than that of the thixoforged AM60B, and both of them always decrease as the temperature rises. However, the elongation of the thixoforged AM60B is always higher than that of the thixoforged composite, and both of them increase as the temperature rises.

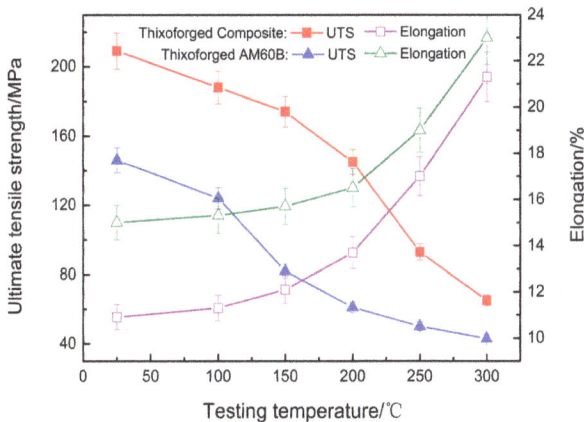

Figure 2. The variations in tensile properties of the thixoforged composite and AM60B with testing temperatures.

Figure 3 presents the fractographs of the thixoforged composite tested at different temperatures. As shown in Figure 3a, the fracture surface is covered by innumerable small dimples and damaged Mg_2Si particles. There are two kinds of damaged Mg_2Si particles: either debonded from the matrix (marked by A), or broken into pieces (marked by B). Figure 4a indicates that the number of the former ones (marked by A) is much fewer than the latter ones (marked by B), owing to the differences of the crystal structures and the lattice constants between the Mg_2Si particles and the α-Mg phases, the interface of which belongs to an incoherent interface [27]. During deformation, the soft α-Mg phases deform plastically and the Mg_2Si particles deform elastically. Thus, a large stress concentration preferentially generates near this interface, and increases to 2–4 times higher than the surrounding matrix [28,29]. Finally, the stress concentration results in the interfacial debonding or/and fragmentation of the Mg_2Si particles. Of course, the stress concentration is relaxed. Thus, it can be expected that the Mg_2Si particles strengthen the matrix through the load transfer mechanism, which results in the excellent UTS of the thixoforged composite at room temperature (Figure 2).

It has been reported that the β phase ($Mg_{17}Al_{12}$) is softened at temperatures above 120 °C [30]. Therefore, the UTS of the thixoforged AM60B decreases at 150 °C owing to the absence of a reinforcing phase for the matrix (Figure 2). As mentioned above, the Mg_2Si particle always locates in the last solidified regions [20], i.e., surrounded by the eutectic phase. Thus, the softening of the eutectic β phase leads to the decrease in the interfacial bonding strength between the Mg_2Si particles and the matrix. As shown in Figure 4b, the interfacial debonding (marked by A) gradually becomes dominant. However, the broken Mg_2Si particles (marked by B) still exist in the fracture surface (Figure 3b). Thus, the Mg_2Si particles still contribute in strengthening the matrix. Therefore, the UTS of the thixoforged composite is much higher than that of the thixoforged AM60B (Figure 2). Moreover, the non-basal slip system of the magnesium alloy is activated at this temperature [31,32], which significantly promotes the plastic deformation ability of α-Mg phase. Correspondingly, the size of the dimples in the fracture surface at the testing temperature of 150 °C becomes large (Figure 3b).

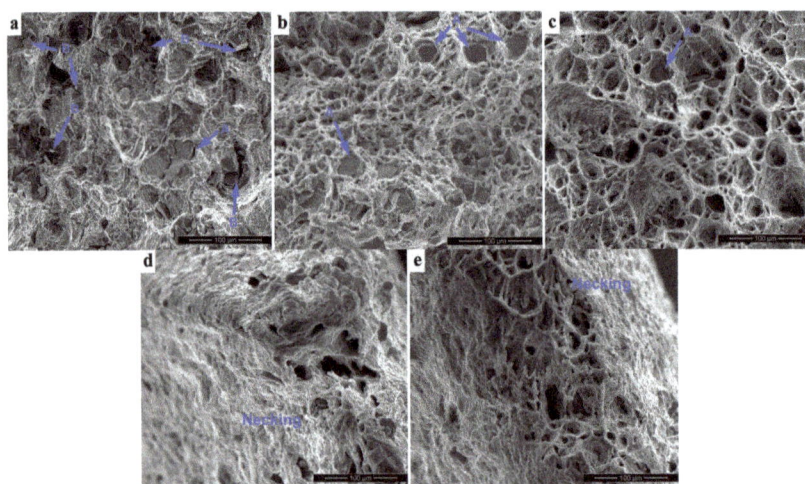

Figure 3. Fracture surfaces of the thixoforged composite tensile tested at different temperatures: (a) 25 °C; (b) 150 °C; (c) 200 °C; (d) 250 °C; (e) 300 °C.

As shown in Figure 3c, those dimples transform into large pits, and deep holes form in the fracture surface at a testing temperature of 200 °C. The Mg_2Si particle always locates to the bottom of these holes (marked by A). As the testing temperature reaches 200 °C, the dislocation motion is promoted from the accelerated atom diffusion ability. Moreover, the non-basal slipping is further active. Both of these lead to the decrease in UTS and the increase in elongation (Figure 2). Therefore, the dimples easily connect with each other, and then the large-sized pits are generated (Figure 3c). In addition, the broken Mg_2Si particles are seldom found in the fracture surface, and the side view of it (Figures 3c and 4c). That is, the Mg_2Si particles mainly debond from the matrix (marked by A in Figure 4c). Subsequently, the surrounding matrix continually plastically deforms as the tensile testing proceeds. Then, the holes, of which the Mg_2Si particle locates to the bottom, are thereby generated (Figure 3c). In addition, as shown in Figure 4c, a kind of texture-like microstructure is formed (marked by C in Figure 4c). Although the critical temperature for recrystallization of the magnesium alloys is

230 °C [33,34], in view of the tensile testing belonging to a dynamic process, it should be suggested that the formation of the texture-like microstructures are attributed to the dynamic recovery and recrystallization regimes, which also result in the decrease of UTS and the increase of elongation of the magnesium alloy. Under the combined effects mentioned above, the UTS of the thixoforged composite apparently decrease and its elongation continually increases (Figure 2).

Figure 4. Side view of fracture surfaces of the thixoforged composite tensile tested at different temperatures: (**a**) 25 °C; (**b**) 150 °C; (**c**) 200 °C; (**d**) 250 °C; (**e**) 300 °C.

When the testing temperature reaches 250 °C, the fracture surface is characterized by the necking feature (as shown in Figure 3d). This implies that the plastic deformation ability of the α-Mg phase is further improved (Figure 2). The combining effects are enhanced as the testing temperature increases, which are responsible for the improved plastic deformation ability. Therefore, the holes' characterization becomes more dominant on the fracture surface instead of those large-sized pits (compare Figure 3d to Figure 3c). Figure 4d displays that these holes are also generated inside the tensile bar (marked by A). This implies that the Mg_2Si particles cannot bear much stress concentration due to the interfacial debonding. Therefore, the load transfer mechanism is decreasingly effective. In this case, the UTS of the thixoforged composite rapidly decreases while its elongation further increases (Figure 2). Even so, the UTS of the thixoforged composite is still higher than that of the AM60B (Figure 2). Thus, it can be expected that the other strengthening mechanisms should take part in reinforcing the matrix and this will be discussed in detail at the Section 3.3.

The fracture surface shown in Figure 3e is characterized by the necking feature as well. However, the debonded Mg_2Si particle is seldom observed and the amount of the texture-like microstructure (marked by C) is increased at the side view of fracture surface (Figure 4e). When the testing temperature further rises to 300 °C, the dynamic recovery and recrystallization mechanisms are further promoted [35]. Thus, the amount of the texture-like microstructure increases. In this case, the stress concentration near the interface between the Mg_2Si particles and the matrix is easily relaxed through the formation of the texture-like microstructures (marked by C in Figure 4e). As a result, the interfacial debonding disappears. Moreover, the non-basal slip system of the magnesium alloys is completely activated at 300 °C [31], which results in the further decrease in UTS and the increase in elongation (Figure 2). The tensile properties of both thixoforged composite and AM60B are at a comparative level (as shown in Figure 2), which implies the strengthening mechanisms from the Mg_2Si particle for the matrix becomes invalid when the testing temperature reaches 300 °C.

3.3. Strengthening Mechanisms of the Mg_2Si Particle

In order to verify the strengthening mechanisms, a typical fracture surface, and a side view of it showing the Mg_2Si particles, are presented in Figure 5. The Mg_2Si particle is broken into pieces (Figure 5a) and there are no visible cracks in the surrounding matrix (Figure 5b). That is, the load transfer mechanism from the Mg_2Si particle effectively protects the matrix from the crack initiation. With the testing temperature rising, the broken degree of the Mg_2Si particle is reduced. As shown in Figure 5c, the Mg_2Si particle only splits into two parts, and even transforms into interfacial debonding as the temperature further increases (Figure 5d). Namely, this mechanism is decreasing, due to the reduced bonding strength of the interface between the Mg_2Si particle and the matrix as the testing temperature rises. However, the UTS of the thixoforged composite is always higher than that of the thixoforged AM60B (Figure 2). It can be expected that the other mechanisms take part in reinforcing the thixoforged composite.

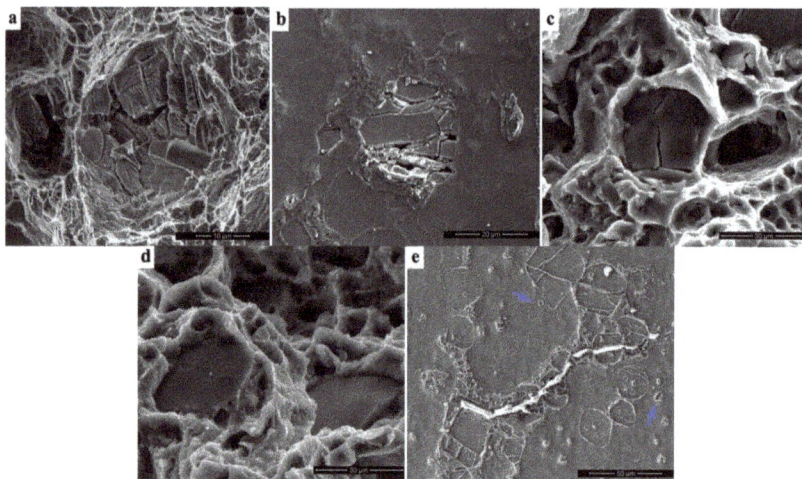

Figure 5. SEM images showing Mg_2Si particles: (**a,b**) fragmentation on fracture surface and side view of it at 25 °C; (**c,d**) fragmentation and interfacial debonding on the fracture surface at 150 °C; (**e**) showing eutectic Mg_2Si particle in the side view of the fracture surface at 25 °C.

As mentioned above, there are two kinds of Mg_2Si particles. After tensile testing, the large-sized ones break into pieces and the small-sized ones maintain their original morphology (marked by arrows in Figure 5e). During tensile testing, the interaction between dislocations and fine particles also contributes in strengthening the matrix through the Orowan looping mechanism [29]. As the testing temperature rises, the dislocation motion is promoted from the accelerated atom diffusion ability. In this case, the Orowan looping mechanism operates increasingly as the testing temperature rises. On the other hand, the mismatch in the coefficient of thermal expansion (CTE) between Mg_2Si particles and the matrix [36] also strengthens the matrix. During thixoforging and tensile testing at elevated temperatures, the dislocations are created near the Mg_2Si/matrix interface due to the relaxation of the thermal expansion mismatch between the Mg_2Si particles and matrix. In this case, the dislocation density increases. The dislocations formed during tensile testing increase as the testing temperature rises. Therefore, this mechanism is gradually in action as the testing temperature rises, can impede the dislocation movement, and also plays a very important role in strengthening the matrix.

Although the contribution of each strengthening mechanism has not been calculated separately, it also suggests that an additive or synergetic effect from the combination of several mechanisms is accountable for strengthening the matrix. In comparison, the load transfer mechanism primarily operates at the testing temperature under 200 °C, and the other two grow dominant as the testing temperature further rises.

3.4. Effect of Initial Strain Rate on Tensile Properties

As mentioned in the Section 3.2, the dynamic recovery and recrystallization regimes influence the tensile properties dramatically. The initial strain rate also has large effects on this regime [37,38]. Thus, the mechanical behavior under 200 °C at different initial strain rates is discussed in detail below.

Figure 6 reveals the tensile properties of the thixoforged composite at different initial strain rates. It indicates that the UTS increases to a peak value at 0.1 s^{-1} and then decreases as the initial strain rate increases from 0.01 to 0.2 s^{-1}. However, the elongation continually decreases.

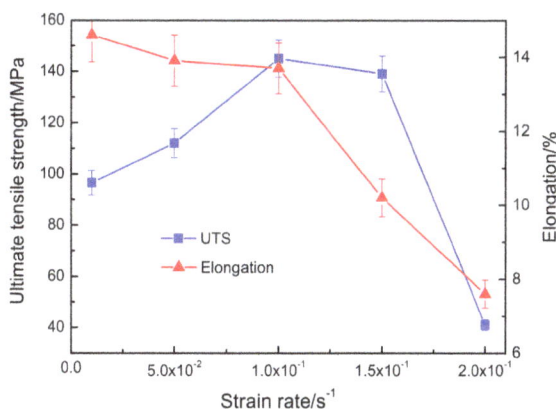

Figure 6. The variations in tensile properties of the thixoforged composite tensile tested at 200 °C under different initial strain rates.

As shown in Figure 7a, the fracture surface is covered by small and uniform dimples. The crack propagates through the α-Mg phases (Figure 8a). The employed initial strain rate 0.01 s^{-1} is the lowest initial strain rate. Thus, the dislocation has enough time to move. Therefore, the stress concentration near the interface between the Mg_2Si particles and the matrix can be easily relaxed through the dislocation motion regime. As the tensile testing proceeds, the microvoids are generated in the α-Mg phase, and then connect with each other to form dimples. Eventually, the connecting leads to the fracture of the composite. Namely, the fracture mode belongs to ductile fracture mechanism and the fracture of the α-Mg phases results in the final fracture of the composite.

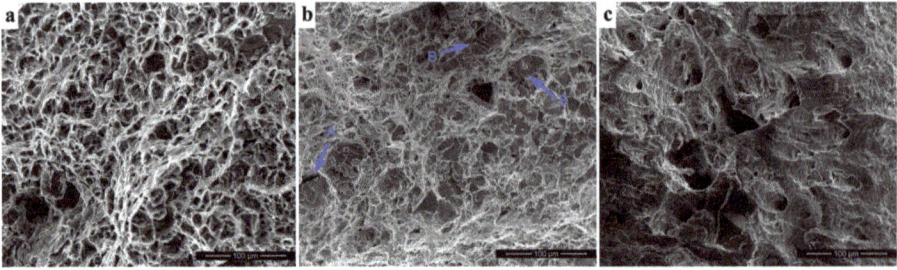

Figure 7. Fracture surface of the thixoforged composite tensile tested at 200 °C under different initial strain rates: (**a**) 0.01 s^{-1}; (**b**) 0.15 s^{-1}; (**c**) 0.2 s^{-1}.

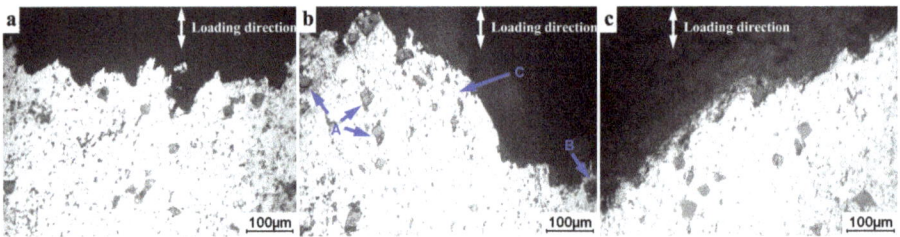

Figure 8. Side view of fracture surface of the thixoforged composite tensile tested at 200 °C under different initial strain rates: (**a**) 0.01 s^{-1}; (**b**) 0.15 s^{-1}; (**c**) 0.2 s^{-1}.

The increase in the initial strain rate leads to the enhanced dynamic recovery and recrystallization mechanisms. Therefore, the texture-like structures are then generated (Figure 4c). This phenomenon results in the softening of α-Mg phases, which leads their improved plastic deformation ability. In this case, these dimples on the fracture surface become large (Figure 3c). However, the stress concentration cannot be relaxed through the dislocation motion owing to the increase of initial strain rate. Thus, the interfacial debonding is obviously observed on the fracture surface (compare Figure 3c with Figure 7a). Moreover, the work hardening mechanism is increasingly promoted as the initial strain rate increases, which takes part in improving the UTS at the cost of the elongation. Consequentially, the UTS is increased while the elongation is decreased as the initial strain rate rises to 0.1 s^{-1} (Figure 6). The fracture mode also obeys to a ductile fracture mechanism.

Figure 7b presents that the cleavage plane appears on the fracture surface, which implies that the fracture is somewhat brittle in nature. In addition, the Mg$_2$Si particles, which debonded (marked by A) and broke into pieces (marked by B), are obviously observed on the fracture surface and the side view of it (Figures 7b and 8b). The dynamic recovery and recrystallization mechanism, and the work hardening mechanism, are at a competitive situation. The increase in the initial strain rate results in the enhancement of both mechanisms. However, when the initial strain rate exceeds a give value, it lacks enough time to realize the dynamic recovery and recrystallization mechanism in action. Therefore, the amount of the texture-like structures decreases. In this case, the work hardening mechanism becomes dominant and leads to the decrease in elongation (Figure 6). On the other hand, the load transfer mechanism from the Mg$_2$Si particle is also enhanced as the initial strain rate rises. Thus, high stress concentration is preferentially formed near the interface between the Mg$_2$Si particles and the matrix, and results in the interfacial debonding or/and fragmentation of the Mg$_2$Si particles (compare Figure 7b with Figure 3c). Then the cracks are formed. Subsequently, the surrounded matrix deforms. For the magnesium alloys, the deformation is achieved through the intergranular slipping accompanied by intragranular dislocation motion [39]. As the initial strain rate rises, the slipping increasingly dominates due to there being insufficient time for the motion of intragranular dislocation. Therefore, the cleavage planes are generated on the fracture surface. The slipping cannot bear much deformation and, thus, the resultant elongation significantly decreases (Figure 6). Correspondingly, the fracture mode transforms into a quasi-cleavage mode.

As the initial strain rate further rises, a high stress concentration is very quickly generated at the interface between the Mg_2Si particles and the matrix. Thus, the interfacial debonding occurs at the early stage of the deformation. Namely, the cracks are thereby formed. Then, the cracks propagate into the matrix, leading to the final fracture. It has been reported that there are three mechanisms responsible for the deformation of the magnesium alloy and the critical resolved shear stress (CRSS) from small to large, in sequence, are: basal slipping, twinning, and non-basal slipping [40]. Generally, the CRSS of the basal slipping and twinning are irrelevant to the initial strain rate while that of the non-basal slipping increases as the initial strain rate rises [41]. Therefore, it can be expected that the deformation of the matrix is completely through the basal slipping of the α-Mg phases. In this case, the fracture surface is characterized by the brittle feature (Figure 7c). Owing to the formation of crack and the deformation mechanism of the α-Mg phase, the resultant UTS and elongation of the composite are rapidly decreased (Figure 6).

Based on the abovementioned, the fracture mode of the composite under different initial strain rates can be summarized by Figure 9. It indicates that the fracture mode gradually transforms from ductile into brittle as the initial strain rate increases from 0.01 to 0.2 s^{-1}. The microvoid coalescence and the sliding are responsible for the fractures under these two conditions, respectively.

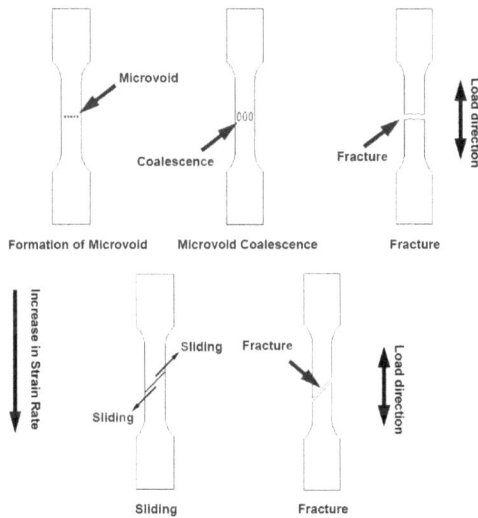

Figure 9. Schematic of the fracture mechanism for the thixoforged composite tensile tested at 200 °C under the low and high initial strain rates.

4. Conclusions

1. The thixoforged in situ Mg_2Si_p/AM60B composite exhibits a higher UTS than that of the thixoforged AM60B at the cost of elongation. The super UTS of the thixoforged in situ Mg_2Si_p/AM60B composite is mainly attributed to the load transfer mechanism and the obstruction for the dislocation motion from the reinforcement Mg_2Si particles.

2. As the testing temperature rises, the UTS of both the thixoforged in situ Mg_2Si_p/AM60B composite and AM60B decrease while their elongation increases. The enhanced dislocation motion

ability, the softened eutectic β phase at 120 °C, the activated non-basal slipping, and the dynamic recovery and recrystallization mechanisms at 150 °C are responsible for the change in tensile properties with testing temperatures.

3. The tensile properties change with the initial strain rate changes tested at 200 °C. The UTS of the composite reaches its peak value at the initial strain rate of 0.1 s^{-1} and then decreases, however, the elongation of the composite continually decreases as the initial strain rate increases. The variations in tensile properties result from the dynamic recovery, recrystallization mechanisms and working hardening.

4. The fracture of the composite transforms from the ductile regime into the quasi-cleavage mode, and finally exhibits the brittle feature as the initial strain rate increases.

Acknowledgments: The authors wish to express thanks to financial support from the National Basic Research Program of China (Grant No. G2010CB635106), the Program for New Century Excellent Talents in University of China (Grant No. NCET-10-0023), and the Program for Hongliu Outstanding Talents of Lanzhou University of Technology.

Author Contributions: Suqing Zhang and Tijun Chen conceived and designed the experiments; Jixue Zhou performed the experiments; Tao Li and Kaiming Cheng analyzed the data; Dapeng Xiu contributed analysis tools; and Suqing Zhang wrote the paper.

Conflicts of Interest: The authors declare no conflict of interest.

References

1. Aghion, E.; Bronfin, B.; Eliezer, D. The role of the magnesium industry in protecting the environment. *J. Mater. Process. Technol.* **2001**, *117*, 381–385. [CrossRef]

2. Luo, A.; Pekguleryuz, M.O. Cast magnesium alloys for elevated temperature applications. *J. Mater. Sci.* **1994**, *29*, 5259–5271. [CrossRef]

3. Zhu, S.; Easton, M.; Abbott, T.; Nie, J.; Dargusch, M.; Hort, N.; Gibson, M. Evaluation of magnesium die-casting alloys for elevated temperature applications: Microstructure, tensile properties, and creep resistance. *Metall. Mater. Trans. A* **2015**, *46*, 3543–3554. [CrossRef]

4. Aghion, E.; Bronfin, B.; Eliezer, D. The art of developing new magnesium alloys for high temperature applications. *Mater. Sci. Forum* **2003**, *419*, 407–417. [CrossRef]

5. Weiss, D.; Kaya, A.A.; Aghion, E.; Eliezer, D. Microstructure and creep properties of a cast Mg–1.7% wt rare earth–0.3% wt Mn alloy. *J. Mater. Sci.* **2002**, *37*, 5371–5379. [CrossRef]

6. Zhang, J.; Leng, Z.; Zhang, M.; Meng, J.; Wu, R. Effect of Ce on microstructure, mechanical properties and corrosion behavior of high-pressure die-cast Mg–4Al-based alloy. *J. Alloys Compd.* **2011**, *509*, 1069–1078. [CrossRef]

7. Hu, H. Squeeze casting of magnesium alloys and their composites. *J. Mater. Sci.* **1998**, *33*, 1579–1589. [CrossRef]

8. Nguyen, Q.B.; Gupta, M. Increasing significantly the failure strain and work of fracture of solidification processed AZ31B using Nano-Al$_2$O$_3$ particulates. *J. Alloys Compd.* **2008**, *459*, 244–250. [CrossRef]

9. Chen, L.; Yao, Y. Processing, microstructures, and mechanical properties of magnesium matrix composites: A review. *Acta Metall. Sin. (Engl. Lett.)* **2014**, *27*, 762–774. [CrossRef]

10. Ye, H.Z.; Liu, X.Y. Review of recent studies in magnesium matrix composites. *J. Mater. Sci.* **2004**, *39*, 6153–6171. [CrossRef]

11. Shen, M.; Ying, T.; Chen, F.; Hou, J. Effect of micro- and nano-SiC particulate reinforcements in magnesium-based metal matrix composites. *J. Mater. Eng. Perform.* **2016**, *25*, 2222–2229. [CrossRef]

12. Guo, W.; Wang, D.; Fu, Y.; Zhang, L.; Wang, Q. Dry sliding wear properties of AZ31-Mg$_2$Si magnesium matrix composites. *J. Mater. Eng. Perform.* **2016**, *25*, 4109–4114. [CrossRef]

13. Lloyd, D.J. Particle reinforced aluminium and magnesium matrix composites. *Int. Mater. Rev.* **1994**, *39*, 1–23. [CrossRef]

14. Nie, K.B.; Wang, X.J.; Wu, K.; Xu, L.; Zheng, M.Y.; Hu, X.S. Fabrication of SiC particles-reinforced magnesium matrix composite by ultrasonic vibration. *J. Mater. Sci.* **2012**, *47*, 138–144. [CrossRef]

15. Zhou, S.; Deng, K.; Li, J.; Shang, S.; Liang, W.; Fan, J. Effects of volume ratio on the microstructure and mechanical properties of particle reinforced magnesium matrix composite. *Mater. Des.* **2014**, *63*, 672–677. [CrossRef]

16. Das, A.; Harimkar, S.P. Effect of graphene nanoplate and silicon carbide nanoparticle reinforcement on mechanical and tribological properties of spark plasma sintered magnesium matrix composites. *J. Mater. Sci. Technol.* **2014**, *30*, 1059–1070. [CrossRef]

17. Xia, K.; Tausig, G. Liquidus casting of a wrought aluminum alloy 2618 for thixoforming. *Mater. Sci. Eng. A* **1998**, *246*, 1–10. [CrossRef]

18. Zhao, Z.D.; Chen, Q.A.; Tang, Z.J.; Hu, C.K. Microstructural evolution and tensile mechanical properties of AM60B magnesium alloy prepared by the SIMA route. *J. Alloys Compd.* **2010**, *497*, 402–411. [CrossRef]

19. Liu, D.; Atkinson, H.V.; Kapranos, P.; Jirattiticharoean, W.; Jones, H. Microstructural evolution and tensile mechanical properties of thixoformed high performance aluminium alloys. *Mater. Sci. Eng. A* **2003**, *361*, 213–224. [CrossRef]

20. Zhang, S.; Chen, T.; Cheng, F.; Li, P. A comparative characterization of the microstructures and tensile properties of as-cast and thixoforged in situ AM60B-10 vol % Mg_2Si_p composite and thixoforged AM60B. *Metals* **2015**, *5*, 457–470. [CrossRef]

21. Zhang, S.; Chen, T.; Cheng, F.; Li, L. Effects of mould temperature on microstructure and tensile properties of thixoforged Mg_2Si_p/AM60B in-situ composites. *J. Alloys Compd.* **2015**, *657*, 582–592. [CrossRef]

22. Zhang, S.; Chen, T.; Li, P. Microstructure and tensile properties of in situ Mg_2Si_p/AM60B composite prepared by thixoforging technology. *J. Mater. Res.* **2016**, *31*, 783–796. [CrossRef]

23. Zhang, S.; Chen, T.; Cheng, F.; Li, L. Microstructural evolution and phase transformation during partial remelting of in-situ Mg_2Si_p/AM60B composite. *Trans. Nonferr. Met. Soc. China* **2016**, *26*, 1564–1573. [CrossRef]

24. Mabuchi, M.; Kubota, K.; Higashi, K. Elevated temperature mechanical properties of magnesium alloys containing Mg_2Si. *Mater. Sci. Technol.* **1996**, *12*, 35–39. [CrossRef]

25. Li, G.H.; Gill, H.S.; Varin, R.A. Magnesium silicide intermetallic alloys. *Metall. Mater. Trans. A* **1993**, *24*, 2383–2391. [CrossRef]

26. Snyder, V.A.; Alkemper, J.; Voorhees, P.W. The development of spatial correlations during Ostwald ripening: A test of theory. *Acta Mater.* **2000**, *48*, 2689–2701. [CrossRef]

27. Yu, B.; Chen, D.; Tang, Q.; Wang, C.; Shi, D. Structural, electronic, elastic and thermal properties of Mg_2Si. *J. Phys. Chem. Solids* **2010**, *71*, 758–763. [CrossRef]

28. Arsenault, R.J. Interfaces in metal matrix composites. *Scr. Metall.* **1984**, *18*, 1131–1134. [CrossRef]

29. Arsenault, R.J. Strengthening mechanisms in particulate mmc. *Scr. Metall.* **1991**, *25*, 2617–2621. [CrossRef]

30. Eliezer, D.; Aghion, E.; Froes, F. The science, technology, and applications of magnesium. *JOM* **1998**, *50*, 30–34.

31. Kang, F.; Li, Z.; Wang, J.; Cheng, P.; Wu, H. The activation of <c + a> non-basal slip in magnesium alloys. *J. Mater. Sci.* **2012**, *47*, 7854–7859. [CrossRef]

32. Yin, D.D.; Wang, Q.D.; Boehlert, C.J.; Ding, W.J. Creep and fracture behavior of as-cast Mg–11Y–5Gd–2Zn–0.5Zr (wt %). *J. Mater. Sci.* **2012**, *47*, 6263–6275. [CrossRef]

33. Biswas, S.; Singh, D.S.; Beausir, B.; Toth, L.S.; Suwas, S. Thermal response on the microstructure and texture of ECAP and cold-rolled pure magnesium. *Metall. Mater. Trans. A* **2015**, *46*, 2598–2613. [CrossRef]

34. Liu, S.; Gao, F.; Zhang, Q.; Li, W. Mechanical properties and microstructures of nano-sized SiC particles reinforced AZ91D nanocomposites fabricated by high intensity ultrasonic assisted casting. *Mater. Sci. Forum* **2009**, *618–619*, 449–452. [CrossRef]

35. Li, Z.; Dong, J.; Zeng, X.Q.; Lu, C.; Ding, W.J. Influence of $Mg_{17}Al_{12}$ intermetallic compounds on the hot extruded microstructures and mechanical properties of Mg–9Al–1Zn alloy. *Mater. Sci. Eng. A* **2007**, *466*, 134–139. [CrossRef]

36. Arsenault, R.J.; Shi, N. Dislocation generation due to differences between the coefficients of thermal expansion. *Mater. Sci. Eng. A* **1986**, *81*, 175–187. [CrossRef]

37. Zhang, D.; Yang, X.; Sun, H.; Li, Y.; Wang, J.; Zhang, Z.; Ye, Y.; Sakai, T. Dynamic recrystallization behaviors and the resultant mechanical properties of a Mg–Y–Nd–Zr alloy during hot compression after aging. *Mater. Sci. Eng. A* **2015**, *640*, 51–60. [CrossRef]

38. Roodposhti, P.S.; Sarkar, A.; Murty, K.L. Microstructural development of high temperature deformed AZ31 magnesium alloys. *Mater. Sci. Eng. A* **2015**, *626*, 195–202. [CrossRef]
39. Fadavi Boostani, A.; Yazdani, S.; Taherzadeh Mousavian, R.; Tahamtan, S.; Azari Khosroshahi, R.; Wei, D.; Brabazon, D.; Xu, J.Z.; Zhang, X.M.; Jiang, Z.Y. Strengthening mechanisms of graphene sheets in aluminium matrix nanocomposites. *Mater. Des.* **2015**, *88*, 983–989. [CrossRef]
40. Watanabe, H.; Ishikawa, K. Effect of texture on high temperature deformation behavior at high strain rates in a Mg–3Al–1Zn alloy. *Mater. Sci. Eng. A* **2009**, *523*, 304–311. [CrossRef]
41. Ulacia, I.; Dudamell, N.V.; Gálvez, F.; Yi, S.; Pérez-Prado, M.T.; Hurtado, I. Mechanical behavior and microstructural evolution of a Mg AZ31 sheet at dynamic strain rates. *Acta Mater.* **2010**, *58*, 2988–2998. [CrossRef]

![metals logo] *metals*

MDPI

Article

Elastic and Plastic Behavior of the QE22 Magnesium Alloy Reinforced with Short Saffil Fibers and SiC Particles

Josef Zapletal [1,*], Zuzanka Trojanová [2], Pavel Doležal [1], Stanislava Fintová [3] and Michal Knapek [2]

[1] Institute of Materials Science and Engineering, Faculty of Mechanical Engineering, Brno University of Technology, Technická 2896/2, 616 69 Brno, Czech Republic; dolezal@fme.vutbr.cz

[2] Department of Physics of Materials, Faculty of Mathematics and Physics, Charles University, Ke Karlovu 5, 121 16 Praha 2, Czech Republic; ztrojan@met.mff.cuni.cz (Z.T.); knapek@karlov.mff.cuni.cz or Michal.Knapek@mff.cuni.cz (M.K.)

[3] CEITEC IPM, Institute of Physics of Materials, Academy of Sciences of the Czech Republic, v. v. i., Žižkova 22, 616 62 Brno, Czech Republic; fintova@ipm.cz

* Correspondence: zapletal@fme.vutbr.cz; Tel.: +420-541-143-197

Received: 30 December 2017; Accepted: 12 February 2018; Published: 15 February 2018

Abstract: Magnesium alloy QE22 (nominal composition 2 wt % Ag, 2 wt % mixture of rare earth elements, balance Mg) was reinforced with 5 vol % Saffil fibers and 15 vol % SiC particles. The hybrid composite was prepared via the squeeze cast technique. The microstructure of the monolithic alloy and composite was analyzed using scanning electron microscopy. Elastic modulus was measured at room temperature and modeled by the Halpin–Tsai–Kardos mathematical model. The strengthening effect of fibers and particles was calculated and compared with the experimentally obtained values. The main strengthening terms were determined. Fracture surfaces were studied via scanning electron microscope. While the fracture of the matrix alloy had a mainly intercrystalline character, the failure of the hybrid composite was transcrystalline.

Keywords: magnesium-alloy-based composite; Halpin-Tsai-Kardos model; deformation behavior; composite strengthening; fracture behavior

1. Introduction

Cast magnesium alloys based on the Mg–Al–Zn–Mn system are designed for applications at ambient temperatures because the mechanical properties of these alloys rapidly deteriorate at elevated temperatures (above 120 °C) [1,2]. Among alloys for application at higher temperatures, the commercial QE22 is one of the most frequently used. Improvement in mechanical properties can be achieved using a proper thermal treatment [2] or, improving thermal stability as well, using various reinforcements that exhibit different mechanical and physical properties. Magnesium matrix composites show better wear resistance, enhanced strength, and creep resistance. Furthermore, they retain low density and good machinability [3,4]. Appropriate materials for the reinforcing phases are ceramics: oxides, nitrides, borides, carbides, carbon nanotubes (CNT), or intermetallic compounds [5–11]. The behavior of composites depends on the content, geometry, and physical and mechanical properties of the reinforcing phases. Metal matrix composites (MMCs) with long fibers, originally used to strengthen a plastic metallic composite matrix, exhibit several disadvantages, such as high anisotropy and low mechanical stability, since fibers are pulled out from the matrix under loading. Extremely high anisotropy is partially suppressed in composites reinforced with short fibers or whiskers [12–17]. This anisotropy may vanish in composites reinforced by particles [18]. On the other hand, mechanical properties of the particle-reinforced composites are inferior compared with fiber-reinforced composites. Using a combination of fibers and particles has proven to be a convenient compromise [19–24]. Reinforcing fibers/particles increases stiffness [25], improves mechanical and creep properties [19–24], and decreases the thermal expansion [26] of the composite. Nanocomposites with a reinforcing phase size of up to 100 nm have been intensively studied in the last decade [27–31]. Magnesium-alloy-based (nano)composites have been considered a material suitable for biomedical applications in the last few years [32]. Al_2O_3 ceramic Saffil fibers and SiC particles (SiCp) are commonly used reinforcements for Mg alloys. A combination of both materials, where Saffil fibers are replaced by cheaper SiC particles, may deliver a material with excellent strength, acceptable plasticity, improved creep properties, and enhanced wear resistance when compared to a mere fiber-reinforced composite with an Mg matrix. If composites are prepared via the hot extrusion technique, mechanical and physical properties are also influenced by the material texture [26]. Bonding between the metallic matrix and the reinforcing phase significantly affects the resulting properties of MMCs [33]. The presence of fibers/particles may also influence the precipitation processes in the metallic matrix during the composite preparation [34]. The chemical composition of magnesium alloys can be chosen according to the required mechanical properties of the composite matrix, considering the strengthening effect of the reinforcing particles/fibers. Rare earth element (RE) additions increase the strength of magnesium alloys at elevated temperatures and reduce weld cracking and casting porosity. The relatively weak tensile properties of Mg–RE–Zr alloys can be improved by an addition of Ag, which results in enhanced thermal treatment possibilities and age hardening processing of the alloy and final composite. QE22 is a widely used Mg–Ag cast alloy, where the content of silver below 2% allows for the formation of Mg–Nd precipitates similar to the Mg–RE alloys; the addition of silver also refines the precipitate size [2].

In this complex study, the elastic modulus and the plastic and fracture properties of the QE22 hybrid composite containing Saffil fibers and SiC particles were investigated with the aim to better understand the factors and processes influencing the resulting composite properties. In addition, the monolithic QE22 alloy was also studied in this work as a reference material.

2. Experimental Material and Procedure

Commercially available cast QE22 magnesium alloy (nominal concentration in wt %: 2 Ag, 2 mischmetal mainly Nd-0.4 Zr, balance Mg) was used as the composite matrix material. The alloy was reinforced with 5 vol % Saffil® (97% δ-Al_2O_3, 3% SiO_2) fibers and 15 vol % SiC particles using a preform consisting of fibers, particles, and a binder system (Al_2O_3 and starch). The orientation of fibers in the preform was planar, with a random orientation of fibers in the plane. Squeeze casting technology (ZFW, Clausthal-Zellerfeld, Germany) was used for the infiltration of liquid Mg alloy into the preform preheated to 1000 °C. Two-stage pressure of 80 and 150 MPa was applied in order to ensure that the resulting hybrid composite contains no pores. Saffil fibers exhibited a mean length of 78 ± 16 μm and were 3 μm in diameter (measured after squeeze casting). Nearly equiaxial SiC particles of an irregular shape were about 9 μm in size.

A scanning electron microscope (SEM) (ULTRA PLUS, Carl Zeiss GmbH, Oberkochen, Germany) equipped with a dispersive X-ray spectrometer (EDS) (X-MAX, Oxford Instruments, Abingdon, England) was used for the microstructure characterization and analysis of the fracture surfaces of the prepared composite. An Everhar–Thornley-type secondary electron (SE) (Oxford Instruments, Abingdon, England) detector and a four-quadrant silicone back-scattered electron (BSE) (Oxford Instruments, Abingdon, England) detector were used to reveal material structural features. Geometry of the samples and the testing procedure corresponded to the EN ISO 6892-1 standard. Samples were not thermally treated preliminarily. Samples for deformation tests were cut from the cast ingot so that the stress axis was parallel to the fiber plane, as indicated in Figure 1. Flat specimens used for the tensile tests had an active length of 30 mm, a width of 8 mm, and a thickness of 3 mm. Cylindrical specimens for compression tests were prepared to have a diameter of 8 mm and a length of 12 mm.

Before the compression tests, the grip heads of the testing machine were lubricated with MoS_2. For both tensile and compression tests, three samples were used. Mechanical tests were carried out at room temperature (23 ± 2 °C) in a Zwick Z250 PC controlled testing device (Zwick GmbH & Co., Ulm, Germany) with a constant cross head speed of 1 mm·min^{-1}, giving the strain rate of 5.5×10^{-4} s^{-1} (tension) and 1.4×10^{-3} s^{-1} (compression). True stress–true plastic strain curves were computed. Characteristic tensile/compression yield stress (*TYS/CYS*) and ultimate tensile/compression strength (*UTS/UCS*) were estimated together with the strain-to-fracture, ε_f.

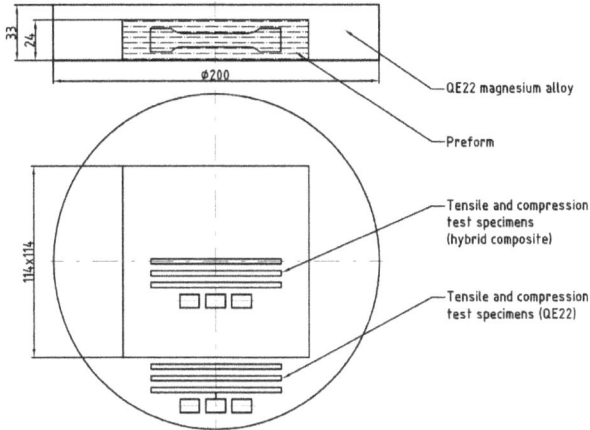

Figure 1. Sample positions in the casting.

The resonant frequency and damping analyzer RFDA HT650 (Integrated Material Control Engineering, Genk, Belgium) was used for the Young's modulus measurements. Samples had a shape of bending beams ($l = 23$ mm, $w = 3.6$, $t = 3.2$ mm) with the longest axis either parallel (L-samples) or perpendicular (T-samples) to the fiber plane. Samples were mechanically excited into vibrations and the resonant frequency was determined. The Young's moduli were calculated for both types of samples using the resonant frequencies.

3. Results and Discussion

3.1. Microstructure Analysis

The QE22 magnesium alloy used in this study contained 2.59 wt % Ag and 1.47 wt % mischmetal (mainly Nd, with Pr, Ce, and La). The microstructure of the QE22 magnesium alloy shown in Figure 2a,b is composed of α grains consisting of a solid solution of alloying elements in Mg and precipitates situated at the grain boundaries. The elements analysis (EDS results) introduced in Figure 2c shows that particles at the grain boundaries are Mg_3(Ag,RE) (63.6 wt % Mg, 17.9 wt % Ag, and 18.5 wt % RE) eutectics [22].

The microstructure of the hybrid composite is shown in Figure 3a,b. It is obvious that the distribution of SiC particles is not uniform; particles formed small clusters in the vicinity of fibers. Bright small needles visible in Figure 3 are the edges of dielectric SiC particles, which were charged with electrons during scanning. Kiehn et al. [34] found that the matrix in the QE22 alloy reinforced with Al_2O_3 fibers changed its chemical composition due to an increased content of Al introduced into the matrix from the inorganic binder in the preform. Al substitutes Ag in the precipitates which remains dissolved in the matrix. The grain size of the QE22 alloy was estimated to be (45.4 ± 0.8) μm and for the composite (6.2 ± 0.8) μm. The grain size was estimated using a linear intercept method from the light micrographs.

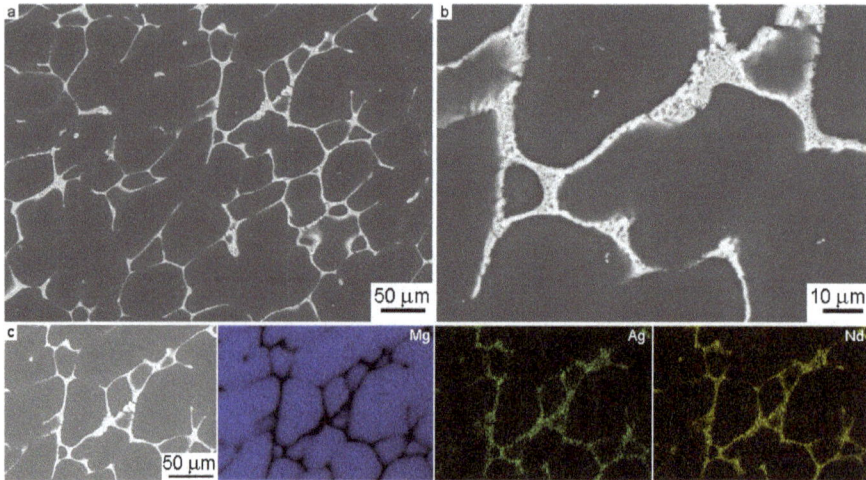

Figure 2. Back-scattered electron (BSE) scanning electron micrograph of the QE22 alloy (**a**); a closer view of the precipitates (**b**); the distribution of main alloying elements in the alloy determined by EDS (**c**).

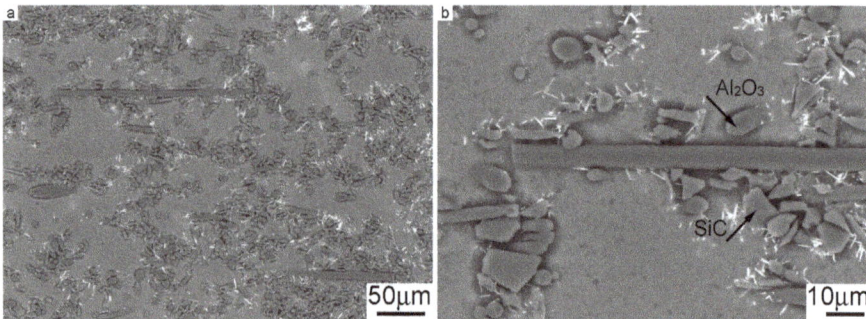

Figure 3. The microstructure (BSE mode) of the hybrid composite (**a**); a closer view of a fiber surrounded by SiC particles (**b**).

3.2. Elastic Properties

Unidirectionally continuous fiber-reinforced MMCs have shown a linear Young's modulus increase with increasing fiber volume fraction in the fibers direction [11]. A Young's modulus increase in the fibers direction is in agreement with the rule of mixtures, while a modulus increase in the transversal direction is very low. Particles in MMCs affect the modulus much less than predicted by the rule of mixtures. Experimentally determined Young's modulus values for the hybrid L-sample, $E_H^L(ex)$ = 74.5 GPa, and the T-sample, $E_H^T(ex)$ = 71.2 GPa, are reported in Table 1. By comparing the experimental modulus values with the values obtained using the rule of mixtures (E_H^{ROM}), we can see that the measured moduli are substantially lower and exhibit certain anisotropy. The modulus measured on the L-sample is higher than that of the T-sample.

Table 1. Experimental moduli values of the matrix (E_m), fibers (E_{Saffil}), particles (E_{SiC}), hybrid L-sample $E_H^L(ex)$, and hybrid T-sample $E_H^T(ex)$. Calculated moduli values for Mg + Saffil in the L direction $E_C^L(th)$, Mg + Saffil in the T direction $E_C^T(th)$, and theoretical values of the hybrid moduli in both directions, $E_H^L(th)$ and $E_H^T(th)$, data from [35–37].

E_m	E_{Saffil}	E_{SiC}	$E_C^L(th)$	$E_C^T(th)$	$E_H^L(th)$	$E_H^T(th)$	$E_H^L(ex)$	$E_H^T(ex)$	$E_{H(ROM)}$
GPa	GPa	GPa	GPa	GPa	GPa	GPa	GPa	GPa	GPa
44.1	300	410	61.42	48.6	76.5	61.5	74.5	71.1	118.5
[35]	[36]	[37]	-	-	-	-	-	-	-

For material exhibiting certain anisotropy, the self-consistent Halpin–Tsai–Kardos model, originally developed for polymer composites, is more suitable [38,39]. Calculation of Young's moduli for the hybrid composite samples according to the Halpin–Tsai–Kardos model can be divided into two steps. In the first step, the Young's modulus, E_C, of the fibers composite QE22 + 5 vol % Saffil (and for both orientations of the fiber plane) is calculated. The Young's modulus of the hybrid composite, E_H, with short fibers and SiC particles is computed in the second step. The modulus in the longitudinal direction (fiber plane is parallel to the longest sample axis) may be expressed as:

$$E_C^L = E_m \frac{1 + \frac{2L}{d}\eta_L v_f}{1 - \eta_L v_f}, \quad \eta_L = \frac{\frac{E_f}{E_m} - 1}{\frac{E_f}{E_m} + \frac{2L}{d}} \tag{1}$$

where E_m is the modulus of the QE22 matrix alloy, E_f the modulus of the Saffil fibers, L the length, and d the diameter of the reinforcing fibers; v_f is the volume fraction of the fibers. The transversal modulus (perpendicular to the fiber plane) can be determined as:

$$E_C^T = E_m \frac{1 + 2\eta_T v_f}{1 - \eta_T v_f}, \quad \eta_T = \frac{\frac{E_f}{E_m} - 1}{\frac{E_f}{E_m} + 2} \tag{2}$$

Calculated values of both moduli, $E_C^L(th)$ and $E_C^T(th)$, are reported in Table 1 together with the experimentally obtained values of the matrix and Saffil moduli. Anisotropy of the material is reflected by different values calculated for longitudinal and transversal moduli. Calculated moduli, $E_C^L(th)$ and $E_C^T(th)$, were used for estimation of hybrid composite moduli according to following equations:

$$E_H^L = E_C^L \frac{1 + A.B^L v_p}{1 - B v_p}, \quad A = \frac{7 - 5v_p}{8 - 10v_p}, \quad B^L = \frac{E_p - E_C^L}{E_p + A.E_C^L} \tag{3}$$

and

$$E_H^T = E_C^T \frac{1 + A.B^T v_p}{1 - B v_p}, \quad A = \frac{7 - 5v_p}{8 - 10v_p}, \quad B^T = \frac{E_p - E_C^T}{E_p + A.E_C^T} \tag{4}$$

where E_p is the Young's modulus of SiC particles, and v_p their volume fraction.

By comparing the calculated moduli of the hybrid composite, $E_H^L(th)$ and $E_H^T(th)$, with the experimentally obtained values, $E_H^L(ex)$ and $E_H^T(ex)$, it is obvious that the agreement is much better than in the case of the rule of mixtures. The agreement of the measured values with the theoretical predictions is not ideal, because the real composite does not fully comply with the theoretical assumptions: fibers are not aligned in one direction (fibers have only planar isotropic distribution), particles are not regularly distributed in the matrix, and fiber length is not uniform. The model, however, depicts the experimentally observed composite anisotropy very well. Generally, it is possible to conclude that the transversal modulus, compared with the longitudinal modulus, is less influenced by the presence of reinforcing fibers. This result is in agreement with the results obtained on Al–Li composites reinforced with short fibers reported by Chawla [11].

3.3. Plastic Deformation

True stress–true strain curves measured in tension for the monolithic QE22 alloy and hybrid composite are presented in Figure 4a. The tensile yield stress and the ultimate tensile strength values are reported in Table 2. A much greater amount of stress is necessary for the plastic deformation of the hybrid composite compared to the monolithic alloy. Furthermore, the work hardening coefficient $\theta = d\sigma/d\varepsilon$ is higher for the composite.

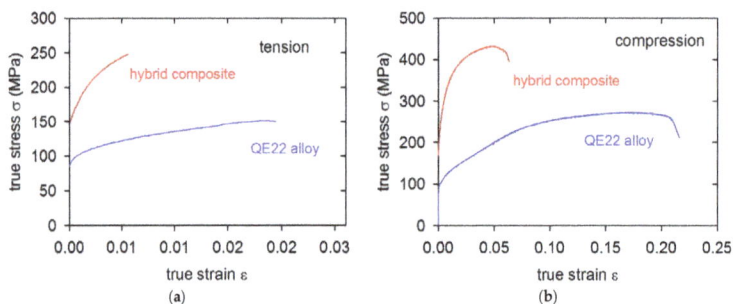

Figure 4. True stress–true strain curves of the QE22 matrix alloy and the hybrid composite in tension (a) and compression (b).

Table 2. Characteristic stresses and the strain-to-fracture ratio obtained in tension and compression.

Material	TYS	CYS	UTS	UCS	ε_f	ε_f
	MPa	MPa	MPa	MPa	Tension	Compression
QE22 alloy	108.3	100.5	151.4	260.3	0.02	0.22
composite	207.6	243.4	247.3	431.4	0.006	0.06

True stress–strain curves obtained in the compression test are given in Figure 4b. Again, a big difference between the hybrid composite and the unreinforced alloy is clearly observed. Characteristic stresses estimated for both materials in compression are shown in Table 2, together with the plastic strain-to-fracture ratio, ε_f.

The plasticity of the alloy and the composite in tension was found to be very low, which is often observed in cast magnesium alloys [40]. Note also the anisotropy observed for both materials: the yield stress found in tension and compression are different. While the *CYS* estimated for the matrix alloy is lower than the *TYS*, the *CYS* in the hybrid composite is higher. These differences may be ascribed to the different deformation and strengthening mechanisms operating in the composite and the matrix alloy in tension and compression. The observed asymmetry of the *TYS* and *CYS* estimated for the matrix alloy is typical for magnesium alloys. Máthis et al. studied acoustic emission signals during the deformation of magnesium samples with different grain sizes. They found strong asymmetry in acoustic emission at temperatures lower than 200 °C corresponding to the activity of twin formation [41]. They concluded that different deformation mechanisms are active in tension and compression. While mainly dislocation glide facilitates plastic deformation in tension, mechanical twinning at the beginning of plastic deformation is the controlling mechanism in compression. This asymmetry depends also on the microstructure: it diminishes with increasing grain size. Twin boundaries that readily form during the compression test are impenetrable obstacles for dislocation motion. Therefore, twin formation contributes to an enhanced strain hardening of the material loaded in compression.

The presence of fibers or particles in a composite material influences the microstructure and straining via different mechanisms, which are summarized in Table 3. Usually, there is a large difference between the thermal expansion coefficient (*CTE*), α, of the metallic matrix and the ceramic reinforcement. The cooling down from manufacturing to room temperature generates thermal stresses in the composite. In this manner, the yield stress can be achieved in the matrix, and new thermal dislocation loops are punched [42]. Because of the incompatibility of the metallic matrix and the ceramic reinforcing phase, dislocations geometrically necessary are created during plastic deformation and contribute to the enhanced dislocation density [43]. The stress increase, $\Delta\sigma_D$, due to increased density of thermal dislocations, ρ_T, and dislocations geometrically necessary, ρ_G, may be calculated using the known Taylor formula. The load transfer mechanism may be considered an important reinforcing mechanism. The stress increase due to load transfer, $\Delta\sigma_{LT}$, was theoretically treated by several authors [44–48]. Nardone and Prewo [48] calculated the stress increase due to load transfer, σ_{LT}, for the aligned short fibers in the matrix, respecting the fact that fibers with the higher aspect ratio L/d contribute more to the composite strengthening; on the other hand, anisotropy of the material increases. Because not all fibers are oriented into the stress axis (but only the fiber plane) in the studied hybrid composite, it is necessary to calculate the load transfer component of the stress, $\beta.\Delta\sigma_{LT}$, where $\beta = 0.6$ is the mean value of the direction cosines. For uniaxial particles, the load transfer value is $\Delta\sigma_{LT} = 0.5\sigma_m v_p$ [49]. Discontinuously reinforced MMCs typically exhibit a finer grain structure compared with the unreinforced matrices [50]. The contribution of this refinement to the yield stress can be estimated using the Hall–Petch relationship [51]. Moreover, Orowan strengthening is also present as a result of the elastic behavior of dislocation segments passing stiff, closely spaced particles [52,53]. In MMCs, where distances between particles are of the order of μm, this mechanism usually plays only a marginal role. Newly created dislocations in the matrix partially accommodate thermal stresses. However, some tensile residual stresses remain in the matrix and either decrease (tension) or increase (compression) the stress necessary for plastic deformation [54].

Table 3. Individual contributions to strengthening following from the presence of the reinforcing phase.

Mechanism	Equation	Symbols
Increased dislocation density due to thermal strain $\Delta\alpha\Delta T$	$\rho_T = \frac{Bv_{f,p}\Delta\alpha\Delta T}{b(1-v_{f,p})}\frac{1}{t}$	b Burgers vector of dislocations, $B = 12$ (p) $B = 10$ (f), t minimum size of (f) or (p)
Dislocation geometrically necessary	$\rho_G = \frac{v_{f,p}8\varepsilon_p}{bt}$	ε_p plastic strain, $v_{f,p}$ volume fraction of fibers/particles
Enhanced dislocation density	$\Delta\sigma_D = \alpha_1 mGb(\rho_T + \rho_G)^{1/2}$	α_1 constant, m Taylor factor, G shear modulus
Load transfer	$\Delta\sigma_{LT} = \sigma_{LT} - \sigma_m = \sigma_m v_f \frac{(L+d)\left(\frac{L}{d}\right)}{4L}$	L fibers length, d fibers diameter
Hall–Petch strengthening	$\Delta\sigma_{GS} = K_y\left(d_2^{-1/2} - d_1^{-1/2}\right)$	d_1, d_2 grain sizes, K_y Hall–Petch constant
Orowan strengthening	$\Delta\sigma_{OR} = \frac{Gb}{\Lambda} + \frac{5}{2\pi}Gv_{f,p}\varepsilon_p$	Λ distance between particles or fibers ends
Residual thermal stresses	$\langle\sigma_m\rangle_{max} = \frac{2}{3}\sigma_m \ln\left(\frac{1}{v_{f,p}}\right)\frac{v_{f,p}}{1-v_{f,p}}$	σ_m yield stress in matrix

Individual strengthening terms were calculated using constants introduced in Table 4 and are reported in Table 5.

Table 4. Constants used for calculations of the strengthening terms, data from [55–60].

α (QE22)	α (Saffil)	α (SiC)	K_y (Mg)	α_1	G	b	m
K^{-1}	K^{-1}	K^{-1}	MPa·mm$^{1/2}$	-	GPa	m	-
26×10^{-6}	6×10^{-6}	6.6×10^{-6}	10	0.35	17	3.2×10^{-10}	4.5
[55]	[56]	[57]	[58]	[59]	-	-	[60]

Table 5. Individual contributions to strengthening of the hybrid composite.

Deformation Mode	YS (A)	YS (M)	$\Delta\sigma_{LT}$	$\Delta\sigma_D$	$\Delta\sigma_{OR}$	$\Delta\sigma_{GS}$	$<\sigma_m>_{max}$	σ_{tot}	YS (ex)
	MPa	MPa	MPa	MPa	MPa	MPa	MPa	MPa	MPa
tension	108.3	144.2	29.2 + 10.8	35.9	2	25.2	-	202.4	207.6
compresion	100.5	136.4	27.6 +10.2	35.9	2	25.2	26.9	229.3	243.4

In Table 5, *YS* (A) represents the yield stress of the unreinforced alloy, *YS* (M) is the yield stress in the matrix of the composite, and σ_{tot} is the calculated theoretical value for the hybrid composite yield stress. From Table 5, it is obvious that the main strengthening contributions in the hybrid composite are the load transfer, the increased dislocation density, and the small grain size of the composite matrix.

The load transfer term was obtained experimentally by Farkas et al. using neutron diffraction in situ measurements [61]. The authors found a load transfer value of about 100 MPa for the AX61 composite containing 26 vol % Saffil fibers. Taking into account the lower volume fraction of the strengthening fibers/particles and the lower strengthening contribution of particles to the load transfer, the calculated value ~30 MPa seems to be a good approximation of the load transfer real value for the studied composite. Mean residual stresses in the matrix were measured in magnesium-based MMCs by Farkas et al. ex situ [62] and in situ using neutron diffraction [61,63]. They found a higher $<\sigma_m>_{max}$ than that of the calculated estimation according to the equation reported in Table 3. In fact, in the first approximation, the difference between the *TYS* and *CYS* can be considered the residual thermal stress (*RTS*), as is schematically depicted in Figure 5. Difficulties with the appropriate combination of individual strengthening terms were solved by Lilhold [64] and Clyne and Whithers [65]. The stress contributions, which act more or less uniformly throughout the matrix, may be superimposed linearly. It follows from Table 5 that a simple sum of the strengthening terms gives a good approximation for the composite offset yield stresses.

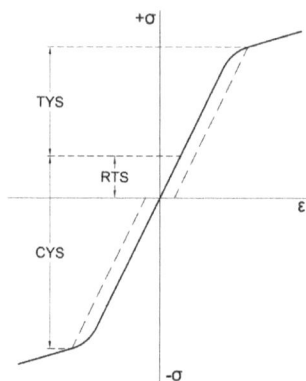

Figure 5. Schematic explanation of the composite tension–compression anisotropy.

3.4. Fractographic Analysis

The fracture surface of the matrix alloy after the tensile test is shown in Figure 6a. The fracture mechanism is intercrystalline; the crack propagated along the grain boundaries. Eutectics situated at the grain boundaries in the naked crystal planes are more visible in Figure 6b (taken in back-scattered electron mode). Secondary cracks along the grain boundaries document that the eutectics present at the grain boundaries reduce their cohesive strength. Broken eutectic particles act as the crack initiation sites and the grain boundaries become a favorable path for the crack propagation. This is also the reason for the low ductility of this alloy. In Figure 6a, the twin boundary fracture is visible in several places. Twin boundary fracture is a typical failure mechanism in the QE22 magnesium alloy [66].

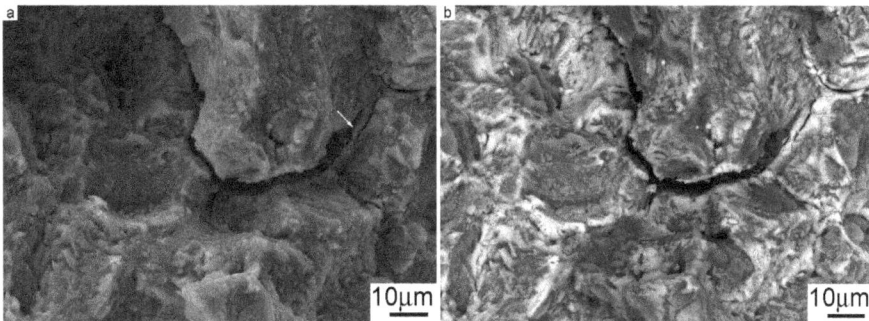

Figure 6. Scanning electron micrograph of the alloy fracture surface after the tensile test (**a**) the same location taken in BSE mode (**b**).

The scanning electron micrographs of the composite fracture surface are reported in Figure 7a,b. Uncovered fibers and particles are characteristically observed. A transcrystalline fracture mechanism can be seen in the case of the hybrid composite. Broken Saffil fibers and SiC particles are visible on the fracture surface shown in Figure 7. The bright needles discernible in the BSE representation are the edges of SiC particles charged in the microscope via the applied electron field. No pulling out of fibers and SiC particles from the matrix alloy due to applied tensile loading was observed on the composite fracture surface after the failure. This gives evidence of a strong bonding between the matrix alloy and the reinforcing fibers and particles. No traces of twinning were observed in the matrix areas of the fracture surface. A smaller grain size of the composite matrix reduces twinning [41], and only a transcrystalline fracture can take place. Besides reinforcement strengthening (presented in the form of increased dislocation density, a load transfer change, and Orowan strengthening), grain boundary strengthening (i.e., the Hall–Petch mechanism) was revealed by the hybrid composite's fracture surface characteristics. The results of the fractographic analysis are, therefore, in agreement with the theoretically obtained results.

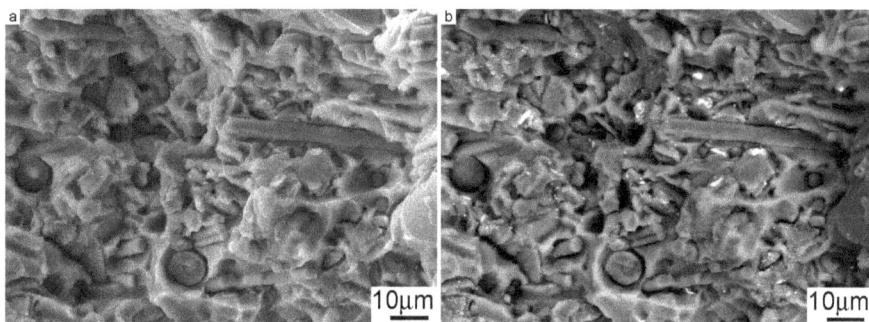

Figure 7. Fracture surface of the hybrid composite after straining in the tension test (**a**); the same location taken in BSE mode (**b**).

4. Conclusions

The elastic, plastic, and fracture properties of a MMC consisting of a QE22 magnesium alloy matrix reinforced with Saffil fibers and SiC particles were studied at room temperature. The following conclusions may be drawn based on the obtained results:

- Young's modulus measurements exhibited anisotropy owing to the 2D fiber distribution.
- The Halpin–Tsai–Kardos self-consistent model was successfully used to model the Young's modulus anisotropy, and the results are in good agreement with the experimentally obtained data.
- The presence of reinforcing fibers and particles substantially increased both tensile and compression deformation flow stresses.
- The ductility of the hybrid composite was radically decreased compared to the cast matrix alloy.
- The load transfer, the increased dislocation density, and the Hall–Petch strengthening are the main reinforcing mechanisms in the case of studied hybrid composite.
- Fracture of the matrix is mainly intercrystalline due to brittle eutectics present at the grain boundaries.
- Fracture of the hybrid composite is transcrystalline; no pulling out of the reinforcing fibers or particles was observed.

Acknowledgments: This research was carried out under the project CEITEC 2020 (LQ1601) with financial support from the Ministry of Education, Youth and Sports of the Czech Republic under the National Sustainability Programme II. Work was also supported by the project NETME Centre Plus (Lo1202), a project of the Ministry of Education, Youth and Sports under the National Sustainability Programme. The research was conducted under IPMinfra, supported through Project No. LM2015069 of MEYS.

Author Contributions: Josef Zapletal, Zuzanka Trojanová, and Pavel Doležal conceived and designed the experiments; Josef Zapletal, Pavel Doležal, and Michal Knapek performed the experiments; Josef Zapletal, Zuzanka Trojanová, Pavel Doležal, and Michal Knapek analyzed the data; Zuzanka Trojanová, Josef Zapletal, and Pavel Doležal contributed reagents/materials/analysis tools; Zuzanka Trojanová and Stanislava Fintová wrote the paper.

Conflicts of Interest: The authors declare no conflict of interest. The founding sponsors had no role in the design of the study; in the collection, analyses, or interpretation of data; in the writing of the manuscript; or in the decision to publish the results.

References

1. Polmear, I.J. *Magnesium Alloys and Their Applications*; Mordike, B.L., Hehmann, F., Eds.; DGM: Oberursel, Germany, 1992; p. 201, ISBN 978-3883551845.
2. Advesian, M.M. *ASM Speciality Handbook: Magnesium and Magnesium Alloys*; Advesian, M.M., Baker, H., Eds.; ASM International: Russell Township, OH, USA, 1999; ISBN 978-0-87170-657-7.
3. Kainer, K.U. *Magnesium Alloys and Their Applications*; Mordike, B.L., Hehmann, F., Eds.; DGM: Oberursel, Germany, 1992; p. 415, ISBN 978-3883551845.
4. Oakley, R.; Cochrane, R.F.; Stevens, R. Recent developments in magnesium matrix composites. *Key Eng. Mater.* **1995**, *104–107*, 387–416. [CrossRef]
5. Dey, A.; Pandey, K.M. Magnesium metal matrix composites—A review. *Rev. Adv. Mater. Sci.* **2015**, *42*, 58–67.
6. Park, Y.; Cho, K.; Park, I.; Park, Y. Fabrication and mechanical properties of magnesium matrix composite reinforced with Si coated carbon nanotubes. *Procedia Eng.* **2011**, *10*, 1446–1450. [CrossRef]
7. Jiang, Q.C.; Wang, H.Y.; Ma, B.X.; Wang, Y.; Zhao, F. Fabrication of B_4C particulate reinforced magnesium matrix composite by powder metallurgy. *J. Alloy. Compd.* **2005**, *386*, 177–181. [CrossRef]
8. Han, G.Q.; Shen, J.H.; Ye, X.X.; Chen, B.; Imai, H.; Kondoh, K.; Du, W.B. The influence of CNTs on the microstructure and ductility of CNT/Mg composites. *Mater. Lett.* **2016**, *181*, 300–304. [CrossRef]
9. Han, B.C.; Dunand, D.Q. Microstructure and mechanical properties of magnesium containing high volume fractions of yttria dispersoids. *Mater. Sci. Eng. A* **2000**, *277*, 297–304. [CrossRef]
10. Yu, W.; Zhao, H.; Hu, X. Anisotropic mechanical and physical properties in textured Ti_2AlC reinforced AZ91D magnesium composite. *J. Alloy. Compd.* **2019**, *732*, 894–901. [CrossRef]

11. Chawla, K.K. *Materials Science and Technology*; Cahn, R.W., Haasen, P., Kramer, E.J., Eds.; Wiley-VCH: Weinheim, Germany, 1993; Volume 13, pp. 121–182.

12. Hassan, S.F.; Gupta, M. Development of high performance magnesium nano-composites using nano-Al_2O_3 as reinforcement. *Mater. Sci. Eng. A* **2005**, *392*, 163–168. [CrossRef]

13. Trojanová, Z.; Gärtnerová, V.; Lukáč, P.; Drozd, Z. Mechanical properties of Mg alloys composites reinforced with short Saffil® fibres. *J. Alloy. Compd.* **2004**, *378*, 19–26. [CrossRef]

14. Trojanová, Z.; Száraz, Z.; Lábár, J.; Lukáč, P. Deformation behaviour of an AS21 alloy reinforced by short Saffil fibres and SiC particles. *J. Mater. Process. Technol.* **2005**, *162–163*, 131–138. [CrossRef]

15. Ho, K.F.; Gupta, M.; Srivatsan, T.S. The mechanical behavior of magnesium alloy AZ91 reinforced with fine copper particulates. *Mater. Sci. Eng. A* **2004**, *369*, 302–308. [CrossRef]

16. Jayalakshmi, S.; Kailas, S.V.; Seshan, S. Tensile behaviour of squeeze cast AM100 magnesium alloy and its Al_2O_3 fibre reinforced composites. *Compos. Part A* **2002**, *33*, 1135–1140. [CrossRef]

17. Trojanová, Z.; Drozd, Z.; Kúdela, S.; Száraz, Z.; Lukáč, P. Strengthening in Mg–Li matrix composites. *Compos. Sci. Technol.* **2009**, *67*, 1965–1973. [CrossRef]

18. Manoharan, M.; Lim, S.C.V.; Gupta, M. Application of a model for the work hardening behavior to Mg/SiC composites synthesized using a fluxless casting proces. *Mater. Sci. Eng. A* **2002**, *333*, 243–249. [CrossRef]

19. Zhang, X.; Zhang, Q.; Hu, H. Tensile behaviour and microstructure of magnesium AM60-based hybrid composite containing Al_2O_3 fibres and particles. *Mater. Sci. Eng. A* **2014**, *607*, 269–276. [CrossRef]

20. Kumar, S.R.; Panigrahi, M.K.; Thakur, S.K.; Kainer, K.U.; Chakraborty, M.; Dhindaw, B.K. Characterization of stress in reinforcements in magnesium based squeeze infiltrated cast hybrid composites. *Mater. Sci. Eng. A* **2006**, *415*, 207–212. [CrossRef]

21. Kumar, S.; Dieringa, H.; Kainer, K.U. Effect of particulate content on the thermal cycling behaviour of the magnesium alloy based hybrid composites. *Compos. Part A* **2005**, *36*, 321–325. [CrossRef]

22. Svoboda, M.; Pahutová, M.; Kuchařová, K.; Sklenička, V.; Kainer, K.U. Microstructure and creep behaviour of magnesium hybrid composites. *Mater. Sci. Eng. A* **2007**, *462*, 220–224. [CrossRef]

23. Arunachaleswaran, A.; Pereira, I.M.; Dieringa, H.; Huang, Y.; Hort, N.; Dhindaw, B.K.; Kainer, K.U. Creep behavior of AE₄₂ based hybrid composites. *Mater. Sci. Eng. A* **2007**, *460–461*, 268–276. [CrossRef]

24. Arunachaleswaran, A.; Dhindaw, B.K.; Dieringa, H.; Hort, N.; Kainer, K.U. Microstructure characterisation and creep properties of AE42 based hybrid composites prepared by squeeze casting process. *Trans. Indian Inst. Met.* **2007**, *60*, 87–91.

25. Roy, S.; Gebert, J.-M.; Stasiuk, G.; Piat, R.; Weidenmann, K.A.; Wanner, A. Complete determination of elastic moduli of interpenetrating metal/ceramic composites using ultrasonic techniques and micromechanical modelling. *Mater. Sci. Eng. A* **2011**, *528*, 8226–8235. [CrossRef]

26. Trojanová, Z.; Drozd, Z.; Minárik, P.; Lukáč, P.; Kasakewitsch, A. Influence of texture on the thermal expansion coefficient of Mg/BN nanocomposite. *Thermochim. Acta* **2016**, *644*, 69–75. [CrossRef]

27. Penchal Reddy, M.; Shakoor, R.A.; Parande, G.; Manakari, V.; Ubaid, F.; Modamed, A.M.A.; Gupta, M. Enhanced performance of nano-sized SiC reinforced Al metal matrix nanocomposites synthesized through microwave sintering and hot extrusion techniques. *Prog. Nat. Sci.* **2017**, *27*, 606–614. [CrossRef]

28. Chen, Y.; Tekumalla, S.; Guo, Y.B.; Shabadi, R.; Shim, V.P.W.; Gupta, M. The dynamic compressive response of a high-strength magnesium alloy and its nanocomposite. *Mater. Sci. Eng. A* **2017**, *702*, 65–72. [CrossRef]

29. Gupta, M.; Wong, W.L.E. Magnesium-based nanocomposites: Lightweight materials of the future. *Mater. Charact.* **2015**, *105*, 30–46. [CrossRef]

30. Casati, R.; Vedani, M. Metal Matrix Composites Reinforced by Nano-Particles—A Review. *Metals* **2014**, *4*, 65–83. [CrossRef]

31. Jayakumar, J.; Raghunath, B.K.; Rao, T.H. Recent Development and Challenges in Synthesis of Magnesium Matrix Nano Composites—A Review. *Int. J. Latest Res. Sci. Technol.* **2012**, *1*, 164–171.

32. Zhao, C.; Wu, H.; Ni, J.; Zhang, S.; Zhang, X. Development of PLA/Mg composite for orthopedic implant: Tunable degradation and enhanced mineralization. *Compos. Sci. Technol.* **2017**, *147*, 8–15. [CrossRef]

33. Trojanová, Z.; Ferkel, H.; Lukáč, P.; Riehemann, W. Two new high-damping magnesium composites. *Phys. Status Solidi* **2002**, *193*, 205–210. [CrossRef]

34. Kiehn, J.; Smola, B.; Vostrý, P.; Stulíková, I.; Kainer, K.U. Microstructure Changes in Isochronally Annealed Alumina Fibre Reinforced Mg-Ag-Nd-Zr Alloy. *Phys. Status Solidi* **1997**, *164*, 709–723. [CrossRef]

35. Magnesium Elektron® QE22 Casting Alloy. Available online: http://www.matweb.com/search/datasheettext.aspx?matguid=05ed4d435e5648058e27e84284aaceae (accessed on 15 January 2018).

36. Saffil—Specific Property Information. Available online: http://www.saffil.com/index/fibre_home/property_information.aspx (accessed on 15 January 2018).

37. Property of Silicon Carbide (SiC). Available online: http://www.qualitymaterial.net/news_list85.html (accessed on 15 January 2018).

38. Halpin, J.C. *Effect of Environmental Factors on Composite Materials*; Technical Report AFML-TR-67-423; US Air Force Material Laboratory: Dayton, OH, USA, 1969.

39. Halpin, J.C.; Kardos, J.L. Halpin-Tsai equations: A review. *Polym. Eng. Sci.* **1976**, *16*, 344–352. [CrossRef]

40. Mendelson, S. Dislocations Dissociations in hcp Metals. *J. Appl. Phys.* **1970**, *41*, 1893–1910. [CrossRef]

41. Máthis, K.; Čapek, J.; Zdražilová, Z.; Trojanová, Z. Investigation of tension-compression asymmetry of magnesium by use of the acoustic emission technique. *Mater. Sci. Eng. A* **2011**, *528*, 5904–5907. [CrossRef]

42. Arsenault, R.J.; Shi, N. Dislocation generation due to differences between the coefficients of thermal expansion. *Mater. Sci. Eng.* **1986**, *81*, 175–187. [CrossRef]

43. Ashby, M.F. The deformation of plastically non-homogeneous materials. *Philos. Mag.* **1970**, *21*, 399–424. [CrossRef]

44. Taya, M.; Arsenault, R.J. A comparison between a shear lag type model and an Eshelby type model in predicting the mechanical properties of a short fiber composite. *Scr. Metall.* **1987**, *21*, 349–354. [CrossRef]

45. Ryu, H.J.; Cha, S.I.; Hong, S.H. Generalized shear-lag model for load transfer in SiC/Al metal-matrix composites. *J. Mater. Res.* **2003**, *18*, 2851–2858. [CrossRef]

46. Zhang, C.Y.; Qiu, Y.P. Modified shear lag model for fibers and fillers with irregular cross-sectional shapes. *J. Adhes. Sci. Technol.* **2003**, *17*, 397–408. [CrossRef]

47. Karbhari, V.M.; Wilkins, D.J. An "engineering" modification to the shear-lag model as applied to whisker and particulate reinforced composites. *Scr. Metall. Mater.* **1991**, *25*, 707–712. [CrossRef]

48. Nardone, V.C.; Prewo, K.M. On the strength of discontinuous silicon carbide reinforced aluminum composites. *Scr. Metall.* **1986**, *20*, 43–48. [CrossRef]

49. Aikin, R.M., Jr.; Christodoulou, L. The role of equiaxed particles on the yield stress of composites. *Scr. Metall. Mater.* **1991**, *25*, 9–14. [CrossRef]

50. Arsenault, R.J.; Wang, L.; Feng, C.R. Strengthening of composites due to microstructural changes in the matrix. *Acta Metall. Mater.* **1991**, *39*, 47–57. [CrossRef]

51. Armstrong, R.W. Theory of the Tensile Ductile-Brittle Behaviour of Polycrystalline h.c.p. Materials with Application to Beryllium. *Acta Metall.* **1986**, *16*, 347–355. [CrossRef]

52. Scattergood, R.O.; Bacon, D.J. The Orowan mechanism in anisotropic crystals. *Philos. Mag. A* **1975**, *31*, 179–198. [CrossRef]

53. Yeh, Y.-H.; Nakashima, H.; Kurishita, H.; Goto, S.; Yoshinaga, H. Absence of Threshold Stress for High-Temperature Creep of Dispersion- and Solution-Hardened Al-3.1 at % Mg-1.3 vol % be Alloy. *Mater. Trans.* **1990**, *31*, 778–785. [CrossRef]

54. Delannay, F. Thermal Stresses and Thermal Expansion in MMCs. In *Comprehensive Composite Materials*; Clyne, T.W., Ed.; Elsevier: Amsterdam, The Netherlands, 2000; Volume 3, ISBN 978-0-08-042993-9.

55. Pekguleryuz, M.O.; Kainer, K.U.; Kaya, A. (Eds.) *Fundamentals of Magnesium Alloy Metallurgy*, 1st ed.; Woodhead Publishing: Cambridge, UK, 2013; ISBN 9780857097293.

56. Rudajevová, A.; Gärtnerová, V.; Jäger, A.; Lukáč, P. Influence of the thermal strain on the dilatation characteristics of Mg8Li and Mg10Li alloys. *Kov. Mater.* **2004**, *42*, 185–192.

57. Lide, D.R. (Ed.) *Handbook of Chemistry and Physics*, 73rd ed.; CRC Press: Boca Raton, FL, USA, 1992; ISBN 978-0849304736.

58. Ono, N.; Nakamura, K.; Miura, S. Influence of Grain Boundaries on Plastic Deformation in Pure Mg and AZ31 Mg Alloy Polycrystals. *Mater. Sci. Forum* **2003**, *419–422*, 195–200. [CrossRef]

59. Lavrentev, F.F.; Pokhil, Y.A. Relation of dislocation density in different slip systems to work-hardening parameters for magnesium crystals. *Mater. Sci. Eng.* **1975**, *18*, 261–270. [CrossRef]

60. Cáceres, C.H.; Lukáč, P. Strain hardening behaviour and the Taylor factor of pure magnesium. *Philos. Mag.* **2008**, *88*, 977–989. [CrossRef]

61. Farkas, G.; Trojanová, Z.; Száraz, Z.; Minárik, P.; Máthis, K. Effect of the fiber orientation on the deformation mechanisms of magnesium-alloy based fiber reinforced composite. *Mater. Sci. Eng. A* **2015**, *643*, 25–31. [CrossRef]

62. Trojanová, Z.; Farkas, G.; Máthis, K.; Lukáč, P. Hardening and softening in an AJ51 magnesium alloy reinforced with Saffil fibres. In *Magnesium Technology 2014*; Alderman, M., Manuel, M.V., Hort, N., Neelameggham, N.R., Eds.; TMS (The Mineral, Metals and Materials Society): Pittsburgh, PA, USA, 2014; pp. 435–440, ISBN 9781118888162.

63. Farkas, G.; Choe, H.; Máthis, K.; Száraz, Z.; Noh, Y.; Trojanová, Z.; Minárik, P. In situ investigation of deformation mechanisms in magnesium-based metal matrix composites. *Met. Mater. Int.* **2015**, *21*, 652–658. [CrossRef]

64. Lilholt, H. Additive strengthening. In *Deformation of Multi-Phase and Particle Containing Materials, Proceedings of the 4th Risø International Symposium on Metallurgy and Materials Science, Risø, Roskilde, Denmark, 5–9 September 1983*; Bilde-Sørensen, J.B., Hansen, N., Horsewell, A., Leffers, T., Lilholt, H., Eds.; Risø National Laboratory: Roskilde, Denmark, 1983; pp. 381–392.

65. Clyne, T.W.; Whithers, P.J. *An Introduction to Metal Matrix Composites*; Cambridge University Press: Cambridge, UK, 1993; ISBN 9780511623080.

66. Khan, F.; Panigradi, S.K. Age hardening, fracture behavior and mechanical properties of QE22 Mg alloy. *J. Magnes. Alloy.* **2015**, *3*, 210–217. [CrossRef]

![metals logo] *metals*

MDPI

Article

Nano-ZnO Particles' Effect in Improving the Mechanical Response of Mg-3Al-0.4Ce Alloy

Sravya Tekumalla [1], Najib Farhan [1], Tirumalai S. Srivatsan [2] and Manoj Gupta [1,*]

[1] Department of Mechanical Engineering, National University of Singapore, 9 Engineering Drive 1, Singapore 117576, Singapore; tvrlsravya@u.nus.edu (S.T.); a0096649@u.nus.edu (N.F.)
[2] Department of Mechanical Engineering, University of Akron, Akron, OH 44325, USA; tss1@uakron.edu
* Correspondence: mpegm@nus.edu.sg; Tel.: +65-6516-6358

Academic Editor: Hugo F. Lopez
Received: 13 October 2016; Accepted: 9 November 2016; Published: 11 November 2016

Abstract: Magnesium based nanocomposites, due to their excellent dimensional stability and mechanical integrity, have a lot of potential to replace the existing commercial Al alloys and steels used in aerospace and automotive applications. Mg-Al alloys are commercially used in the form of AZ (magnesium-aluminum-zinc) and AM (magnesium-aluminum-manganese) series in automobile components. However, the $Mg_{17}Al_{12}$ phase in Mg-Al alloys is a low melting phase which results in a poor creep and high temperature performance of the alloys. Rare earth additions modify the phase and hence improve the properties of the materials. In this paper, Ce and nano ZnO particles were added to Mg-Al alloys to attain a favorable effect on their properties. The developed materials exhibited promising properties in terms of thermal expansion coefficient (CTE), hardness, and tensile strength. Further, the ZnO addition refined the microstructure and helped in obtaining a uniform distribution, however without grain size refinement. The increased addition of ZnO and the improvement in the distribution led to an enhancement in the properties, rendering the materials suitable for a wide spectrum of engineering applications.

Keywords: Mg-3Al-0.4Ce alloy; nano ZnO particles; uniform distribution; strength

---

1. Introduction

The extraction and utilization of magnesium (Mg) in significant amounts started only in the 20th century, despite the fact that the discovery of Mg by Sir H. Davy dates back to 1808 [1]. Mg, with a density of 1.738 g/cm^3, is one of the lightest engineering and structural metals available on the earth, with its density being about two-thirds that of aluminium and one-fourth that of steel [2]. Therefore, there is an increasing interest in the usage of Mg in certain engineering industries, specifically in the automobile, electronic, and aviation industries. In the above mentioned sectors, for material selection, weight of the components is one of the most crucial criteria, thus giving light metals such as Mg a lot of potential for applications. Besides low density, Mg based materials also display other beneficial properties such as specific mechanical strength, excellent damping capacity, good castability, machinability, thermal stability, weldability, and resistance to electromagnetic radiation [2]. However, despite these advantages, Mg has limited ductility under tensile deformation loads, poor elastic modulus, and low resistance to creep at high temperatures. As Mg has a Hexagonal Close Packed (HCP) crystal structure, at ambient temperatures, deformation is, thus, confined to a few modes such as basal slip and twinning. Due to fracture (transgranular and intergranular) at basal planes or twin zones, pure Mg shows a tendency for low ductility [1]. There is also a tension-compression yield asymmetry, and a significant difference in tensile and compressive yield strengths exhibited by Mg, which narrows the opportunities of Mg in structural applications. To circumvent these limitations,

Mg is alloyed with other elements, such as aluminium, zinc, zirconium, rare earths, etc., to improve strength, ductility, corrosion resistance, and other properties [1].

Among different alloying constituents, Al addition seems to have favourable effects on magnesium as it leads to increases in strength and ductility. Despite several extensive research studies and characterizations on magnesium-aluminium alloys, the main focus was laid on specific materials, such as AZ31, AZ61, and AZ91 [1]. Other potential alloying elements with Mg would be rare earth (RE) metals, which have been known for many years [1]. Furthermore, RE metals also are promising for increasing strength and ductility [3] when alloyed to Mg. However, one limitation of RE metals is the high cost, limiting their usage. Furthermore, the solubility of RE metals in Mg is not very high, hence, increase in the content of solute RE concentration reduces strength [4]. The results of literature studies showed that the addition of 0.4% Ce [5] yields the best properties of tensile strength and ductility and no research has been done to study the effect of reinforcement on Mg-Al-Ce alloys. Furthermore, reinforcement with nano-particulates is also another possibility to improve the properties of Mg. Previously reported literature shows that the improvement in strength and ductility was highest with the addition of nano-scale ZnO [6]. Furthermore, it may also be noted that the ZnO nanoparticle addition helped in reducing the tension-compression yield asymmetry (tensile yield strength/compressive yield strength) [6].

Accordingly, the aim of the current work was to develop Mg-Al-RE nanocomposites through the integration of aluminium, nano zinc oxide, and a rare earth element (Ce) using the Disintegrated Melt Deposition (DMD) technique followed by hot extrusion. Microstructure and mechanical properties are studied and critically analysed. The developed materials are expected to be promising in many engineering applications that include high performance automobile components such wheels (by replacing currently used AM60 alloys), transmission case, bumper beam, etc.

2. Materials and Methods

2.1. Materials

Mg was obtained in the form of turnings from Acros Organics, NJ, USA, with 99% purity. Al was obtained in powder form with a size of 7–15 µm from Alfa Aesar, with 99.5% purity. Ce was incorporated by using Mg-30 wt. % Ce master alloy procured from Sunrelier Metal Co., Limited, Pudong, Shanghai, China. ZnO was obtained in the form of powder from Nanostructured and Amorphous Materials (Houston, TX, USA). The powder exhibited a particle size of less than 200 nm and an average particle size of 90 nm.

2.2. Primary Processing

Synthesis of the Mg-3Al-0.4Ce alloy using magnesium turnings, Al powder, and Mg-Ce master alloy was carried out using the disintegrated melt deposition (DMD) technique [7]. Synthesis of the alloy involved heating the Mg turnings with the addition of respective weights of Al powder and Mg-Ce master alloy in a graphite crucible to 750 °C using an electrical resistance furnace and an atmosphere of inert argon gas. The superheated melt was then stirred (at 465 rpm for 5 min) and bottom-poured into a steel mold (following disintegration by two jets of argon gas oriented normal to the melt stream). This technique displays the advantages of spray processing and conventional casting i.e., utilizing higher superheat temperatures and lower impinging gas jet velocities to produce the bulk composite material. The advantage of using this technique is to obtain a uniform distribution of reinforcements, fine equiaxed grains, and low porosity in the materials due to the rapid solidification of the atomized melt leading to an improvement in properties. An ingot of 40 mm diameter was obtained following the deposition stage. This technique was repeated for the other compositions, which contained ZnO nanoparticles.

2.3. Secondary Processing

2.3.1. Pre-Extrusion

The 40 mm diameter ingots, obtained from casting, were machined down to a diameter of 36 mm. Further, billets with lengths of approximately 45 mm were cut and machined.

2.3.2. Extrusion

The billets were, first, soaked at 400 °C for 1 h in a furnace, before extrusion. Extrusion was done at 350 °C on a 150 tonne hydraulic press. An extrusion ratio of 20.25:1 was used to produce rods of 8 mm diameter.

2.3.3. Post-Extrusion

After extrusion, the rods were machined to the required sample dimensions: dog bone shaped tensile test specimens with a gauge diameter of 5 mm. Samples were also prepared for other characterisation tests.

2.4. Characterisation

2.4.1. Density and Porosity

Density measurements were performed in accordance with the Archimedes' principle on eight randomly selected samples taken from the extruded rods. The experimental density measurements involved weighing samples in air and in distilled water. Theoretical densities of materials were calculated assuming the samples are fully dense to measure the volume percentage of porosity in the end materials. The porosity was calculated by using the theoretical and experimental densities. An electronic balance (A&D HM-202, Bradford, MA, USA) with an accuracy of ±0.0001 g was used for all the measurements.

2.4.2. Microstructure

Microstructural characterization studies were conducted on metallographically polished extruded samples to investigate morphological characteristics of grains, reinforcement distribution, and interfacial integrity between the matrix and reinforcement. Picric acid was used as the etching solution. The sample was then analysed using the Leica DM2500 M metallographic optical microscope equipped with a Leica E3 digital colour camera (procured from Baldock, UK). Scion image analysis software was used to determine the grain size and morphology of the samples. The presence and distribution of the secondary phases as well as reinforcement was studied using the JEOL JSM-5600LV Scanning Electron Microscope (SEM) (procured from Tokyo, Japan).

2.4.3. X-ray Diffraction

X-ray diffraction (XRD) analysis of the samples was performed using the automated Shimadzu LAB-X XRD-6000 X-ray diffractometer (procured from Tokyo, Japan). The specimens, approximately 5 mm in height, were exposed to CuKα radiation (λ = 1.54056 Å) at a scanning speed of 2°/min. The scanning range was 20°–80° for all samples. The graph intensity against 2θ (θ represents the Bragg angle) was obtained, illustrating peaks at different Bragg angles. The Bragg angles corresponding to different peaks were noted, and the values of interplanar spacing (d-spacing) obtained from the computerized output were compared with the standard values from the Powder Diffraction File (PDF).

2.4.4. Coefficient of Thermal Expansion

Coefficient of Thermal Expansion (CTE) values were measured using a thermo mechanical analysis instrument "INSEIS TMA PT 1000LT" (procured from Tokyo, Japan). A heating rate of 5 °C/min was maintained with an argon flow rate of 0.1 lpm. Using an alumina probe, the displacement of the test samples was measured as a function of temperature.

2.4.5. Microhardness

Microhardness measurements were obtained using the Shimadzu HMV automatic digital microhardness tester (procured from Tokyo, Japan) with a Vickers indenter (square-based pyramidal-shaped diamond indenter with a face angle of 136°). Cylindrical samples were prepared and polished. Tests were conducted in compliance with ASTM test method E384-11e1. A 25 g force load was applied with a dwell time of 15 s. Three samples were prepared per composition and 10 readings were taken per sample for an accurate and consistent result. Indentations and the microhardness readings were recorded in Vickers Hardness number (H_V).

2.4.6. Tensile Test

Tensile tests were performed using the MTS 810 machine (procured from Eden Prairie, MN, USA). Specimens of 5 mm diameter and 25 mm gauge length were prepared. Tests were conducted in compliance with ASTM test method E8M-13a. The crosshead speed was 2.54×10^{-4} m/min^{-1} (strain rate: 0.01 min^{-1}). Five samples were tested per composition to obtain more accurate and consistent results. Fractured specimens were studied under SEM to analyse the fracture features.

3. Results and Discussion

3.1. Density and Porosity

The results of addition of ZnO nano-particulates show an increase in the density, but only marginally. While porosity values do not manifest any particular trend, the values remained relatively insignificant. The results for density and porosity can be found in Table 1. Within the synthesised materials, a positive trend was noticed with the addition of ZnO nano-particulates in terms of density. This is expected to be due to the high density of ZnO (5.61 g/cc) as compared to that of Mg. Furthermore, the addition of increasing amounts of ZnO nano-particulates increased the porosity of the alloy. The same trend was noticed in other studies on nano-composites [7]. The highest porosity was recorded for the Mg-3Al-0.4Ce-2.5ZnO nanocomposite. This is due to the possible clustering effects caused by the increased addition of nano-reinforcements (a limitation of the casting technique). Nevertheless, the absolute porosity values are relatively low, with the highest recorded porosity value being about 1%. Thus, obtaining nanocomposites with <1% porosity was achieved and is accredited to the secondary processing (extrusion and the extrusion ratio).

Table 1. Density, Porosity, and Grain size results.

Composition	Experimental Density (g/cm³)	Theoretical Density (g/cm³)	Porosity (%)	Grain Size (μm)
Mg-3Al-0.4Ce	1.759 ± 0.013	1.762	0.201	6.5 ± 1.5
Mg-3Al-0.4Ce-1.5ZnO	1.779 ± 0.012 (↑ 1.13%) [1]	1.781	0.120	6.4 ± 1.5
Mg-3Al-0.4Ce-2ZnO	1.783 ± 0.013 (↑ 1.36%) [1]	1.787	0.235	6.1 ± 1.5
Mg-3Al-0.4Ce-2.5ZnO	1.774 ± 0.028 (↑ 0.85%) [1]	1.794	1.080	6.0 ± 1.6

[1] (↑ x%) corresponds to an increase in the density by x% as compared to the monolithic alloy.

3.2. Microstructure/Phase Analysis

3.2.1. Grain Size Analysis

Nearly similar grain morphology was noticed when the samples were studied under the optical microscope. The results and images of the grain size and morphology are shown in Table 1 and

Figure 1a–d. The addition of ZnO did not affect the grain size of the monolithic alloy, considering the standard deviation.

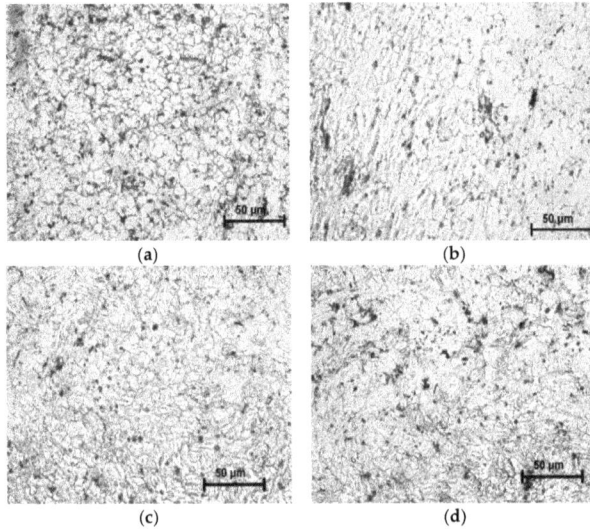

(a)

(b)

(c)

(d)

Figure 1. Images observed under an optical microscope for: (a) Mg-3Al-0.4Ce; (b) Mg-3Al-0.4Ce-1.5ZnO; (c) Mg-3Al-0.4Ce-2ZnO and (d) Mg-3Al-0.4Ce-2.5ZnO.

3.2.2. Microstructure Analysis

Figure 2a–d shows the secondary phase distribution in the alloy and the nanocomposites. The images indicate the reduced size of the secondary phases and the improved phase and reinforcement distribution with an increase in the amount of ZnO particles.

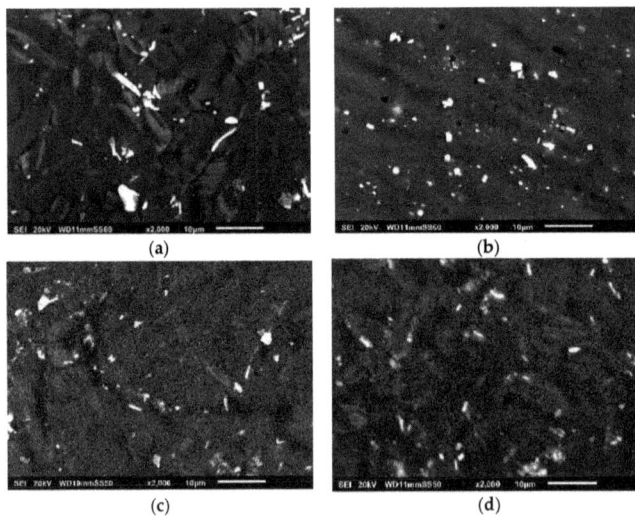

(a)

(b)

(c)

(d)

Figure 2. Images observed under SEM for: (a) Mg-3Al-0.4Ce; (b) Mg-3Al-0.4Ce-1.5ZnO; (c) Mg-3Al-0.4Ce-2ZnO and (d) Mg-3Al-0.4Ce-2.5ZnO.

3.2.3. Phase Analysis

The XRD patterns confirmed the existence of secondary phases as shown in Figure 3. The peaks seen in the obtained patterns are correlated with the standard diffraction patterns obtained from the Powder Diffraction File data (PDF). The following phases were confirmed to be present in all the synthesized compositions, in varying intensities: α-Mg, $Mg_{17}Al_{12}$, $Al_{11}Ce_3$, $(Al,Mg)_2Ce$.

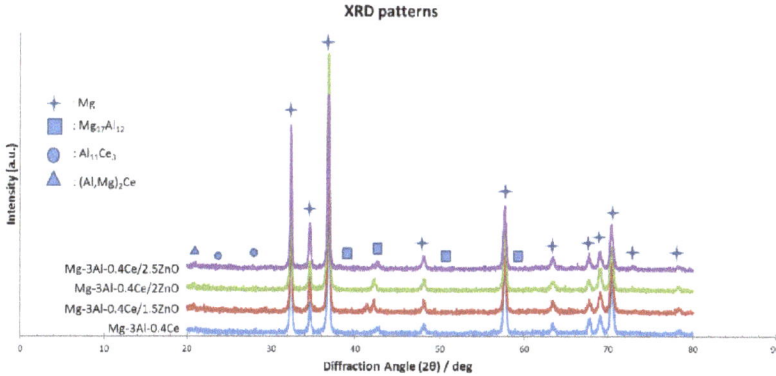

Figure 3. X-ray diffraction patterns.

No peaks corresponding to the Mg-Ce intermetallic phases were observed from the XRD patterns. This is ascribed to be due to the greater affinity of Ce towards Al [8], therefore, forming a $Al_{11}Ce_3$ phase instead of a Mg-Ce phase. From the SEM images in Figure 2, for alloy and nanocomposites, blocky shaped structures and rod-like structures were observed in the matrix. Spherical structures were also observed only in the case of nanocomposites.

From the energy dispersive spectroscopy (EDS) results in Figure 4, the following inferences are made: (i) The matrix correspond to Mg-Al solid solution; (ii) blocky shaped structures correspond to the $Mg_{17}Al_{12}$ phase (iii) rod-like structures are a combination of a phase containing Mg, Al, and Ce. While this could be either the $Al_{11}Ce_3$ phase or $(Al,Mg)_2Ce$ phase, previous research has shown that $Al_{11}Ce_3$ phase has a needle-like structure [9] and is the more dominant phase compared to $(Al,Mg)_2Ce$ [10]. Furthermore, XRD results also suggests that more peaks were observed for $Al_{11}Ce_3$ phase. Thus, this rod-like structure could be the $Al_{11}Ce_3$ phase. Further; (iv) the spherical structures are noted to be ZnO nano-particles. Also, these structures were detected only on the materials containing the ZnO particles. Furthermore, no evident peaks of ZnO were observed in the XRD patterns. This is due to the inability of XRD to detect nano-length scale particles in low volume fractions.

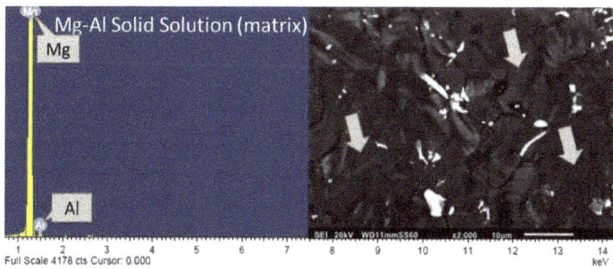

(a)

Figure 4. *Cont.*

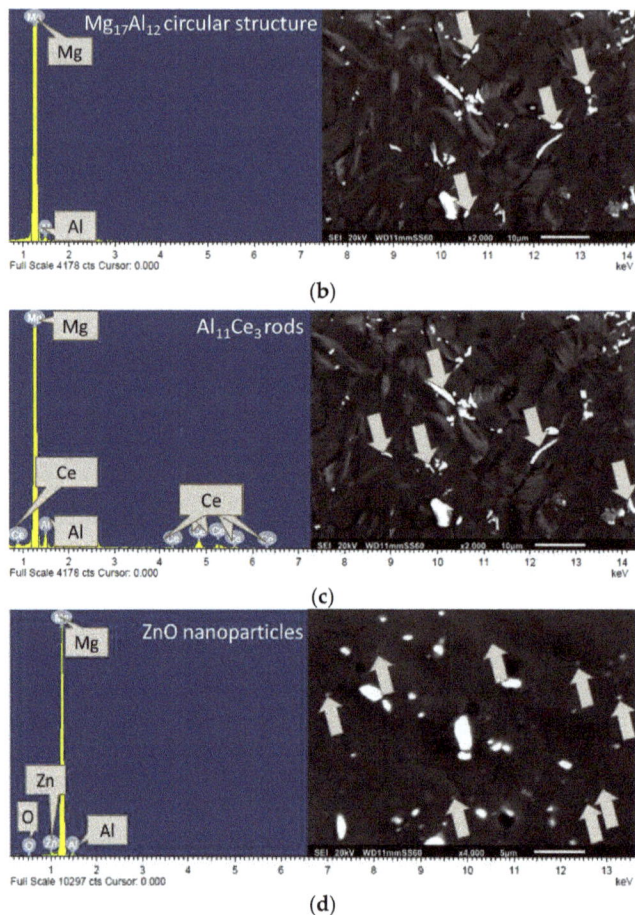

Figure 4. EDS analysis of the phases (**a**) Mg-Al matrix; (**b**) circular $Mg_{17}Al_{12}$ phase; (**c**) rod shaped $Al_{11}Ce_3$ phase and (**d**) ZnO nanoparticles (in a high magnification image of Mg-3Al-0.4Ce-1.5ZnO showing nano-particle distribution).

Thus, from the microstructural characterization of the Mg-3Al-0.4Ce alloy and its ZnO reinforced nanocomposites, the following points are to be factored in: (a) no drastic change in the grain morphology or grain size among the alloy and nanocomposites and (b) a more uniform distribution of the hard secondary phases and ZnO nano-particulates in the nanocomposites.

3.3. Coefficient of Thermal Expansion (CTE)

Thermal expansion refers to the tendency to change shape in response to a change in temperature, occurring through heat transfer. The lesser the CTE value, the lesser the tendency for the material to change shape when the temperature is increased. Thus, a material with lower CTE values indicates a thermally stable material. The addition of Al and Ce corresponded to a drop in the CTE value of Mg. Comparing Mg-Al-0.4Ce with Mg-3Al-0.4Ce-1.5ZnO, a further decrease in CTE value was noticed. However, when the wt % of ZnO nano-particles increased, the CTE values increased. Nevertheless, the addition of ZnO particles resulted in overall lower CTE values and thereby, more thermally stable materials in contrast with the monolithic alloy. Results of the CTE are shown Table 2. The obtained results are due to the lower CTE value of ZnO as compared to that of Mg.

Table 2. CTE results.

Composition	CTE (μK^{-1})
Mg-3Al-0.4Ce	25.52 ± 0.8
Mg-3Al-0.4Ce-1.5ZnO	22.03 ± 1.0
Mg-3Al-0.4Ce-2ZnO	24.16 ± 2.0
Mg-3Al-0.4Ce-2.5ZnO	24.36 ± 0.6

3.4. Mechanical Properties

3.4.1. Microhardness

The microhardness of Mg-3Al-0.4Ce alloy is seen to be much higher as compared to that of pure Mg [5] and commercially available AZ31B [11]. This is due to the addition of a rare earth element. Nayyeri et al. [12] reported an increase in hardness when RE metals were added to Mg-Al alloys, due to the Al-RE intermetallic phase formation. Further, the increasing addition of ZnO nanopartices progressively increased the microhardness of the nanocomposites (Table 3). This is due to the existence of harder ZnO particles that can resist deformation under indentation loads [13].

Table 3. Tensile test results.

Composition	Microhardness (Hv)	0.2% Offset Tensile Yield Strength (MPa)	Ultimate Tensile Strength (MPa)	Fracture Strain (%)
Mg-3Al-0.4Ce	105 ± 4	144 ± 8	232 ± 11	17 ± 1
Mg-3Al-0.4Ce-1.5ZnO	113 ± 3 ($\uparrow 7.6\%$) [1]	157 ± 8 ($\uparrow 9\%$) [1]	263 ± 8 ($\uparrow 13\%$) [1]	17 ± 5
Mg-3Al-0.4Ce-2ZnO	144 ± 4 ($\uparrow 37.1\%$) [1]	173 ± 11 ($\uparrow 20\%$) [1]	288 ± 11 ($\uparrow 24\%$) [1]	18 ± 2
Mg-3Al-0.4Ce-2.5ZnO	161 ± 5 ($\uparrow 53.3\%$) [1]	180 ± 7 ($\uparrow 25\%$) [1]	294 ± 9 ($\uparrow 27\%$) [1]	15 ± 2
Mg-3Al-0.2Ce [8]	-	120	235	-
Mg-3Al-0.5Ce [8]	-	125	230	-

[1] ($\uparrow x\%$) corresponds to an increase in the respective property by $x\%$ as compared to the monolithic alloy.

3.4.2. Tensile Properties

A trend similar to that of the microhardness results was observed for the results of the tensile tests. An increase in the 0.2% offset yield strength and ultimate tensile strength was observed with the progressive addition of ZnO to the Mg-3Al-0.4Ce alloy. The addition of ZnO displayed no significant effect on the ductility. The overall tensile response assessed in terms of the energy absorption capability prior to failure remained superior in the case of the nanocomposites. The results are found in Table 3 and Figure 5. When compared to the previously reported results on Mg-3Al-0.2Ce and Mg-3Al-0.5Ce alloys [8] (similar processing conditions), it is observed that the materials synthesized in this study possess superior strengths. Within the nanocomposites, the fracture strain and energy absorption values increased from Mg-3Al-0.4Ce to Mg-3Al-0.4Ce-2ZnO and decreased for Mg-3Al-0.4Ce-2.5ZnO. This could be due to a clustering effect of ZnO nanoparticles [14], as well as the increased porosity as seen in Table 1 for the Mg-3Al-0.4Ce-2.5ZnO nanocomposite. Furthermore, an interesting observation from the curves is that all the materials fractured at (or before realizing) their ultimate tensile strengths, i.e., they had not undergone any deformation after reaching the peak values. This shows that the ultimate failure has occurred at (or before) the peak stress. This occurs when the plastic strain at any point in the material reaches the maximum value. The common stress concentration sites in all the materials, in this study, are the rod shaped $Al_{11}Ce_3$ structures. These rods seemed to act as the stress concentration sites, i.e., sites for continued local stresses. When the local stress at these sites reach the threshold stress, the materials begin to fail at these sites. Thus, the rod shaped $Al_{11}Ce_3$ is responsible for the materials exhibiting this phenomenon.

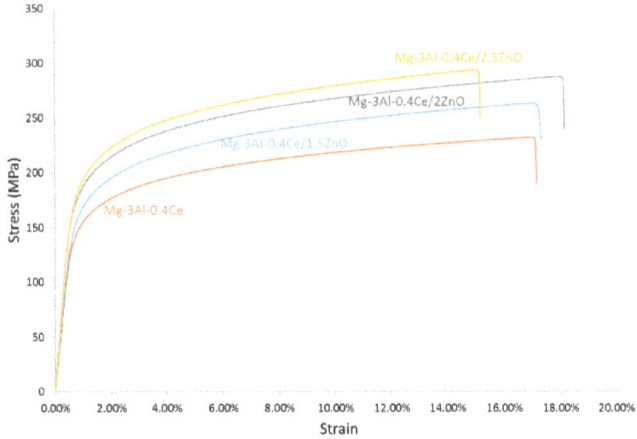

Figure 5. Engineering tensile stress strain curves.

3.4.3. Fractography

The microscopic fracture features of the samples fractured in the tensile test are shown in Figure 6. These images revealed pronounced ductile features due to plastic deformation. Typically, magnesium materials exhibit cleavage features which are suggestive of their inability to deform plastically (due to the HCP crystal structure). However, in the current study, the developed Mg-3Al-0.4Ce alloy and its ZnO reinforced nanocomposites showed noticeable ductile features representing the occurrence of pronounced plastic deformation, thus validating the tensile test results.

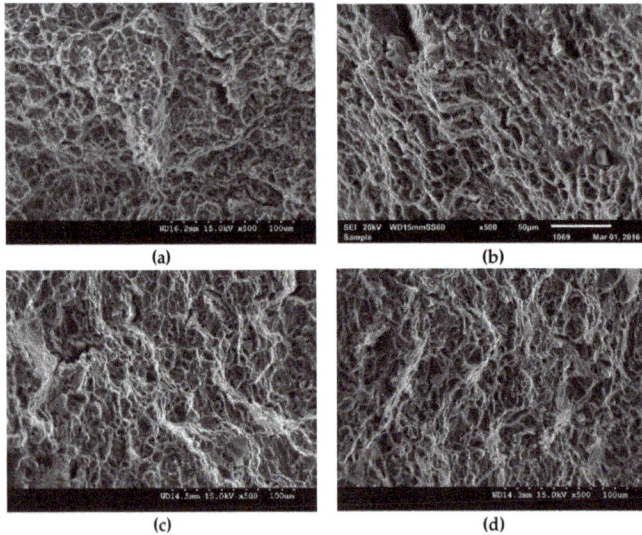

Figure 6. Fracture surfaces of the (**a**) Mg-3Al-0.4Ce alloy; (**b**) Mg-3Al-0.4Ce-1.5ZnO; (**c**) Mg-3Al-0.4Ce-2ZnO and (**d**) Mg-3Al-0.4Ce-2.5ZnO nanocomposites.

4. Conclusions

Synthesis of the Mg-3Al-0.4Ce alloy and its ZnO nano-particulate reinforced nanocomposites was carried out using the DMD technique followed by hot extrusion. The microstructural and mechanical properties were studied. With the current findings, the following can be concluded:

1. The processing methodology used in this study is capable of synthesizing the Mg-3Al-0.4Ce alloy and its nanocomposites with porosities restricted to ~1%.
2. The addition of ZnO nanoparticles reduced the CTE of the Mg-3Al-0.4Ce alloy, resulting in more dimensionally stable nanocomposites.
3. The addition of ZnO nano-particulates corresponded with the increase in the microhardness of the Mg-3Al-0.4Ce alloy. A maximum microhardness was realized in the case of the Mg-3Al-0.4Ce/2.5ZnO nanocomposite, with a value of 161 Hv.
4. The addition of an increasing amount of ZnO nanoparticles led to an increase in the 0.2% offset yield strength and ultimate tensile strength, while strain to fracture remained unaffected. A maximum yield strength of 180 MPa and ultimate tensile strength of 294 MPa was achieved for the Mg-3Al-0.4Ce/2.5ZnO nanocomposite.
5. These superior mechanical properties, achieved with the addition of the ZnO nanoparticles, are attributed to the uniform distribution of the secondary phases and the presence of ZnO reinforcement.

Acknowledgments: This work was supported by the Singapore Ministry of Education Academic Research Funding (grant number WBS# R-265-000-498-112).

Author Contributions: Sravya Tekumalla and Manoj Gupta conceived and designed the experiments; Sravya Tekumalla and Najib Farhan performed the experiments, Sravya Tekumalla, Manoj Gupta, and Tirumalai S. Srivatsan analyzed the data, and all the authors wrote the paper.

Conflicts of Interest: The authors declare no conflict of interest.

References

1. King, J.F. Magnesium: Commodity or exotic? *Mater. Sci. Technol.* **2007**, *23*, 1–14. [CrossRef]
2. Alam, M.E.; Han, S.; Nguyen, Q.B.; Hamouda, A.M.S.; Gupta, M. Development of new magnesium based alloys and their nanocomposites. *J. Alloy. Compd.* **2011**, *509*, 8522–8529. [CrossRef]
3. Tekumalla, S.; Seetharaman, S.; Almajid, A.; Gupta, M. Mechanical Properties of Magnesium-Rare Earth Alloy Systems: A Review. *Metals* **2015**, *5*, 1. [CrossRef]
4. Rokhlin, L.L. Structure and properties of alloys of the Mg-REM system. *Met. Sci. Heat Treat.* **2006**, *48*, 487–490. [CrossRef]
5. Tekumalla, S.; Seetharaman, S.; Bau, N.Q.; Wong, W.L.E.; Goh, C.S.; Shabadi, R.; Gupta, M. Influence of Cerium on the Deformation and Corrosion of Magnesium. *J. Eng. Mater. Technol.* **2016**, *138*, 031011. [CrossRef]
6. Sankaranarayanan, S.; Nayak, U.P.; Sabat, R.; Suwas, S.; Almajid, A.; Gupta, M. Nano-ZnO particle addition to monolithic magnesium for enhanced tensile and compressive response. *J. Alloy. Compd.* **2014**, *615*, 211–219. [CrossRef]
7. Gupta, M.; Wong, W.L.E. Magnesium-based nanocomposites: Lightweight materials of the future. *Mater. Charact.* **2015**, *105*, 30–46. [CrossRef]
8. Luo, A.A.; Wu, W.; Mishra, R.K.; Jin, L.; Sachdev, A.K.; Ding, W. Microstructure and Mechanical Properties of Extruded Magnesium-Aluminum-Cerium Alloy Tubes. *Metall. Mater. Trans. A* **2010**, *41*, 2662–2674. [CrossRef]
9. Song, Y.L.; Liu, Y.H.; Wang, S.H.; Yu, S.R.; Zhu, X.Y. Effect of cerium addition on microstructure and corrosion resistance of die cast AZ91 magnesium alloy. *Mater. Corros.* **2007**, *58*, 189–192. [CrossRef]
10. Zhang, J.; Leng, Z.; Zhang, M.; Meng, J.; Wu, R. Effect of Ce on microstructure, mechanical properties and corrosion behavior of high-pressure die-cast Mg-4Al-based alloy. *J. Alloy. Compd.* **2011**, *509*, 1069–1078. [CrossRef]

11. Nguyen, Q.B.; Gupta, M. Increasing significantly the failure strain and work of fracture of solidification processed AZ31B using nano-Al$_2$O$_3$ particulates. *J. Alloy. Compd.* **2008**, *459*, 244–250. [CrossRef]
12. Nayyeri, M.J.; Ganjkhanlou, Y.; Kolahi, A.; Jamili, A.M. Effect of Ca and Rare Earth Additions on the Texture, Microhardness, Microstructure and Structural Properties of As-Cast Mg–4Al–2Sn Alloys. *Trans. Indian Inst. Met.* **2014**, *67*, 469–475. [CrossRef]
13. Yahya, N.; Puspitasari, P.; Latiff, N.R.A. *Characterization and Development of Biosystems and Biomaterials*; Springer: Berlin/Heidelberg, Germany, 2013; pp. 9–32.
14. Zhaoxu, D.; Daining, F.; Kah, S.A. On the effect of nano-particle clustering on toughening of nano-composite ceramics. *Acta Mech. Sin.* **2002**, *18*, 146–158. [CrossRef]

Article

Magnesium-β-Tricalcium Phosphate Composites as a Potential Orthopedic Implant: A Mechanical/Damping/Immersion Perspective

Gururaj Parande [1], Vyasaraj Manakari [1], Harshit Gupta [2] and Manoj Gupta [1,*]

[1]   Department of Mechanical Engineering, National University of Singapore, 9 Engineering Drive 1, Singapore 117576, Singapore; gururaj.parande@u.nus.edu (G.P.); mbvyasaraj@u.nus.edu (V.M.)

[2]   Department of Mechanical Engineering, Indian Institute of Technology (BHU), Varanasi 221005, India; harshit.gupta.mec14@itbhu.ac.in

*   Correspondence: mpegm@nus.edu.sg; Tel.: +65-651-635-8

Received: 6 April 2018; Accepted: 8 May 2018; Published: 11 May 2018

Abstract: The design and development of novel magnesium-based materials with suitable alloying elements and bio-ceramic reinforcements can act as a possible solution to the ever-increasing demand of high performance bioresorbable orthopedic implant. In the current study, Mg-β-tricalcium phosphate composites are synthesized using the hybrid powder metallurgy technique, followed by hot extrusion. The influence of addition of (0.5, 1, and 1.5) vol % β-tricalcium phosphate on the mechanical, damping, and immersion characteristics of pure magnesium are studied. The addition of β-tricalcium phosphate enhanced the yield strength, ultimate compressive strength, and compressive fracture strain of pure magnesium by about ~34%, ~53%, and ~22%, respectively. Also, Mg 1.5 vol % β-tricalcium phosphate composite exhibited a ~113% enhancement in the damping characteristics when compared to pure magnesium. A superior ~70% reduction in the grain size was observed by the addition of 1.5 vol % β-tricalcium phosphate particles to pure Mg. The response of Mg-β-tricalcium phosphate composites is studied under the influence of chloride environment using Hanks' balanced salt solution. The dynamic passivation was realized faster for the composite samples as compared to pure Mg, which resulted in decreased corrosion rates with the addition of β-tricalcium phosphate particles to pure Mg.

Keywords: magnesium; tricalcium phosphate; compression; damping; corrosion; powder metallurgy

---

1. Introduction

Bone is a natural composite that is made up of hydroxyapatite and type I collagen [1]. The collagen prevents brittle failure of the bones and makes it elastic whereas hydroxyapatite provides the necessary mechanical strength. In the past few years, noteworthy headways have been made in the research community to keep introducing novel materials that are targeting various biomedical applications to the market. Ceramic and polymer-based biomaterials have superior biocompatibility and bioactivity, thereby finding applications in tissue regeneration and drug delivery [2]. However, insufficient strength and non-biodegradability hinder their application in fixation devices like pins, screws, and plates targeting load-bearing orthopaedic applications. Metal-based biomaterials, like 316 L stainless steel, Ti6Al4V alloy, and Co-Cr biomedical alloy, amongst others, have been long used as commercial orthopaedic implants. Although these implant materials perform the suitable function of assisting in bone remodelling and resorption, a mismatch in elastic modulus between these materials and the bone induces several stress-shielding effects on the bone/implant interface, inducing severe pains to the patient [1]. Either these implants materials are in the human body throughout leading to apparent toxicity at later stages or may require a secondary surgery to remove the implant adding to the long facing trauma of the patient. In order to solve this issue, the need is to design and develop a biodegradable metal-based orthopaedic implant with superior strengths and optimum degradation rates to serve the purpose of bone remodelling and disintegration into the human body without causing any ill-effects to the patient. Mg, being the lightest structural metal, is also biodegradable, non-toxic, and is abundantly available, and hence it can be a viable solution to be used as orthopedic implant. However, low room temperature ductility and inferior corrosion resistance in biological environments are the reasons why it is not already commercially used in the biomedical sector [3]. However, low room temperature ductility and reduced corrosion resistance in biological environments are the reasons why it is not already commercially used in the biomedical sectors [4–6]. Degradation occurs faster than the bone remodelling process, and hence there is minimum retention of mechanical integrity in the bone chips. There are several ways to retard the degradation rate of Mg, namely alloying, composite technology, and in form of coatings. Although, alloying and coating technology have been effective in improving the biocompatibility and corrosion response of Mg, additional processing steps add to the cost of the eventual product, thus making it less feasible in mass production of implants. Hence, the development of biocompatible materials using cost effective processing techniques is crucial. Therefore, the addition of biocompatible reinforcements using the composite technology to control the degradation rates without adversely affecting the strength properties is the key. Further, Mg has prime importance in the metabolism process, being the second most abundant cation in the human body and it helps in the formation of antibodies maintaining the required wall tension in blood vessels and aids in muscle contraction regulation [7]. Any sort of deficiency in Mg may lead to the change in bone structure, the reduction in osteoblast/osteoclast activity, and may also result in cardiovascular issues, leading to death [8,9].

Calcium (Ca) and Phosphorous (P) are the main minerals in the human bone. Hence, using Ca-P as a reinforcement or coating may be a viable option for enhanced corrosion protection of magnesium-based materials in order to promote bone growth, absorption, and osseointegration [10]. Incorporation of bio-ceramics, like hydroxyapatite (HA) and tricalcium phosphates (TCP), both forms of calcium phosphates, into the Mg matrix seems very promising in the field of bone regeneration. Not only do they exhibit superior biocompatibility and no visible signs of systemic and local toxicity, but also their crystal structure and chemical composition are close to the mineral parts of bone, which may help in tailoring the desired biological properties. Several studies, like the development of Mg-HA composites, AZ91-HA composites, and ZK60-calcium polyphosphate (CPP) composites have been carried out by means of the powder metallurgy technique in the recent past [11–13]. The HA composites reported a decrease in the strength properties although biocompatibility and cell viability of the HA composites were good [14]. The CPP containing composites reported faster pH stabilization, and hence an increased corrosion resistance, however CPP containing composites responded poorly in the tensile mode as compared to the ZK60 base alloy. WE43/HA composites exhibited superior corrosion resistance when immersed in 1% NaCl solution with an appreciable compromise in the compressive strength properties [15]. Hence, superior sintering capability is required for better densification, leading to better strength properties in the Mg bio-ceramic composites. Although, HA possesses superior bioactivity, the better wettability of β-TCP with Mg matrix, and higher dissolution in human physiological environments makes it an ideal candidate for Mg-composite technology targeting orthopaedic implants [16]. β-tricalcium phosphate (β-TCP) is an excellent bio-ceramic with superior biocompatibility, chemical stability, and osteointegration behaviour in the body environment, with the resorption rate being better than HA ceramics, thereby finding numerous applications in skeletal and dental prosthetics [14]. Sheng Ying He and coworkers studied the influence of β-TCP nanoparticles on the strength and corrosion properties of the Mg-3Zn-0.8Zr alloy. Both tensile strength and corrosion resistance of the alloy was improved with the addition of β-TCP particles [17]. Liu et al. synthesized Mg-2Zn-0.5Ca-1 β-TCP composite using equal channel extrusion processing and noted that the addition of β-TCP improves the mechanical and corrosion properties of Mg-2Zn-0.5Ca alloy significantly [18]. β-TCP was also investigated as a coating material for surface modification of Mg alloys, with positive results [19]. However, its influence as reinforcement on the mechanical, damping, and immersion response of Mg matrix that is synthesized using solid-state blend-press-sinter powder metallurgy technique is not available in the public domain, which is the novelty of the current study.

2. Materials and Methods

2.1. Materials and Processing

Magnesium powder of purity ≥98.5% (assay > 98.5%, Fe < 500 ppm, substances that are insoluble in HCl < 0.005) with a size range of 60–300 μm, supplied by Merck, Germany was used as the base material. β-tricalcium phosphate (reinforcement) with a size range of 0.7–4.6 μm and a purity of ≥96% (assay > 96%, Cl < 0.05%, Fe < 0.1%, K < 0.005%, Na < 0.005%, Ni < 0.005%, Pb < 0.005%, Zn < 0.005%) was supplied by Sigma-Aldrich, St. Louis, MO, USA. Pure Mg and Mg 2 vol % β-tricalcium phosphate (β-TCP) composite was synthesized using the powder metallurgy technique, incorporating hybrid microwave sintering [20]. The as-sintered billets were homogenized at 400 °C for 1 h and were then hot extruded at 350 °C to obtain cylindrical rods of 8 mm diameter at an extrusion ratio of 20.25:1. Samples that were cut from the rods were then characterized for physical and mechanical properties.

2.2. Material Characterization

2.2.1. Density Measurements

Density measurements were performed on both monolithic and composite samples using the Archimedes principle. Four samples were cut from different parts of the extruded rods and were tested ten times for conformance. The samples were weighed separately in air and water using an A&D ER-182A electronic balance (Bradford, MA, USA) with an accuracy of 10^{-4} g. The theoretical density was calculated using the densities and weight percentages of the constituents by means of the rule of mixtures. From the experimental and theoretical densities, the porosity values of the samples were determined.

2.2.2. Microstructural Characterization

Cylindrical samples were finely polished and etched according to the conventional techniques of metallography to obtain a clear distinction between the grain boundaries with the help of a LEICA-DM 2500M metallographic light microscope (Singapore). Four representative micrographs were analyzed for each composition in order to obtain accurate grain sizes. The OLYMPUS metallographic microscope (Singapore) and JEOL JSM-5800 LV Scanning Electron Microscope (SEM, Kyoto, Japan) was used for the microstructural characterization studies.

X-ray diffraction studies were carried out on extruded samples in the direction along the axis of extrusion. The studies were performed using an automated Shimadzu LAB-XRD-6000 (Cu Kα; λ = 1.54056 Å, Kyoto, Japan) using a scan speed of 2°/min. The studies were conducted to identify the possible formation of any impurities/secondary phases. The XRD analysis was also conducted on the post-corroded samples to identify the corrosion products that were formed.

2.2.3. Damping and Elastic Modulus

Damping characteristics and elastic modulus of the cylindrical samples (7 mm diameter and 60 mm length) were analyzed using the resonant frequency and the damping analyzer (RFDA) equipment from IMCE, Genk, Belgium. Recordings of the vibration signal were obtained in terms of amplitude vs. time. Damping capacity, loss rate, and elastic modulus values for both pure Mg and Mg (0.5, 1.0, and 1.5) vol % β-TCP composite sample were recorded.

2.2.4. Mechanical Properties

Microhardness tests were performed on the composite samples using Vickers microhardness tester Matsuzawa MXT 50 (Kyoto, Japan) with an indenter phase angle ~136°; in conformance with ASTM standard E384-11-1 [21]. Fifteen readings were taken to arrive at an average representative value.

Compression testing in the quasi-static mode was performed on cylindrical samples having 8 mm diameter and 8 mm length, utilizing a fully automated servo-hydraulic mechanical testing machine (Model-MTS 810; in conformance with ASTM test method E9-09, Eden Prairie, MN, USA) at a strain rate 8.33×10^{-5} s^{-1} [22]. Four specimens each for both of the compositions were tested to ensure reproducibility. Fracture surface analysis of the samples failed under compression was done using SEM.

2.2.5. Immersion Studies

Cylindrical samples of (5 mm diameter and 5 mm length) were immersed for 96 h in Hanks balanced salt solution (HBSS) procured from Lonza Chemicals Pte Ltd. Singapore. The setup was immersed in a water bath that was maintained at 37 °C to simulate the temperature of the human body. The sample dimensions of 5 mm diameter and 5 mm length was used. Solution to sample ratio was maintained at 20 mL:1 cm^2. The solution was changed every 24 h. Weight loss and pH measurements were measured after every 24 h. A solution containing 20 g CrO$_3$ and 1.9 g AgNO$_3$ dissolved in 100 mL of de-ionized water was used for removing the corrosion products. The samples post-corrosion were observed under the SEM in order to gain further information about the nature of corrosion products that were formed.

3. Results and Discussion

3.1. Density and Porosity

Table 1 shows the density and porosity levels of pure Mg and Mg-β-TCP composites. The experimental density of Pure Mg slightly increased with the incorporation of β-TCP, and Mg-1.5 TCP composite exhibited an experimental density value of 1.7449 g·cm^{-3}. The slight increase (0.2%) in the density can be attributed to the fact that there is a density difference between the matrix (1.74 g·cm^{-3}) and reinforcement (3.14 g·cm^{-3}). Porosity levels marginally increased with the addition of the β-TCP and the highest porosity value of ~0.28% was observed for the Mg-1.5 TCP composite. The observed porosity is less than 1% porosity, which is an advantage when compared to conventional sintering processes that can achieve only up to ~95% densification [23]. Microstructural examination of the extruded rod revealed the absence of blowholes, defects and a superior surface finish indicated the suitability of the powder metallurgy technique to generate near dense composites [3].

Table 1. Density, porosity, grain size and microhardness measurements of pure Mg and Mg-β-tricalcium phosphate (β-TCP) composite.

Material	Theoretical Density (g cm^{-3})	Experimental Density (g cm^{-3})	Porosity (%)	Grain Size (μm)	Hardness (Hv)
Pure Mg	1.74	1.7363 ± 0.002	0.21	34 ± 2	46 ± 3
Mg-0.5 TCP	1.7412	1.7371 ± 0.0147	0.23	18 ± 2 (↓47%)	52 ± 2 (↑13.04%)
Mg-1.0 TCP	1.7424	1.7381 ± 0.0067	0.24	13 ± 1 (↓61%)	54 ± 3 (↑17.39%)
Mg-1.5 TCP	1.7449	1.7387 ± 0.0048	0.28	10 ± 1 (↓70%)	54 ± 1 (↑17.39%)

3.2. Microstructural Characterisation

Table 1 shows the average grain size values of pure Mg and Mg (0.5, 1.0, and 1.5) vol % β-TCP composites. The grain size of pure Mg in as-extruded form was observed to be ~34 μm. The addition of 0.5 vol % β-TCP particles resulted in superior grain refinement of up to ~18 μm, which is ~47% finer than that of pure Mg. Increased addition of 1.0 and 1.5 vol % β-TCP particles resulted in a further refinement in the grain size of up to ~13 and ~10 μm. Near equiaxed grain morphology was observed with the addition of β-TCP particles, as observed in Figure 1. This superior grain refinement can be attributed to mainly two aspects namely (a) Particle stimulated nucleation phenomenon that promotes the nucleation of grains, hence restricting the grain growth; (b) the ability of the β-TCP particles to pin the grain boundaries resulting in finer microstructure [24]. As the size of the reinforcement is predominantly in micron length scale, simulated dynamic recrystallization phenomenon can be expected during the extrusion process [25]. Distribution of the β-TCP particle in the Mg matrix is shown in Figure 2. The efficient extrusion process has managed to break down the large β-TCP particles and clusters, leading to a reasonable distribution pattern. Hence, hot extrusion can be considered as a suitable secondary process to promote the near uniform distribution of reinforcement and simultaneously reducing the spatial heterogeneity of the mechanical properties of the Mg-based composites [16]. The superior grain refinement also aids in the strengthening of the composites by means of Hall-Petch mechanism activation.

High wettability of β-TCP particles with the Mg matrix leads to the easy densification under sintering, and hence showing a superior interfacial integrity between the particle and the matrix. The near-uniform distribution of β-TCP throughout the Mg matrix can also be attributed to the suitable primary and secondary processing parameters that are optimized for the primary processing of Mg-β-TCP composites. Energy Dispersive Spectroscopy (EDS) analysis of Mg-1.0 TCP composite is also shown in Figure 2. The EDS spectra are studied at the reinforcement (A) and matrix (B) location. The analysis of the matrix reveals predominantly Mg phases with traces of O due to surface oxidation during the processing of the material. The β-TCP particles have settled at the grain boundaries of the composite, hence confirming the grain boundary pinning mechanism that is responsible for grain refinement as quantitatively confirmed by the predominant Ca, O, P peaks in the spectrum. The EDS also confirms the absence of any sign of impurities or secondary phases in the composite.

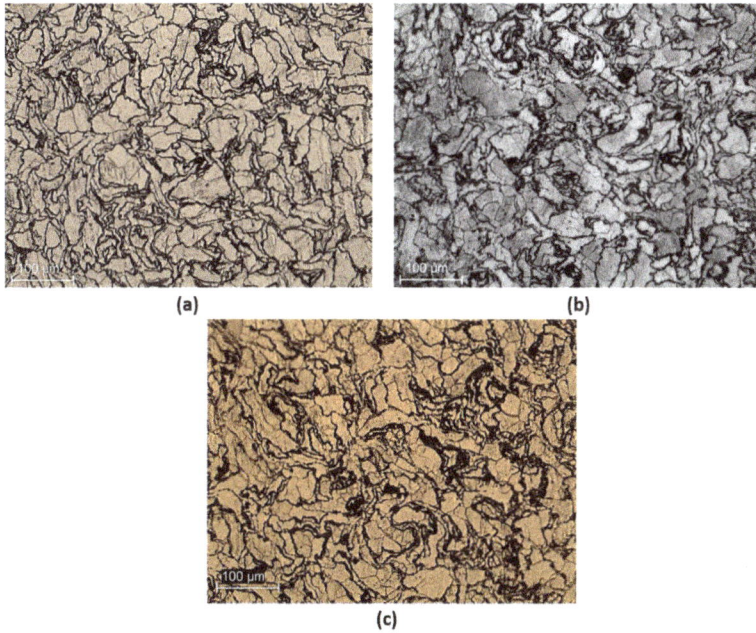

Figure 1. Optical micrography images of Mg-β-TCP composites: (**a**) Mg-0.5 TCP; (**b**) Mg-1.0 TCP; and, (**c**) Mg-1.5 TCP.

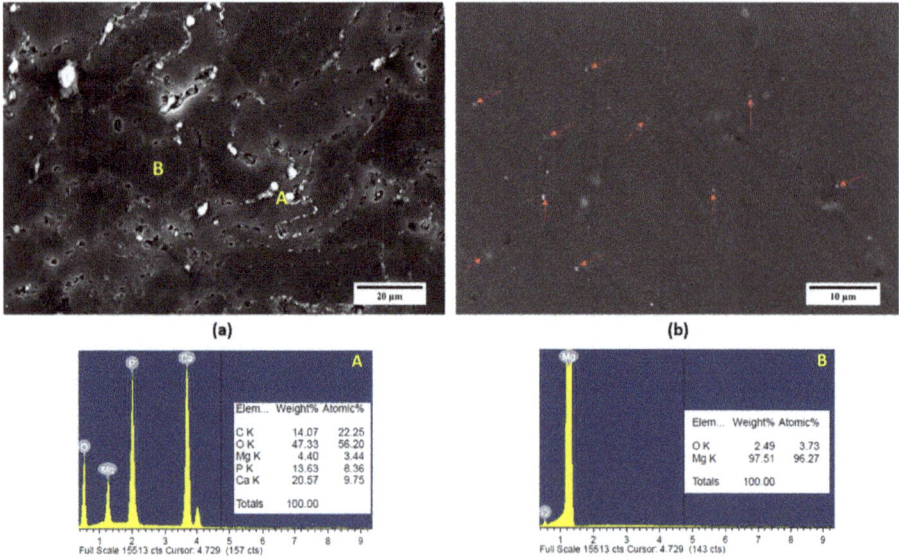

Figure 2. (**a**) Grain boundary pinning mechanism of Mg-1.0 TCP composite; (**b**) β-TCP particle distribution within the Mg matrix in Mg-1.5 TCP composite. Energy Dispersive Spectroscopy (EDS) analysis of the Mg-1.0 TCP composite at the matrix and reinforcement location.

The X-ray diffraction analysis of the developed composites was performed along the extruded direction and shown in Figure 3. The X-ray diffraction peaks of pure Mg and Mg-β-TCP composites reveal mainly Mg peaks. The reinforcement peaks are not visible, as the amount of reinforcement in the Mg matrix is low (<2 vol %), which might go undetected in the X-ray analysis. No MgO peaks, secondary phase, or impurity peaks were observed, which suggests that the surface oxidation during the compaction, sintering, or extrusion process is minimum, and the higher densification of the composite can be realized.

The ratio of the respective intensities to the maximum intensity of the composites (I/I_{max}) is also shown in Table 2. The addition of β-TCP to Mg matrix has resulted in a texture randomization with all of the developed composites having dominant intensities corresponding to the pyramidal plane. From Table 2, it can be seen that the initial addition of the reinforcement results in a decrease in the intensity corresponding to the basal plane, which reaches a maximum value with the addition of 1 vol % β-TCP and decreases with further addition up to 1.5 vol % β-TCP. The high I/I_{max} values corresponding to the basal plane assists in the corrosion protection of the material. The texture modification due to the presence of β-TCP particles may result in a particle stimulated nucleation mechanism, and it assists in systematic recrystallization of randomly oriented grains along the extrusion axis [16].

3.3. Microhardness

The results of the microhardness tests that were performed on pure Mg and Mg (0.5, 1.0, and 1.5) vol % β-TCP composite are presented in Table 1. The addition of 0.5 vol % β-TCP particle increased the microhardness of pure Mg (~46 Hv) by ~13% to an average value of ~52 Hv. Further addition of 1.0 and 1.5 vol % β-TCP particles show minimal enhancement in the microhardness of the composite. The microhardness results of Mg 1.5 vol % β-TCP composite reveal the improvement of ~17% when compared to pure Mg, hereby indicating the increased resistance to indentation. This increase in the hardness value can be attributed to the nearly uniform distribution of β-TCP particles throughout the

Mg matrix (Figure 2) and the reduced grain size (Table 1) of the Mg-β-TCP composite, leading to the increased resistance to localized plastic deformation [3].

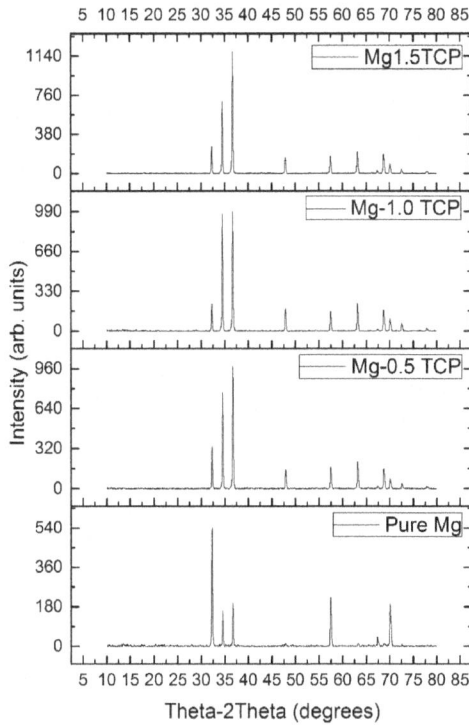

Figure 3. X-ray diffractograms of pure magnesium and Mg (0.5, 1.0, and 1.5) vol % β-TCP composites along the longitudinal direction.

Table 2. The ratio of intensities w.r.t the maximum intensity for the as-extruded Mg-β-TCP composites.

Material	Plane	I/I_{max}
Pure Mg	Prismatic	1.00
	Basal	0..29
	Pyramidal	0.36
Mg-0.5 TCP	Prismatic	0.34
	Basal	0.76
	Pyramidal	1.00
Mg-1.0 TCP	Prismatic	0.22
	Basal	0.98
	Pyramidal	1.00
Mg-1.5 TCP	Prismatic	0.23
	Basal	0.58
	Pyramidal	1.00

3.4. Damping Characteristics and Elastic Modulus

Figure 4 and Table 3 shows the damping characteristics of pure Mg and Mg (0.5, 1.0, and 1.5) vol % β-TCP composites. The time taken for pure Mg to absorb vibration is ~0.37 s. The gradual addition of β-TCP particles resulted in a linear decrease in the time that is taken by the material to absorb vibration with Mg-1.5 TCP absorbing the vibration as quick as ~0.28 s. Table 3 discusses the damping loss rate, damping capacities, and elastic modulus of the composite samples. The damping loss rate (L) and damping capacity (Q^{-1}) of pure Mg enhanced with the increased presence of β-TCP with Mg-1.5 TCP composite exhibiting the best value of ~17.7 and ~7.59 × 10^{-4}, respectively. The enhancement in damping loss rate and damping capacity for Mg-1.5 TCP composite was ~109% and ~15.7% when compared to that of pure Mg, respectively.

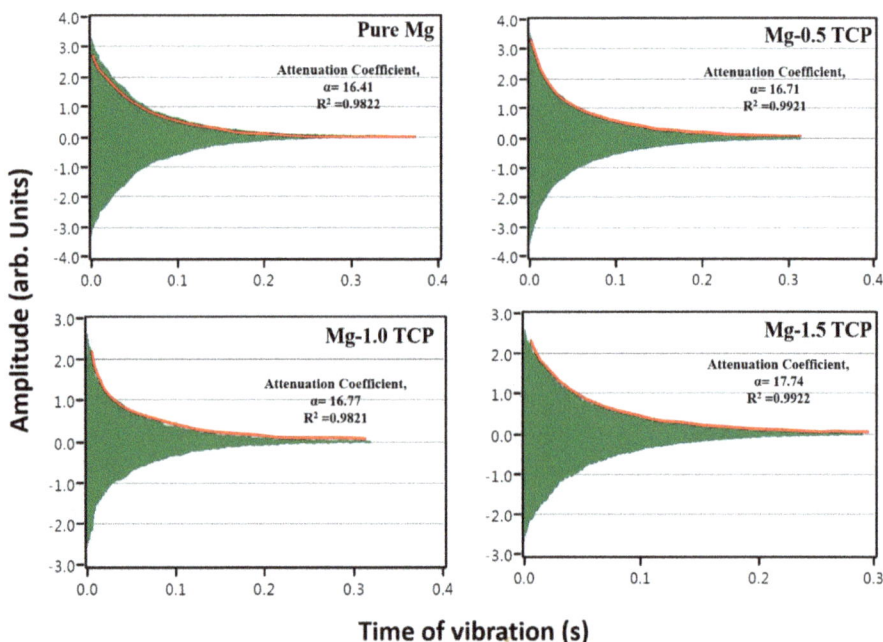

Figure 4. Damping characteristics of pure magnesium and Mg (0.5, 1.0, and 1.5) vol % β-TCP composites.

Table 3. Damping characteristics of pure Mg and Mg (0.5, 1.0, and 1.5) vol % β-TCP composites.

Material	Damping Loss Rate (L)	Damping Capacity (Q^{-1}) (× 10^{-4})	Elastic Modulus (GPa)
Pure Mg	8.3 ± 0.2	6.56 ± 0.2	44.7 ± 0.2
Mg-0.5TCP	15.7 ± 0.9 (↑89%)	6.94 ± 0.2 (↑5.7%)	43.7 ± 0.4
Mg-1.0 TCP	17.4 ± 0.7 (↑109%)	6.96 ± 0.3 (↑6.0%)	43.5 ± 0.08
Mg-1.5 TCP	17.7 ± 0.5 (↑113%)	7.59 ± 0.2 (↑15.7%)	43.7 ± 0.6

The superior enhancement in the damping characteristics of pure Mg with the addition of β-TCP may be attributed to the contribution of several damping mechanisms, namely (a) damping behavior at particle/matrix interface; (b) dislocation density owing to Mg and TCP thermal mismatch; and, (c) bulk texture modifications [26]. Other properties, like porosities and micro-defects, also affect the damping response of magnesium. However, the effect of this phenomenon on the damping characteristics of the material is a combined effect of their interactions and is not monotonically related to a particular mechanism. Attenuation coefficient calculations provide interesting insights into the qualitative understanding of the damping mechanisms of Mg-based materials. The amplitude versus time plots are shown in Figure 4. In theory, with the passage of time, the amplitude of free vibration decreases in Mg-based materials. Attenuation coefficient quantifies this difference and it is represented as the steepness of the curve in the amplitude-time plot. The composite samples excited during the study vibrate at their natural resonant frequency before attaining equilibrium. This amplitude of the dampening vibration ($A(t)$) can be expressed as Equation (1).

$$A(t) = A_0 \cdot \exp[-at] + C \tag{1}$$

where, A_0 is the amplitude at t = 0, a is the apparent attenuation coefficient, t is the time after the removal of impulse, and C is the fitting coefficient.

The attenuation coefficients of pure Mg increased with the presence of β-TCP particles. The attenuation coefficients of Mg-0.5 TCP and Mg-1.0 TCP are very similar to the attenuation coefficient of pure Mg. The addition of 1.5 vol % β-TCP particle has further enhanced the attenuation coefficient. The trend that was observed in this case is very similar to the Q^{-1} values of the composites with respect to pure Mg. This behavior is majorly dependent on the size, shape, density, and elastic modulus of the specimen that was used for measurement. The change in vibrational amplitude and resonant frequency influenced the overall improvement of the composite material when compared to that of pure Mg.

In case of Mg-based materials, dislocation pinning that is caused by increased dislocation density is favorable for achieving superior damping characteristics. This increased dislocation density owing to the mismatch in the coefficient of thermal expansions of the Mg matrix and the ceramic reinforcement results in a high volume of energy dissipation agents. This mismatch can also cause several plastic deformation zones at the matrix/reinforcement interface. The higher the plastic deformation zones and strain amplitudes, the higher the damping capacities that can be realized using the composite technology. In order to obtain high volume of plastic deformation zones, high amount of reinforcement is desirable. However, in the current case the addition of reinforcement in micron-length scale is <2 vol %. Hence, the damping capacities of the composites are close to each other [27,28].

Elastic modulus and damping capacity are two crucial properties in order to qualify any material as an orthopedic implant [3]. Currently used biomaterials, like Ti6Al4V (113 GPa), 316 L stainless steel (193 GPa), and Co-Cr alloy (230 GPa), amongst others, have high elastic modulus when compared to that of the natural bone (2–20 GPa). This huge elastic modulus mismatch might lead to stress shielding effects, hence decreasing the stimulated bone growth resulting in the failure of the implant [29]. Mg-β-TCP composites display elastic modulus (~45 GPa) that is closer to that of the natural bone when compared to the commercially available implants, and hence could be highly effective in decreasing the effect of stress shielding. Superior damping capacity values assists in the mitigation of the vibrations caused by the patient movement by suppressing the developed stresses at the implant-bone interface to result in superior osseointegration [30].

3.5. Immersion Studies

The pH, weight change, and corrosion rate measurements with respect to the time of immersion of Mg and Mg-β-TCP composites are shown in Figure 5 and Table 4. The pH of Mg showed a sudden increase post 24 h of immersion. Further immersion of up to 96 h led to a relatively slow and uniform increase in the pH, as shown in Figure 5a. The addition of β-TCP particles to the Mg matrix resulted in a lower pH reading at the end of every 24 h as compared to pure Mg. The sudden increase in the pH during the initial stage is characteristic for Mg-based materials due to high anodic Mg^{2+} dissolution [31]. However, post 24 h, the stabilization of pH is observed to be quicker for the composites when compared to pure Mg. Lower pH values and the corrosion rate values in the case of composites suggest that the amount of Mg^{2+} dissolution is lesser for the Mg-β-TCP composites when compared to pure Mg. It is well known in Mg-biomaterial community that the maximum hydrogen evolution happens in the first 12–24 h of immersion in salt solutions and biofluids [32]. Hence, the presence of β-TCP particle in the magnesium matrix helps in faster pH stabilization, thereby a controlled degradation can be achieved.

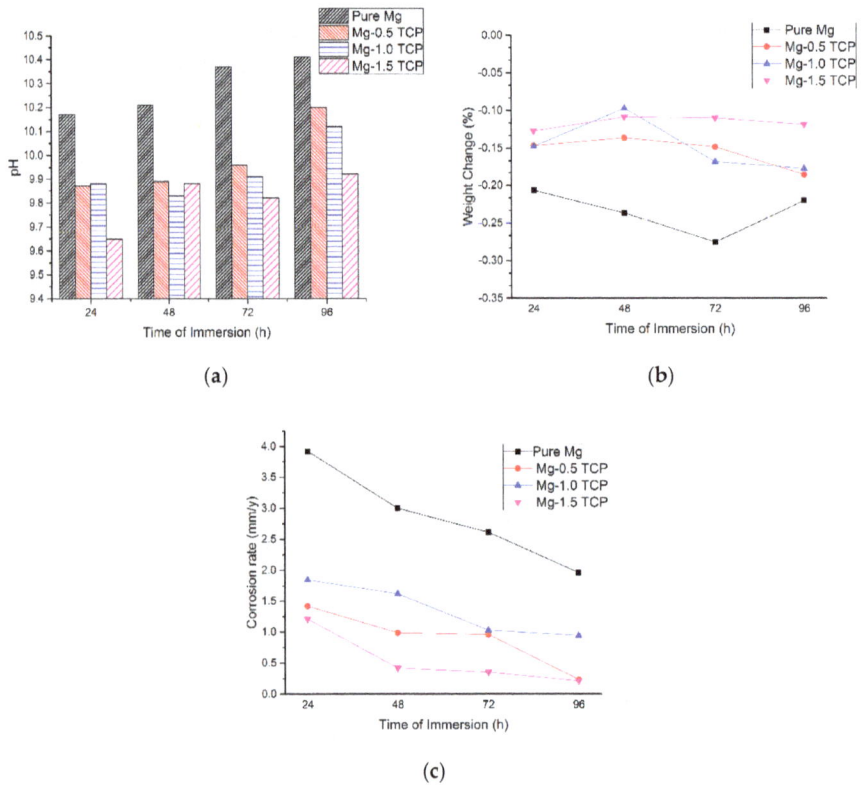

(a)

(b)

(c)

Figure 5. (**a**) pH vs. Time of immersion; (**b**) Weight change (%) vs. Time of immersion (h); and, (**c**) Corrosion rate vs. Time of immersion (h).

Table 4. Corrosion rate and pH measurements of the composite samples after 96 h of Hanks balanced salt solution (HBSS) immersion. A comparison is made with existing Mg-based alloys.

Material	Corrosion Rate (mm/Year)	pH
Pure Mg	1.95	10.41
Mg-0.5 TCP	0.23	10.23
Mg-1.0 TCP	0.92	10.11
Mg-1.5 TCP	0.21	9.92
Pure Mg [32]	2.08	
Mg1Ca [32]	3.16	
Mg1Ca1Zn [32]	2.13	
Mg1Ca3Zn [32]	2.92	
Mg5Zn [33]	2.25	-
Mg5Zn0.2Sr [33]	1.75	
Mg3Sr [32]	0.75	
ZE41 [34]	2.04	
AZ91 [34]	3.56	

From Figure 5b, it can be observed that the weight loss of the material due to immersion is gradually increasing for pure Mg with an increased time of immersion of up to 72 h. In the case of Mg-β-TCP composites, the weight changes as compared to pure Mg are observed to be lesser in all of the cases. In the case of Mg-1.5 TCP composite, the weight change is almost constant through the period, which is desirable in order to achieve a uniform dissolution of a material in vitro and in vivo conditions. Mg-0.5 TCP and Mg-1.0 TCP composites also exhibit near uniform weight loss with the increase in immersion time. Figure 5c represents the change in corrosion rate (mm/year) with the time of immersion, and is calculated by Equation (2) [35].

$$CR = \frac{(K \times W)}{(A \times T \times D)} \tag{2}$$

where, time conversion coefficient, $K = 8.76 \times 10^4$, W is the change in weight pre and post-immersion (g), A is the surface area of the cylinder exposed to the immersive medium (cm^2), T is the time of immersion (h), and D is the experimental density of the material (g·cm^{-3}).

The corrosion rate is a factor of the change in weight, surface area exposed to immersion liquid, and the time of immersion. The corrosion rates of the composites are observed to be lesser than that of pure Mg with Mg-1.5 TCP composite exhibiting the least corrosion rates for all of the conditions. The corrosion rate values owing to the weight loss of the composites mainly depend on two factors, namely (a) an initial period of incubation from the protective layer formation and breaking and (b) subsequent increase in the volume of hydrogen evolved with respect to the immersion time [34]. The corrosion rates of the composites in the current study are compared to the corrosion rates of several potential Mg-based orthopedic materials in HBSS. The corrosion rate of Mg-0.5 TCP composite (Table 4) is either better or as good as potential Mg-based alloys that are suited for orthopedic applications [32]. This justifies the suitability of Mg-β-TCP composite as a potential candidate for orthopedic applications and encourages the scientific community to further the research in this domain.

The corrosion of pure Mg and Mg-β-TCP composites are mainly governed by the way that Mg responds in aqueous environments. The Mg samples once immersed in the HBSS leads to severe anodic dissolution during the initial immersion, which leads to a a sudden increase in the pH of the developed composites. However, post-initial anodic dissolution, Mg^{2+} from the anodic metal will react with OH$^-$ in the HBSS to form a protective porous Mg(OH)$_2$ layer [4,31]. The formation of this layer will thereby protect the material by covering the surface and discouraging attack from the immersive medium. This formation of Mg(OH)$_2$ layer might stabilize the corrosion rate and the pH values of the developed composites that are observed in the current study as well. The pH and corrosion rate of the composites stabilized around a time range from 24–72 h. However, the immersive medium will slowly permeate into the sample, thereby increasing the weight loss of the sample. It is common knowledge that the corrosion resistance of Mg in chloride environments is low and the presence of Cl$^-$ ions in Hank's solution promotes the corrosion rate by forming more resoluble MgCl$_2$, thereby increasing the concentration of OH$^-$ ions in the solution and hence rupturing the protective hydroxide layer [36]. The Cl$^-$ attack on the sample leads to the pitting corrosion of the material, which is the most commonly seen mechanism for Mg-based materials. Post this behaviour, a dynamic stabilization is observed in the pH and corrosion rate values and a near uniform dissolution of the material is observed. However, with the passage of time, these reactions might continue to have a deeper impact along the matrix/particle interface, allows for the swelling of the composites due to electrolyte penetration, leading to the eventual failure of the composite [37]. Figure 6 shows the optical micrographs of the samples under immersion. The formation of corroded pits is very evident with pure Mg and the number of corroded pits that were formed on the composites samples are much lesser than that of pure Mg.

By theory, in addition to pitting corrosion, corrosion at the grain boundaries is a major form of corrosion in the Mg-based materials [4,5,38]. However, due to the high chemical activity of Mg in chloride environments, realizing a near uniform corrosion rate is ideal. Hence, altering Mg microstructurally can assist in realizing this. In the current study, the refinement in the grain size of pure Mg with the addition of β-TCP particles strongly aids in the corrosion protection of Mg. The increased number of grain boundaries due to their higher energies and chemical activities increase the surface reactivity, realising the faster formation of the protective layer by relieving the extent of localized corrosion [5,39]. The reduction of the number of cathodic sites can also enhance the corrosion resistance, owing to the dissolution of impurities at the grain boundaries [40]. Further, the absence of secondary phases/impurities in the material, which might serve as a stress concentration sites leading to stress corrosion cracking, also supports the enhancement of the corrosion protection. In addition to the superior grain refinement, the high intensities corresponding to the basal plane of the Mg-β-TCP composites also assists in increasing the corrosion resistance of the material [40]. The lower porosities that were observed for the composite samples can also assist in mitigating the effect of corrosion. The cumulative effect of lower grain size coupled with higher basal intensities of the composites is responsible for the lower corrosion rates of the composites when compared to that of pure Mg [40]. Further, the presence of Fe impurity (<500 ppm) in the magnesium powder might be responsible for a high corrosion rate value of pure Mg. The addition of β-TCP tends to override the effect of the Fe impurity, as observed by the lower corrosion rates of the composites. Although, the presence of the Fe impurity cannot be confirmed by the XRD and EDS analysis of the composites, its effect on the corrosion rate could still be acknowledged to be significant. In order to understand the effect of any alloying element/secondary reinforcement on the corrosion performance of pure Mg, ultra-pure Mg must be used so that the Fe content can be restricted to <45 ppm in the extruded form and the direct effect of the secondary reinforcement could be more accurately discussed [9].

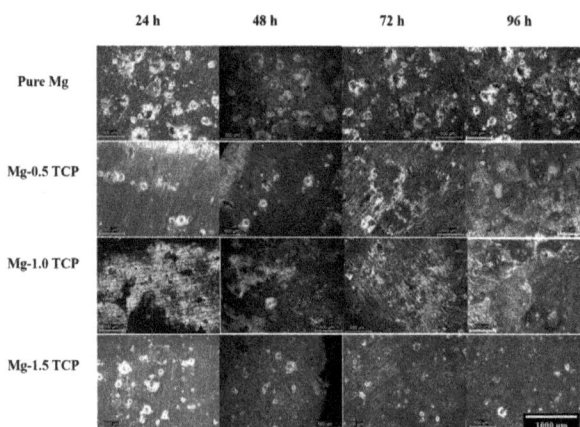

Figure 6. Optical micrographs of pure magnesium and Mg (0.5, 1.0, and 1.5) vol % β-TCP composites after 24, 48, 72, and 96 h of immersion.

The SEM analysis of the corroded surface of the Mg-1.0 TCP composite samples post 96 h of immersion is shown in Figure 7. The removal of the composite samples from the immersive medium and drying in the air before undergoing SEM observation shrinks the protective film on the surface, owing to dehydration [41]. This leads to severe cracking on the sample surface with two types of layers namely quasi-adherent layer and cracked layer being formed [4]. Several uneven corrosion pits with varying sizes and cracks were observed throughout the sample surface [41]. Apart from this, many tiny cracks were observed through the sample surface. Figure 8 shows the XRD analysis of the samples post-corrosion. The XRD analysis of the composites reveals the presence of corrosion products like $Mg(OH)_2$, $(Ca_2Mg)_3(PO_4)_2$ and Hydroxyapatite (HA). These corrosion products fill these corrosion pits, forming a protective layer and hence delaying the corrosion attack to the surface [36]. Hence, the behaviour of the Mg-β-TCP composites can be observed to be better than that of pure Mg. The corrosion product layer that was formed in pure Mg destabilizes faster, leading to the HBSS to penetrate the matrix influencing the corrosion along the grain boundaries and the impurities, if any.

The HA corrosion products can encourage the formation of apatite layer at the implant/bone interface, which therefore assists in the osteoclast activity and resulting in faster bone formation, as compared to monolithic Mg [42].

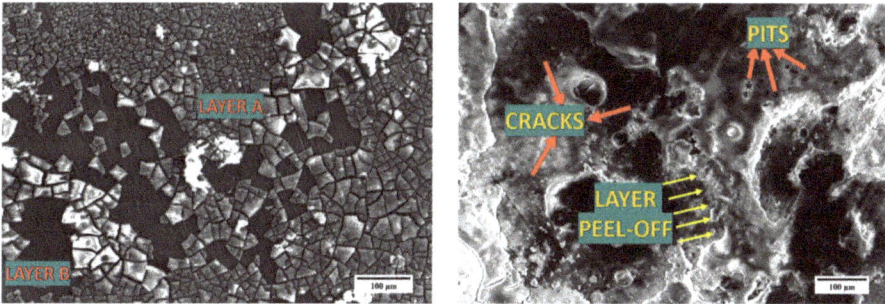

Figure 7. Scanning Electron Microscope (SEM) analysis of Mg-1.0 TCP composite post 96 h of Hanks balanced salt solution (HBSS) immersion.

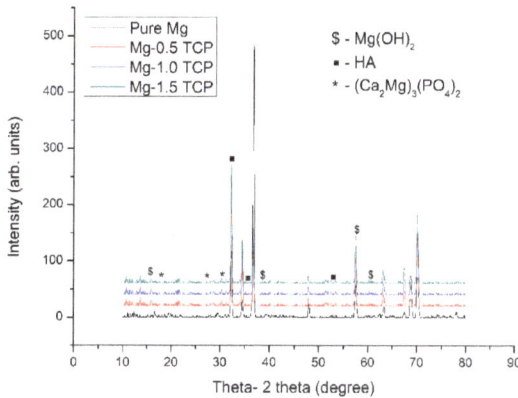

Figure 8. X-ray diffraction studies on the corroded samples post 96 h immersion in HBSS.

3.6. Compression Properties

Table 5 and Figure 9 shows the compressive properties and the stress-strain behavior of the composites. Addition of β-TCP enhances the overall compressive properties of pure Mg. The compressive yield strength (0.2% CYS) of Mg (0.5, 1.0 and 1.5) vol % TCP was observed to be ~92 MPa, ~96 MPa and ~103 MPa, respectively, which is ~19%, ~24%, and ~34% greater than that of pure Mg (~77 MPa). The addition of 0.5 vol % TCP increases the ultimate compressive strength (UCS) to the value of ~258 MPa, which is a ~65% enhancement when compared to pure Mg (~156 MPa). Further addition of TCP particles resulted in a decrease in the UCS values, with both Mg-1.0 TCP and Mg-1.5 TCP composites remaining lower than that of Mg-0.5 TCP. However, the UCS values of the composites still exhibited a ~42% and ~53% enhancement, respectively, with respect to pure Mg. The maximum fracture strain and energy absorbed under compressive loading was observed for Mg-1.5 vol % TCP composite with ~19.3% and ~29.2 MJ/m^3, which is ~22% and ~64% greater than that of pure Mg.

Table 5. Room temperature compressive testing results.

Material	0.2 CYS (MPa)	UCS (MPa)	Fracture Strain (%)	Energy Absorbed (MJ/m^3)
Pure Mg	77 ± 5	156 ± 7	15.8 ± 0.3	17.7 ± 0.7
Mg-0.5 TCP	92 ± 1 (↑19%)	258 ± 4 (↑65%)	18.5 ± 0.6 (↑17%)	28.3 ± 1.3 (↑59%)
Mg-1.0 TCP	96 ± 2 (↑24%)	223 ± 7 (↑42%)	17.2 ± 0.7 (↑9%)	23.2 ± 2.9 (↑31%)
Mg-1.5 TCP	103 ± 7 (↑34%)	240 ± 7 (↑53%)	19.3 ± 0.5 (↑22%)	29.2 ± 2.4 (↑64%)

The significant increase in the strengths of Mg-β-TCP composite may be due to: (a) superior grain refinement (Table 1), leading to activation of Hall-Petch strengthening mechanism [3]; (b) uniformly distributed β-TCP particles that are acting as an obstacle to dislocation movement through the Orowan strengthening mechanism [16]; (c) forest strengthening of the composite owing to the thermal coefficient mismatch between the matrix (Mg) and the ceramic reinforcement (β-TCP); (d) combined effects of texture randomization and deformation twinning creating additional barriers to the crack path leading to enhanced compressive strengths [43]; and, (e) effective load transfer from the ductile matrix to the brittle ceramic, owing to the good interface with each other. Due to high interfacial integrity between the matrix and the reinforcement, the crack propagates preferentially through the soft matrix phase under compressive loading. When the crack encounters the soft matrix, the matrix deforms, thereby creating a bridging mechanism, leading to crack closure and resistance to crack initiation and growth [36,44]. Hence, the sample fracture under compression is mainly driven by the matrix deformation. The fracture strain of the composite samples remained as marginally superior to that of pure Mg, unlike in the case of submicron and micron size ceramic particles where it is adversely affected. Figure 9 shows the fractured surface analysis of the composite samples. The fractured surfaces show typical shear band formation, which corroborates the fact that the composite fracture is mainly matrix driven.

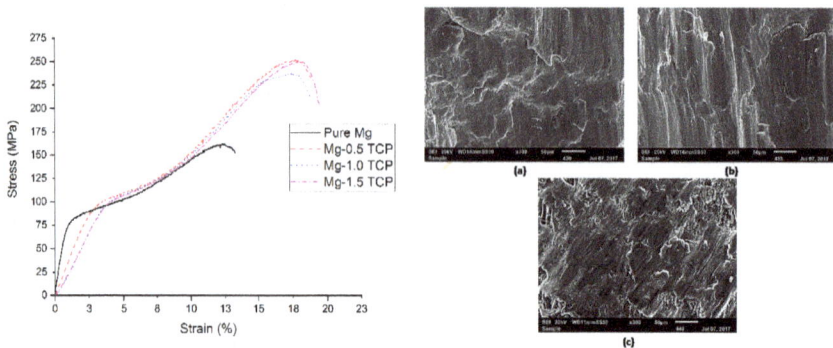

Figure 9. Compressive stress-strain relationship and fractured surface behavior of Mg ((a) 0.5, (b) 1.0, and (c) 1.5) vol % β-TCP composites.

The cumulative effect of enhanced damping, compression, and corrosion properties is key to qualify a certain material to be a potential orthopaedic implant. In addition to the mitigation of stress-shielding effects and the bioresorbable nature of magnesium, enhanced structural properties are also equally important when considering that Mg-based implants and stents may witness a 15–20% decrease in strength and ductility when being immersed in a physiological environment [45]. With the bone remodelling process taking 10–12 weeks for an average human being, incorporating novel biocompatible reinforcements in the magnesium matrix that can protect pure Mg from expedited degradation and simultaneously perform the task of load-bearing. Further, to understand and to critically evaluate the possibility of using Mg-β-TCP composites in orthopaedic applications, in vivo immersion and cytotoxicity tests must be conducted.

4. Conclusions

- Near dense Mg-β-TCP composites were successfully synthesized with blend-press-sinter powder metallurgy technique with a porosity of less than 1%.
- The microhardness of pure Mg increased due to the presence of β-TCP particles with ~17.39% enhancement realized in the case of Mg-1.5 TCP composite.
- Mg-1.5 TCP composite exhibited a compressive yield strength, ultimate compressive strength, compressive fracture strain, and total energy absorbed under compression loading of ~103 MPa (↑~34%), ~240 MPa (↑~53%), ~19.3% (↑~22%), and ~29.2 MJ/m^3 (↑~64%), respectively. The enhancements with respect to the base pure Mg are significant in all of the cases.
- The damping response of pure Mg enhanced with the addition of β-TCP particles, with Mg-1.5 TCP composite exhibiting the best damping capacity (~15.7% increase as compared to pure Mg) and damping loss rate (~113% increase compared to pure Mg) values.
- The presence of β-TCP particles assisted in the corrosion protection of pure Mg. The pH values stabilized earlier for the composites as compared to pure Mg and displayed lower corrosion rate values, which a superior ~9 times protection displayed by the Mg-1.5 TCP composite as compared to pure Mg.

Author Contributions: M.G. and G.P. proposed the original project and supervised the investigation. H.G. and G.P. performed the experiments. G.P., V.M. and H.G. analyzed the data and wrote the paper with assistance from all authors. All authors contributed to the discussions in the manuscript.

Conflicts of Interest: The authors declare no conflict of interest.

References

1. Kuśnierczyk, K.; Basista, M. Recent advances in research on magnesium alloys and magnesium–calcium phosphate composites as biodegradable implant materials. *J. Biomater. Appl.* **2017**, *31*, 878–900. [CrossRef] [PubMed]
2. Boccaccini, A.R.; Ma, P.X. *Tissue Engineering Using Ceramics and Polymers*; Elsevier: Amsterdam, The Netherlands, 2014.
3. Parande, G.; Manakari, V.; Meenashisundaram, G.K.; Gupta, M. Enhancing the hardness/compression/damping response of magnesium by reinforcing with biocompatible silica nanoparticulates. *Int. J. Mater. Res.* **2016**, *107*, 1091–1099. [CrossRef]
4. Song, G.L.; Atrens, A. Corrosion mechanisms of magnesium alloys. *Adv. Eng. Mater.* **1999**, *1*, 11–33. [CrossRef]
5. Atrens, A.; Song, G.-L.; Liu, M.; Shi, Z.; Cao, F.; Dargusch, M.S. Review of recent developments in the field of magnesium corrosion. *Adv. Eng. Mater.* **2015**, *17*, 400–453. [CrossRef]
6. Atrens, A.; Song, G.-L.; Cao, F.; Shi, Z.; Bowen, P.K. Advances in mg corrosion and research suggestions. *J. Mages. Alloys* **2013**, *1*, 177–200. [CrossRef]
7. Tan, L.; Yu, X.; Wan, P.; Yang, K. Biodegradable materials for bone repairs: A review. *J. Mater. Sci. Technol.* **2013**, *29*, 503–513. [CrossRef]
8. Witte, F. The history of biodegradable magnesium implants: A review. *Acta Biomater.* **2010**, *6*, 1680–1692. [CrossRef] [PubMed]
9. Atrens, A.; Liu, M.; Abidin, N.I.Z. Corrosion mechanism applicable to biodegradable magnesium implants. *Mater. Sci. Eng. B* **2011**, *176*, 1609–1636. [CrossRef]
10. Lu, Y.; Tan, L.; Xiang, H.; Zhang, B.; Yang, K.; Li, Y. Fabrication and characterization of Ca–Mg–P containing coating on pure magnesium. *J. Mater. Sci. Technol.* **2012**, *28*, 636–641. [CrossRef]
11. Witte, F.; Feyerabend, F.; Maier, P.; Fischer, J.; Störmer, M.; Blawert, C.; Dietzel, W.; Hort, N. Biodegradable magnesium–hydroxyapatite metal matrix composites. *Biomaterials* **2007**, *28*, 2163–2174. [CrossRef] [PubMed]

12. Feng, A.; Han, Y. The microstructure, mechanical and corrosion properties of calcium polyphosphate reinforced ZK60A magnesium alloy composites. *J. Alloys Compd.* **2010**, *504*, 585–593. [CrossRef]
13. Gu, X.; Zhou, W.; Zheng, Y.; Dong, L.; Xi, Y.; Chai, D. Microstructure, mechanical property, bio-corrosion and cytotoxicity evaluations of Mg/Ha composites. *Mater. Sci. Eng. C* **2010**, *30*, 827–832. [CrossRef]
14. Koepp, H.E.; Schorlemmer, S.; Kessler, S.; Brenner, R.E.; Claes, L.; Günther, K.P.; Ignatius, A.A. Biocompatibility and osseointegration of β-TCP: Histomorphological and biomechanical studies in a weight-bearing sheep model. *J. Biomed. Mater. Res. Part B* **2004**, *70*, 209–217. [CrossRef] [PubMed]
15. Dieringa, H.; Fuskova, L.; Fechner, D.; Blawert, C. Mechanical and corrosion behaviour of a hydroxyapatite reinforced magnesium alloy WE43. In Proceedings of the 17th International Conference on Composite Materials, ICCM, Edinburgh, UK, 27–31 July 2009.
16. Yan, Y.; Kang, Y.; Li, D.; Yu, K.; Xiao, T.; Deng, Y.; Dai, H.; Dai, Y.; Xiong, H.; Fang, H. Improvement of the mechanical properties and corrosion resistance of biodegradable β-Ca$_3$(PO$_4$)$_2$/Mg-Zn composites prepared by powder metallurgy: The adding β-Ca$_3$(PO$_4$)$_2$, hot extrusion and aging treatment. *Mater. Sci. Eng. C* **2017**, *74*, 582–596. [CrossRef] [PubMed]
17. He, S.-Y.; Sun, Y.; Chen, M.-F.; Liu, D.-B.; Ye, X.-Y. Microstructure and properties of biodegradable β-TCP reinforced Mg-Zn-Zr composites. *Trans. Nonferr. Met. Soc. China* **2011**, *21*, 814–819. [CrossRef]
18. Liu, D.B.; Huang, Y.; Prangnell, P.B. Microstructure and performance of a biodegradable Mg–1Ca–2Zn–1TCP composite fabricated by combined solidification and deformation processing. *MatL* **2012**, *82*, 7–9. [CrossRef]
19. Famery, R.; Richard, N.; Boch, P. Preparation of α- and β-tricalcium phosphate ceramics, with and without magnesium addition. *Ceram. Int.* **1994**, *20*, 327–336. [CrossRef]
20. Matli, P.R.; Ubaid, F.; Shakoor, R.A.; Parande, G.; Manakari, V.; Yusuf, M.; Mohamed, A.M.A.; Gupta, M. Improved properties of Al–Si$_3$N$_4$ nanocomposites fabricated through a microwave sintering and hot extrusion process. *RSC Adv.* **2017**, *7*, 34401–34410. [CrossRef]
21. ASTM Standard. *E384, Standard Test Method for Microindentation Hardness of Materials*; ASTM International: West Conshohocken, PA, USA, 2000.
22. ASTM Standard. *E9-09. Standard Test Methods of Compression Testing of Metallic Materials at Room Temperature*; ASTM International: West Conshohocken, PA, USA, 2009.
23. Wong, W.; Gupta, M. Using microwave energy to synthesize light weight/energy saving magnesium based materials: A review. *Technologies* **2015**, *3*, 1–18. [CrossRef]
24. Parande, G.; Manakari, V.; Meenashisundaram, G.K.; Gupta, M. Enhancing the tensile and ignition response of monolithic magnesium by reinforcing with silica nanoparticulates. *J. Mater. Res.* **2017**, *32*, 2169–2178. [CrossRef]
25. Wang, X.; Wu, K.; Zhang, H.; Huang, W.; Chang, H.; Gan, W.; Zheng, M.; Peng, D. Effect of hot extrusion on the microstructure of a particulate reinforced magnesium matrix composite. *Mater. Sci. Eng. A* **2007**, *465*, 78–84. [CrossRef]
26. Kujur, M.S.; Mallick, A.; Manakari, V.; Parande, G.; Tun, K.S.; Gupta, M. Significantly enhancing the ignition/compression/damping response of monolithic magnesium by addition of SM$_2$O$_3$ nanoparticles. *Metals* **2017**, *7*, 357. [CrossRef]
27. Anilchandra, A.R.; Surappa, M.K. Microstructure and damping behaviour of consolidated magnesium chips. *Mater. Sci. Eng. A* **2012**, *542*, 94–103. [CrossRef]
28. Carreño-Morelli, E.; Urreta, S.E.; Schaller, R. Mechanical spectroscopy of thermal stress relaxation at metal–ceramic interfaces in aluminium-based composites. *Acta Mater.* **2000**, *48*, 4725–4733. [CrossRef]
29. Walker, J.; Shadanbaz, S.; Woodfield, T.B.; Staiger, M.P.; Dias, G.J. Magnesium biomaterials for orthopedic application: A review from a biological perspective. *J. Biomed. Mater. Res. B Appl. Biomater.* **2014**, *102*, 1316–1331. [CrossRef] [PubMed]
30. Tsai, M.-H.; Chen, M.-S.; Lin, L.-H.; Lin, M.-H.; Wu, C.-Z.; Ou, K.-L.; Yu, C.-H. Effect of heat treatment on the microstructures and damping properties of biomedical Mg–Zr alloy. *J. Alloys Compd.* **2011**, *509*, 813–819. [CrossRef]
31. Wang, B.; Xu, D.; Dong, J.; Ke, W. Effect of corrosion product films on the in vitro degradation behavior of Mg-3% Al-1% Zn (in wt %) alloy in hank's solution. *J. Mater. Sci. Technol.* **2018**. [CrossRef]
32. Gupta, M.; Meenashisundaram, G.K. *Insight into Designing Biocompatible Magnesium Alloys and Composites: Processing, Mechanical and Corrosion Characteristics*; Springer: Berlin, Germany, 2015.

33. Cheng, M.; Chen, J.; Yan, H.; Su, B.; Yu, Z.; Xia, W.; Gong, X. Effects of minor sr addition on microstructure, mechanical and bio-corrosion properties of the Mg-5Zn based alloy system. *J. Alloys Compd.* **2017**, *691*, 95–102. [CrossRef]

34. Taltavull, C.; Shi, Z.; Torres, B.; Rams, J.; Atrens, A. Influence of the chloride ion concentration on the corrosion of high-purity Mg, Ze41 and Az91 in buffered hank's solution. *J. Mater. Sci. Mater. Med.* **2014**, *25*, 329–345. [CrossRef] [PubMed]

35. Meenashisundaram, G.K.; Nai, M.H.; Gupta, M. Effects of ti and TIB$_2$ nanoparticulates on room temperature mechanical properties and in vitro degradation of pure mg. In *Magnesium Technology 2015*; Manuel, M.V., Singh, A., Alderman, M., Neelameggham, N.R., Eds.; Springer International Publishing: Cham, Switzerland, 2016; pp. 413–418.

36. Ma, X.; Dong, L.; Wang, X. Microstructure, mechanical property and corrosion behavior of co-continuous β-tcp/mgca composite manufactured by suction casting. *Mater. Des.* **2014**, *56*, 305–312. [CrossRef]

37. Wang, X.; Li, J.; Xie, M.; Qu, L.; Zhang, P.; Li, X. Structure, mechanical property and corrosion behaviors of (ha+ β-TCP)/Mg–5Sn composite with interpenetrating networks. *Mater. Sci. Eng. C* **2015**, *56*, 386–392. [CrossRef] [PubMed]

38. Mostaed, E.; Hashempour, M.; Fabrizi, A.; Dellasega, D.; Bestetti, M.; Bonollo, F.; Vedani, M. Microstructure, texture evolution, mechanical properties and corrosion behavior of ecap processed ZK60 magnesium alloy for biodegradable applications. *J. Mech. Behav. Biomed. Mater.* **2014**, *37*, 307–322. [CrossRef] [PubMed]

39. Ahmadkhaniha, D.; Järvenpää, A.; Jaskari, M.; Sohi, M.H.; Zarei-Hanzaki, A.; Fedel, M.; Deflorian, F.; Karjalainen, L. Microstructural modification of pure Mg for improving mechanical and biocorrosion properties. *J. Mech. Behav. Biomed. Mater.* **2016**, *61*, 360–370. [CrossRef] [PubMed]

40. Pu, Z.; Song, G.-L.; Yang, S.; Outeiro, J.; Dillon, O., Jr.; Puleo, D.; Jawahir, I. Grain refined and basal textured surface produced by burnishing for improved corrosion performance of AZ31B Mg alloy. *Corros. Sci.* **2012**, *57*, 192–201. [CrossRef]

41. Zainal Abidin, N.I.; Atrens, A.D.; Martin, D.; Atrens, A. Corrosion of high purity Mg, Mg$_2$Zn$_{0.2}$Mn, ZE41 and AZ91 in hank's solution at 37 °C. *Corros. Sci.* **2011**, *53*, 3542–3556. [CrossRef]

42. Geng, F.; Tan, L.; Jin, X.; Yang, J.; Yang, K. The preparation, cytocompatibility, and in vitro biodegradation study of pure β-TCP on magnesium. *J. Mater. Sci. Mater. Med.* **2009**, *20*, 1149–1157. [CrossRef] [PubMed]

43. Parande, G.; Manakari, V.; Kopparthy, S.D.S.; Gupta, M. Utilizing low-cost eggshell particles to enhance the mechanical response of Mg-2.5Zn magnesium alloy matrix. *Adv. Eng. Mater.* **2017**, *1700919*. [CrossRef]

44. Tun, K.; Zhang, Y.; Parande, G.; Manakari, V.; Gupta, M. Enhancing the hardness and compressive response of magnesium using complex composition alloy reinforcement. *Metals* **2018**, *8*, 276. [CrossRef]

45. Hermawan, H. Biodegradable metals: State of the art. In *Biodegradable Metals*; Springer: Berlin, Germany, 2012; pp. 13–22.

MDPI

St. Alban-Anlage 66

4052 Basel

Switzerland

Tel. +41 61 683 77 34

Fax +41 61 302 89 18

www.mdpi.com

Metals Editorial Office

E-mail: metals@mdpi.com

www.mdpi.com/journal/metals